THE NEXT GENERATIONS OF ARMY AVIATION SYSTEMS

Setting the Record Straight on Development of the Next Generations of Army Aviation Systems

Book 2 in the trilogy

A Full Lifetime Career of Seeking Perfection Driven by Family and Mentors

Daniel P. Schrage

Publish Authority

"In Praise For Daniel P. Schrage"

"Dan's operational, acquisition and academic prowess establishes him as the preeminent rotorcraft developer. An outstanding leader and technical mentor." —Dr. William D. Lewis Former Director for U.S. Army Aviation Development

"As a decorated combat Army Aviator, accomplished aeronautical engineer, advanced systems innovator and past Director of the Army Rotorcraft Research Center of Excellence, Dr. Schrage brings an expert perspective in this important publication." – George T Singley III, Former Deputy Assistant Secretary of the Army (Research & Technology)

"Dan Schrage was an excellent athlete, my teammate, and captain of the 1966-67 Army Basketball Team with Bob Knight as Coach. He was our leader and our hardest worker and competitor" –Mike Krzyzewski (Ret.) Duke University Basketball Coach.

"Dan Schrage provides first-hand experiences and valuable lessons learned from the Cold War in Europe and the Vietnam War." --Paul Fardink, LTC USA (Ret), Vertical Flight Society History Committee

The Next Generations of Army Aviation Systems: Setting the Record Straight on Development of the Next Generations of Army Aviation Systems - Book 2 in the trilogy A Full Lifetime Career of Seeking Perfection Driven by Family and Mentors

ISBN 978-1-954000-44-5 (Paperback)
ISBN 978-1-954000-45-2 (eBook)

Editor: Bob Laning
Interior design: Teresa Evans
Cover design lead: Raeghan Rebstock

Published 2024, by Publish Authority
300 Colonial Center Parkway, Suite 100
Roswell, GA 30076-4892 USA
PublishAuthority.com

Printed in the United States of America

First edition

To my family, friends, and mentors.

TABLE OF CONTENTS

PREFACE

DEVELOPMENT OF NEXT GENERATIONS OF ARMY AVIATION SYSTEMS

After I returned from Vietnam in February 1971, I attended the nine-month Field Artillery Officers Advanced Course at Fort Sill, OK, starting in March 1971. It was an excellent time to spend with my growing family and for Nancy and me to re-acquaint with some classmates and their spouses from our days at West Point.

This was followed by my attending graduate school at the Georgia Institute of Technology (GIT)/Georgia Tech. This provided me with the transition from an active duty operational officer to an active duty aerospace engineer with a subsequent assignment to the Army Aviation Systems Command (AVSCOM). This assignment was timely as Army Aviation was initiating the development of the 3rd generation of Army Aviation Systems, Figure 1. (Ref. 1). First developed by Mike Hirschberg, Vertical Flight Society (VFS), then modified by me to account for generations of Army Aviation Systems.

My AVSCOM assignment was an excellent opportunity to apply lessons learned to new Army aircraft from my experience as an Army Aviator in planning and conducting airmobile operations in South Vietnam and Cambodia.

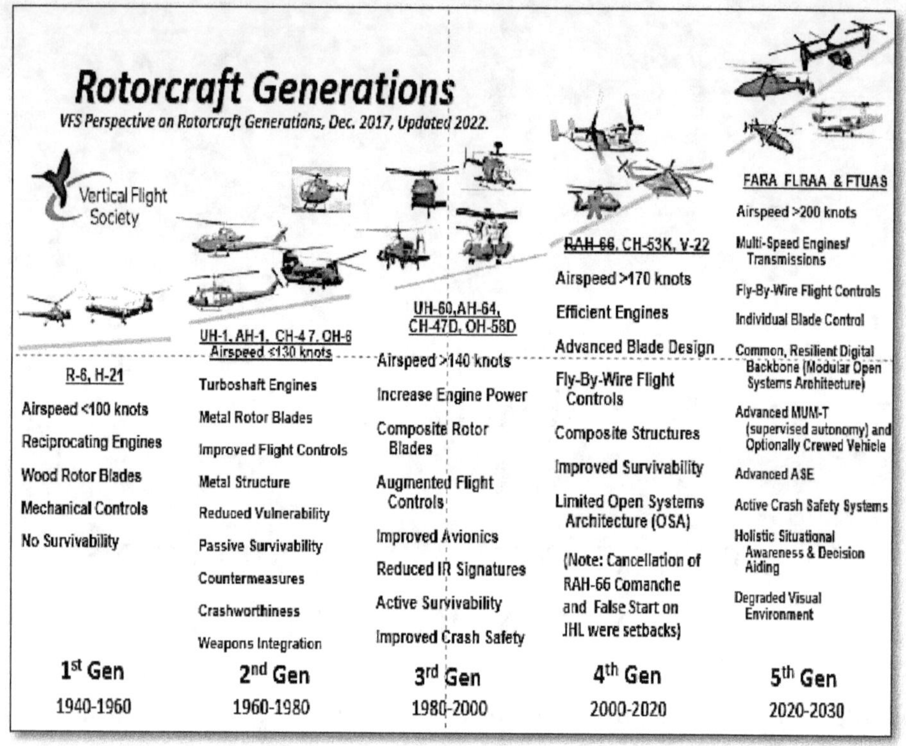

Figure 1. Rotorcraft Generations and Key Technologies Developed (Ref.1)

Illustrated in Figure 2 is a summation of Army 3rd Rotorcraft Generation development. It began as part of the Army Big Five Program in the early 1970s with the Army Utility Tactical Transport Aircraft System (UTTAS) and Advanced Attack Helicopter (AAH) Programs initiated to replace the Vietnam helicopters, e.g., UH-1 Huey Utility Helicopter and AH-1 Cobra Attack Helicopter.

Another program, Heavy Lift Helicopter (HLH) Program, was also initiated in the early 1970s but was canceled in 1973. The need for an HLH continued to exist in the following years, with attempts for a Joint Heavy Lift (JHL) in the 2000s, but again, it never materialized. While the CH-47D, E, and F Chinook Programs have partially solved a medium to heavy Army need, there is still an Army need for heavy lift today. As illustrated in Figure 2, four Government-Industry competitions were held in the 1970s. In addition to the UTTAS and AAH Competitions in the early and mid-1970s, modification programs were also started in the late 1970s, e.g., the Army Helicopter Improvement Program (AHIP) to develop a near-term scout helicopter (NTSH) for the replacement of the OH-58A/C Scout Helicopter. It also included the CH-47D Cargo Helicopter Modernization Program to upgrade the CH-47C Chinook helicopters.

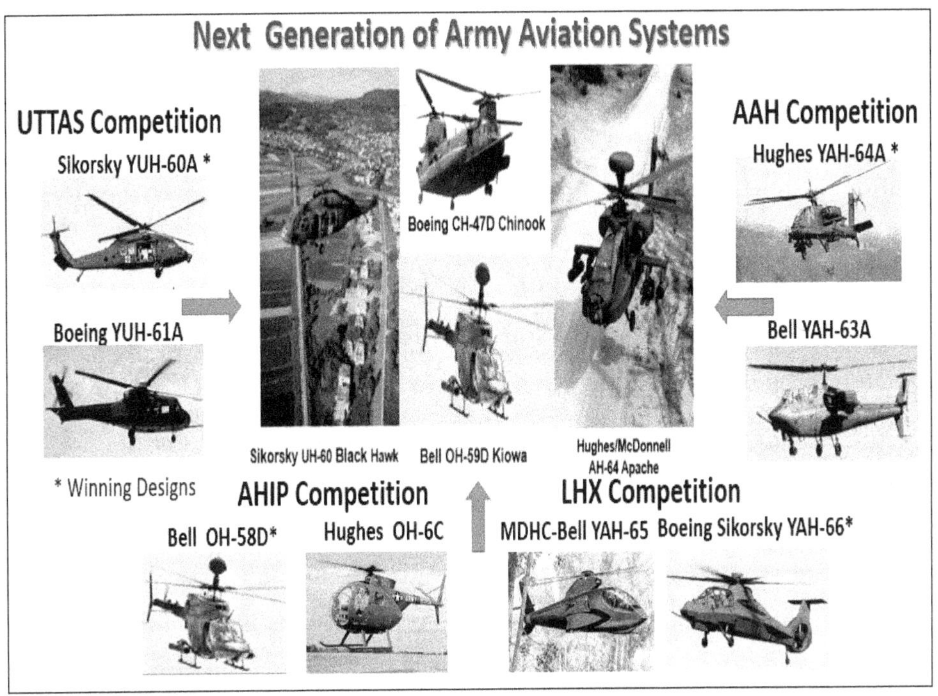

Figure 2. 3rd Generation of Rotorcraft

In the early 1980s, the Light Helicopter Experimental (LHX) Program began Concept Formulation for the new development of light Scout-Attack (SCAT) and Light Utility Helicopter (LUH) aircraft. The initial plan was to replace approximately 4500 Vietnam helicopters, e.g., OH-58/OH-6 scout helicopters, UH-1 utility helicopters, and AH-1 attack helicopters. During their prototype developments, I played significant engineering evaluation roles in the development of the UTTAS and AAH programs, as the Aeroelasticity, Dynamics, and Vibrations (ADV) evaluation engineer.

I also served on both competitive Source Selection Evaluation Boards (SSEBs) for these aircraft from 1974 to 1978. In addition, during this time, I completed advanced degrees in Business Administration and Mechanical Engineering at Webster College and Washington U. in St. Louis, MO, respectively.

The first half of Book 2 will also address setting the record straight on developing the Third Generation of Army Aviation Systems. These developments led to key government-sponsored industry competitions and winning designs, as illustrated in Figure 2.

As a baseline, I will build off the Program Executive Officers (PEO) Aviation Report, *Army Aviation Foundations of the Modern Fleet* Report (Ref.2). It was prepared by and can be obtained from, the Program Executive Office, Aviation Public Affairs Office, 1ˢᵗ Addition, Published August 2013. I was a major contributor to this effort and will try and set the record straight on why it has been so difficult for the Army to develop and field the next generations of Aviation Systems. My interviews were with Mr. Tom House, a friend and former Army Aviation and Troop Command (ATCOM) Technical Director, who served as the interviewer. My recorded interviews are included on videos under Book 2 on my website, DanielSchrage.com. The 3ʳᵈ Generation of Rotorcraft, e.g., UH-60A. AH-64A, plus the OH-58D and the CH-47D, are addressed in the *Army Aviation Foundations of the Modern Fleet Book*. (Ref.2)

However, this Book 2 will also address setting the record straight for the development programs for both the 3ʳᵈ and 4ᵗʰ Generation, e.g., the V-22 Osprey, Light Helicopter Experimental (LHX), and the RAH-66 Comanche.

I will also refer to the currently projected 5ᵗʰ Generation Future Vertical Lift (FVL) aircraft, e.g., the Army Future Long Range Assault Aircraft (FLRAA) and the Future Attack and Reconnaissance Aircraft (FARA) aircraft. Previous and next generations of these rotorcraft developments are shown in Figure 3.

Figure 3. Army Advanced Rotorcraft Concepts leading to Potential Fifth Generation

The Army has not had a successful completion of a new concept development program since the LHX Concept Formulation in the early 1980s, which led to the Demonstration and Validation (DEMVAL) of the RAH-66 Comanche in the 1990s. In June 2000, a 6-year engineering and manufacturing development (EMD) contract was awarded to the Boeing-Sikorsky First Team. However, on 23 February 2004, the US Army announced the termination of the Comanche program, stating they had determined that the RAH-66 would require numerous upgrades to be viable on the battlefield. Also, there was a need to fund the existing fleet.

These issues will be addressed in a later chapter in Book 2 and will review a fairly recent book published on the Comanche, *The RAH-66 Comanche Helicopter: Technical Accomplishment, Program Frustration*, by Arthur W. Linden and the Comanche Team (Ref.3). While this book provides detailed information about the Comanche Development and the First Team's frustrations, it ignores their poor selection of a Bearing-less Main Rotor (BMR) and their lack of design experience in BMRs. It also shows their insufficient knowledge of Systems Engineering, Integrated Product and Process Development (IPPD), and the use of Integrated Product Teams (IPTs).

The initiation of the Joint Multi-Role (JMR) Program in the 2013-2015 timeframe led by Dr. Bill Lewis, Director of the Army Aviation Development Directorate (AAD), my former colleague and PhD student, was the first new Army development since the late 1970s and early 1980s. Under his leadership, JMR resulted in the successful demonstration of two advanced rotorcraft configurations, the Bell V-280 tiltrotor, and the Lockheed-Sikorsky Boeing Defiant X. This, in turn, has led to successful prototypes flown and demonstrated over the past few years. Recently, the Bell V-280 was selected for Future Long Range Assault Aircraft (FLRAA) Full-Scale Development (FSD) in the 2020s and for production beginning in 2030s.

This will lead to a 5th Generation of Rotorcraft, as illustrated in Figure 1. If successful, the FLRAA will be the first tiltrotor aircraft purchased by the Army. As shown on the right side of Figure 3, the Army under the Army Aviation Research and Development Command (AVRADCOM), with the National Aeronautics and Space Administration (NASA) support, led to tiltrotor aircraft research and development of the XV-15 aircraft since the 1970s. The Army-NASA led Joint Technical Assessment (JTA) resulted in the Navy-led V-22 Program. (Ref.4)

However, the Bell LHX tiltrotor aircraft concept evaluated in the early 1980s was deemed unacceptable by the Army due to size, weight, and cost for the ~ 4000+ light attack and utility aircraft it was to replace. Several Army LHX Cost and Operational Effectiveness Analyses (COEAs) determined the tilt-rotor was not a viable concept for the LHX, e.g., too heavy, too large, and too expensive. Interestingly, the Sikorsky preferred concept for LHX was a coaxial rotor concept illustrated on the left side of Figure 3. However, the Army dictated that LHX be a conventional helicopter, with the Sikorsky Boeing being selected, as shown in Figure 2.

The second set of Fifth Generation Rotorcraft, in addition to the FLRAA, is the Future Attack and Reconnaissance Aircraft (FARA) that is being pursued to include a smaller version of the Lockheed-Sikorsky Boeing FARA Defiant Aircraft, called the Raider X. The Bell Helicopter FARA Concept is the Bell 360 Invictus which is producing an advanced scout helicopter to fulfill requirements set out by the Army FARA program. The Bell 360 Invictus is an advanced helicopter with some wing lift designed to provide improved lethality, survivability, and some extended reach for Army Aviation. The winner of this FARA Competition will possibly be declared in 2023/ 2024.

However, the Army canceled the FARA Program in early 2024 due to concerns about weight, performance, cost, and evolving threats. The Epilogue to this book illustrates how these concerns could have and should have been addressed.

INTRODUCTION

BACKGROUND ON ARMY AVIATION SYSTEMS DEVELOPMENT AND READINESS ORGANIZATIONS

Army Aviation Development Following the Vietnam Way

During the 1970s and into the early 1980s, the US Army undertook a series of studies and ensuing actions that essentially reinvented the service in light of past experience, current developments, and future global threats. The end of the service's involvement in Southeast Asia, in conjunction with the introduction of the all-volunteer Armed Forces, posed a variety of questions and challenges for the Army that were not fully resolved for another twenty years. An upsurge in anti-militarism and isolationism among some segments of the US populace impacted Army recruiting efforts, as well as leading to constrained budgets. The service's concentration on fighting in Vietnam also hampered its first efforts to deal effectively with changing threats throughout the world, particularly in Europe, where the buildup of Warsaw Pact forces and more extensive deployment of technologically superior weapons increasingly alarmed the nation's North Atlantic Treaty Organization (NATO) allies.

Once the service committed itself to the reality of the all-volunteer Army, it was able to make considerable headway by 1974 in recruiting both male and female "quality enlistees," who, in turn, helped to revitalize the standards and pride of the nation's oldest military institution. In addition, Army leaders began to identify, define, and refine new doctrines to guide how it would prepare for, as well as fight, the differing levels of conflict threatened by possible foes closer to home or overseas. Also of major concern was the failure to adequately modernize in light of the extensive technological developments that had been altering the battlefield while the Army was engaged in Vietnam.

Confronted with hostile forces able to commit larger numbers of troops equipped with superior firepower, Army leaders looked for ways to counter these unfavorable odds successfully. This it did by reorganizing its major commands, revising Army Field Manual (FM) 100-5, Operations, and maximizing its available arsenal with technologically superior weapon systems, such as the so-called "big five" development programs for the M-1 Abrams tank, the Bradley Fighting Vehicle, the AH-64A Apache attack helicopter, the UH-60A Blackhawk utility helicopter, and the Patriot air defense missile. The latter three items were managed by the Army Material Command (AMC) major subordinate commands (MSCs) in St. Louis and at Redstone Arsenal, which would eventually be merged into today's US Army Aviation and Missile Command (AMCOM).

New concepts for how the Army was to fight and quickly win any future war were also introduced and reflected evolving technology. The 1976 revision of FM 100-5 introduced the theory of Active Defense and emphasized the pivotal role of heavy armor and concentrated artillery fire in a primarily ground war. Various aspects of this concept proved to be unacceptable or difficult to apply outside Europe, which led to the 1982 edition of the Operations Field Manual (FM)and the introduction of the Air Land Battle doctrine. Further revisions in 1986 and subsequent refinements of the concept called for simultaneously aggressive close, deep, and rear operations, with greater reliance on tactical air power. Air Land Battle was the doctrine that eventually helped guide the Army's actions during Operation Desert Storm in 1991.

One of the most significant changes involved the reorganization of the Army in 1973, which led to the division of the multifunctional US Army Continental Army Command (CONARC) into the US Army Forces Command (FORSCOM) and the US Army Training and Doctrine Command (TRADOC). FORSCOM exercised jurisdiction over all the service's operational units in the continental United States (CONUS) and concentrated on readiness. TRADOC combined authority over most of the Army's schools with combat development functions formerly assigned to the Combat Developments Command.

In some respects, this division was similar to the separation of the US Army Material Command (AMC) into readiness and research and development organizations. Readiness-oriented commands were responsible for current operations and sustaining the force, while the future-oriented doctrine and development commands provided the training and technology essential to prepare for potential conflicts.

Army Aviation Systems Command History and Evolution in the Army Material Command (AMC)

(Ref.5. Dr. Kaylene Hughes, AMC Historian, 2018)

The Army Aviation Systems Command (AVSCOM) evolved from the U.S. Army Aviation Material Command (AVCOM) which was re-designated on 23 September 1968 the U.S. Army Aviation Systems Command (AVSCOM). It was located in the Mart Building in downtown St. Louis, MO. also shown below. The AVSCOM Headquarters was located in the Mart Building also shown.

Transition from AVCOM to AVSCOM

It was, according to a former Army Aviation historian, "the evolutionary offspring of four years of functional alteration and geographic reorganizations involving activities scattered coast to coast." The newly realigned and renamed command consisted of 300 military personnel and 4,000 DA civilians (DACs) employed at AVSCOM headquarters in St. Louis and across the nation in research and development laboratories, plant cognizance activities and test flight activities assigned to AVSCOM's jurisdiction. When AVRADCOM was formed in 1977 it was located at Goodfellow Blvd with Troop the Army Troop Support and Aviation Materiel Readiness Command (TSARCCOM) near the St. Louis Airport

AMC Reorganizations, Realignments, and Acquisition Restructuring, 1966-1973

Like the rest of the Army, AMC also experienced several major organizational changes during this period. The command's original structure when it stood up in Aug. 1962 included five major commodity commands, one each for electronics, missiles, munitions, mobility, and weapons. There also were two functional MSCs covering supply and maintenance as well as test and evaluation. In addition, AMC initially established thirty-six Project Management Offices (PMOs) responsible for the development of major weapons and equipment.

The first modifications to this organizational structure began not long afterward. In 1966 and 1969, for example, AMC implemented two major realignments of its headquarters (HQ), the main objectives of which were to provide better control over assigned missions and functions, streamline the HQ, and reduce the number of commands, agencies, and individuals reporting to the AMC command group, especially to the commanding general. These significant changes were followed in the period between 1970 and 1973 by the command's loss of almost 35,000 authorized civilian manpower spaces; additional cuts also decreased AMC's authorized military spaces.

The command's assigned depots experienced even greater cutbacks because of their primary focus on supporting the war in Vietnam and their more extensive reliance on the use of temporary employees. Consequently, Headquarters Department of the Army's (HQDA's) 1969 decision to accelerate the termination of temporary and part-time positions, in conjunction with the subsequent manpower cuts from 1970 to 1973, seriously eroded the morale, productivity, and efficiency of the entire depot workforce. Also, in this period, AMC had to deal with the announced closure of thirty-six installations under its jurisdiction.

Further reorganizations undertaken in 1973 transformed several of AMC's original commodity commands. Designed to improve materiel readiness and enhance management, AMC restructured, merged, and consolidated many of its subordinate commands and other elements in accordance with guidance provided by the Total Optimum Army Materiel Command initiatives, DA's Baseline Development and Utilization Planning Project, and the Army Reorganization of 1973. Despite all these changes, however, AMC faced an even more extensive transformation based on the findings of the Army Materiel Acquisition Review Committee (AMARC), established by the Secretary of the Army in Dec. 1973.

AMARC and the Creation of Development and Readiness Command (DARCOM)

The Secretary of the Army assigned AMARC the task of analyzing the Army's materiel acquisition process and recommending improvements, with special emphasis on organization and procedures, particularly those of AMC.

One significant committee finding was the fact that the AMC mission of maintaining the readiness of fielded weapon systems had taken precedence over the command's responsibility for materiel acquisition. AMC's intense focus on readiness, AMARC members concluded, was a major reason for the many problems confronting post-Vietnam Army acquisition and had put the service behind the power curve in staying abreast of technological advances in weapon systems development.

Gen. John R. Deane, Jr., then the AMC Commander, acknowledged that by not concentrating sufficient management attention on the acquisition process, Army procurement, at times, did not incorporate the latest technology. Further compounding the problem was the Army's failure to find effective contractor incentives designed to encourage companies to push state-of-the-art technologies, to develop more effective, cost-effective, and better manufacturing techniques early enough in system development to drive down costs once it reached production.

In response to these concerns, one of the key recommendations of the AMARC study involved the creation of separate centers within the AMC focused on either development or logistics. The development centers were to be formed from the existing laboratories, research and development elements, and other separate acquisition activities. The new organizations would be independent commands reporting directly to the AMC Commander and would have no logistics support missions. Once established, the new development centers would also initiate programs and policies to strengthen the Army's working relationship with industry. The logistics centers would provide follow-on procurement and support to fielded systems. In addition, these commands would cooperate with the development centers throughout the initial acquisition process to ensure that new systems could be maintained in a high state of readiness once developed and fielded.

The AMARC study resulted in an extensive Department of the Army (DA) assessment of the AMC organizational and management structure, which led to a major reorganization of AMC headquarters and its MSCs. On Jan. 23, 1976, AMC became the US Army Materiel Development and Readiness Command (DARCOM). Designed to function as a smaller, corporate-style headquarters, with 30 percent fewer personnel than the former AMC, the objective of the command's restructuring was maximum decentralization via the transfer of operational functions to the field. DARCOM Headquarters retained responsibility for developing broad programs and policies, establishing priorities, allocating resources, and evaluating performance.

A major feature of the new DARCOM HQ was the creation of two three-star deputy commanding general (DCG) positions. The DCG for Materiel Development supervised the R&D commands and those organizations' assigned project managers, corporate laboratories, the Test and Evaluation Command, all-Army research, and standardization offices, as well as the Foreign Science and Technology Center. The DCG for Materiel Readiness was in charge of the materiel readiness commands, their assigned project managers, the International Logistics Command, the Depot System Command, arsenals, logistics management activities, and logistic assistance offices.

Subsequently, DARCOM HQ added a third DCG for Resources and Management to oversee management and control of total command resources by formulating and maintaining systems and procedures to develop and execute fully balanced and integrated appropriation budgets.

A vital part of this responsibility was the provision of a consistent resources management framework for development, testing, procurement, production, and integrated logistic support planning. This became the basis for the Planning Program Budget Execution System (PPBES). The PPBES evolved from the Planning, Programming, and Budgeting System (PPBS), which was introduced into the Department of Defense (DoD) in the early 1960s by Robert McNamara during his tenure as Secretary of Defense (SECDEF). PPBS was a cyclic process consisting of three distinct but interrelated phases: planning, programming, and budgeting. PPBS established the framework and provided the mechanisms for decision-making for the future, and provided the opportunity to annually re-examine prior decisions in light of the existing environment at that particular time (e.g., evolving threat, changing economic conditions, etc.

Equal emphasis on both readiness and development was to be achieved by separating these two functions within the DARCOM MSCs, including the US Army Aviation Systems Command (AVSCOM), US Army Troop Support Command (TROSCOM), and US Army Missile Command (MICOM).

Consequently, DARCOM restructured its former commodity commands into eleven MSCs, six of which were dedicated to development. In January 1979, the eleven MSCs expanded to thirteen. This was in keeping with DARCOM's initial plans for reorganizing into a completed organizational structure with eight R&D commands, five readiness commands, a Test and Evaluation Command, a Depot System Command, and an International Logistics Command.

The Creation of AVRADCOM

Troop Support and Aviation Materiel Readiness Command (TSARCOM)

On July 1, 1977, in compliance with AMARC recommendations, the R&D components of AVSCOM evolved into the US Army Aviation Research and Development Command (AVRADOM). TROSCOM became the US Army Troop Support and Aviation Materiel Readiness Command (TSARCOM) as a result of the merging of its own as well as the logistics support and materiel readiness functions. TSARCOM and AVRADCOM Insignia are shown in Figure 4.

Figure 4. TSARCOM and AVRADCOM with the Same Insignia

AVRADCOM's basic mission was to manage and control the Army's aviation materiel development and the initial acquisition portion of the aviation materiel life cycle.

TSARCOM's primary responsibility was the management of logistics for assigned troop support and aviation materiel, which included additional acquisition and materiel readiness support for the life cycle of various items once they were developed and field tested. TSARCOM also served as the "host command" for AVRADCOM in keeping with DARCOM policy that the newly created R&D commands not be burdened with installation support type functions. Consequently, TSARCOM provided common service support to its "tenants" in areas such as finance and accounting, morale and welfare activities, and installation safety.

In 1979, both the TSARCOM and AVRADCOM headquarters underwent further realignments and transfers of workload in several of their subordinate elements to achieve "more efficient and productive organizational and position structures." A primary action of the realignment involved the reallocation of seventy-one engineering and technical spaces from TSARCOM to AVRADCOM and the move of twenty-two equipment specialists and supply cataloger spaces from AVRADCOM to TSARCOM. These refinements in the alignment of the commands' engineering activities, operational concepts, organization, and manpower allocations occurred because not enough engineering spaces had been given to AVRADCOM in the original transfer of functions. Also, TSARCOM failed to receive enough space to handle the functions of configuration management, data planning and acquisition, type classification, or new equipment training.

Other reorganization manpower issues also confronted both commands in 1979. These included the reassignment of employees made excess by the AMARC reorganization, the abolition of spaces due to a high-grade reduction action, the placement of employees from the Iranian Project Management Office (PMO), which had been curtailed and substantially reduced by the overthrow of the Shah, and an unscheduled manpower adjustment imposed by changes at DARCOM and the Department of the Army (DA).

AMARC Revisited

AMC's two MSCs in St. Louis were not the only organizations encountering difficulties in operating under the AMARC-mandated realignment. Previously in 1978, the AMC Commander, Gen. John Guthrie, initiated a baseline study known as "AMARC Revisited," designed to rejoin its severed commodity commands.

The first of AMC's separated MSCs to undergo such a realignment were the US Army Missile Readiness Command (MIRCOM) and the US Army Missile Research and Development Command (MIRADCOM) located at Redstone Arsenal, Alabama. On Apr. 25, 1979, DA approved the merger of MIRCOM and MIRADCOM personnel, missions, and assets into a single organization to reduce duplication, improve efficiency, eliminate interface, and transition problems, and optimize the use of dwindling resources.

To minimize personnel turbulence, the merger was implemented in two phases. MICOM was reactivated at Redstone Arsenal on Jul. 1, 1979. Concurrently, the organizational elements of MIRCOM and MIRADCOM were transferred in place to a single command. During the 90-day stable period that followed, MICOM consisted of a dual command structure, with former MIRCOM elements grouped under the DCG for Readiness and the former MIRADCOM elements under the DCG for R&D.

Phase II of the MIRCOM/MIRADCOM merger into MICOM began on Oct. 1, 1979. The dual command structure was realigned in a two-step process. The first step entailed the consolidation of functionally similar or duplicate missions into the merged command structure, which became operational on Oct. 7.

This process continued in the fiscal year (FY) 1980-81 period by other realignments to streamline and optimize the consolidated command in accordance with DARCOM-approved organizational concepts. Although the re-establishment of MICOM fueled speculation that DARCOM planned to reunite all of its readiness and R&D commands immediately, such was not the case. Instead, AMC undertook a series of evaluations at its other MSCs to determine how well the DARCOM reorganization, in line with the AMARC study, was working, and whether it was achieving the expected objectives. Nonetheless, from 1979 to 1984, AMARC Revisited "resulted in the reconciliation of the commodity commands and the elimination of the many problems created by AMARC."

In St. Louis, starting in March 1983, command representatives began devising a detailed step-by-step plan to consolidate control of the existing aviation and troop system-related organizations into two unified, life cycle-oriented commands called AVSCOM and TROSCOM. In a prepared statement released on Mar. 11, DARCOM announced the return to the prior MSC structures and explained that the realignment would reduce the complexity of a single command (TSARCOM) managing two such dissimilar commodities as aviation and troop support. During a re-designation ceremony held in St. Louis on Oct. 3, 1983, DARCOM's newest major subordinate commands (MSCs) were provisionally implemented by Gen. Donald R. Keith, the DARCOM Commanding General. According to Keith, "Making our Army a better one is what these two commands are all about. We have retained the best of a strengthened research and development structure and combined it with an improved readiness organization. We will be able to meet the challenges of modernization and readiness and fulfill the promise of attaining and sustaining quality equipment and support for an excellent Army."

In Aug.1984, DARCOM resumed its original name of Army Materiel Command. The command expected the change to remove a perceived boundary between development and logistics support implied in the

DARCOM name. Additionally, the re-designation features brevity and clarity and will be better understood by allies and the general public.

AVRADCOM and TSARCOM Accomplishments, 1977-1983

Despite the organizational and manpower challenges of the six years during which AVSCOM and TROSCOM were divided into separate R&D and readiness commands, both AVRADCOM and TSARCOM made notable headway in several of the programs assigned to their respective commands. Work on most of these programs had actually begun prior to the AMARC reorganization.

During this period, AVRADCOM supported three major aviation programs, including the Black Hawk utility helicopter, which in 1978 began maturity flight testing, continuing through a second production year contract with Sikorsky Aircraft, acceptance of the first production T-700 engine, the establishment of the first pilot ground school, and the start of Army mechanic training.

The command also managed the AH-64A Apache attack helicopter program initiated in 1972. Full-scale production began in 1982, and the Army officially accepted the first of 515 Apache helicopters in Feb. 1984.

The Aquila Remotely Piloted Vehicle (RPV), the third major program assigned to AVRADCOM, began development in 1975, and made its initial flight on Jul. 16, 1982. Being developed by the Army as an unmanned aerial platform designed to carry a complement of mission payloads, the successful flight was the first of a series to prove the RPV system's effectiveness as a battlefield support system. A predecessor of the unmanned aerial vehicles (UAVs) now being used in combat, the operational control of the Aquila RPV transferred to MICOM in Aug. 1985 because of its more extensive laboratory resources. Later, MICOM also sought and achieved consolidation with AVSCOM at MICOM.

Other AVRADCOM accomplishments included delivery of the initial production model CH-47D Chinook helicopter in May 1982; the first of 436 CH-47s to be modernized to the "D" configuration. Begun in 1976, the effort was designed to extend the service life, increase the operational capabilities, and lower the operating costs of the Army's medium-lift workhorse cargo helicopter.

The Opportunities Provided From the Breakout of AVRADCOM

In 1977, in principle, the Research and Development (R&D) elements of the Aviation Systems Command (AVSCOM) were broken out and became the Aviation Research and Development Command (AVRADCOM). This provided a significant opportunity to advance Army Aviation Research and Development leading to the next generations of rotorcraft.

On July 1, 1978, I transferred from Army Active Duty to the Army Reserves and converted to Civil Service. I became the Chief of the Aeromechanics Branch and was the technical lead for the Army Helicopter Improvement Program (AHIP), which led to the OH-58D Kiowa Development. I was then made the Technical Chief for the AHIP SSEB. In 1979, I became the Chief of the Structures & Aeromechanics Division in the Development and Qualification Directorate, AVRADCOM. This also led to technical development and airworthiness qualification for the CH-47D Chinook Modernization Program. My participation and evaluation in these two programs are addressed later in the chapters in Book 2.

In 1981 I was selected for Senior Executive Service (SES) Level 3 and became the Director of Advanced Systems (DAS) and Associate Technical Director for the AVRADCOM Science and Technology (S&T) Program. At age 37, I was the youngest SES Level 3 in DARCOM. In this position, I led the Concept Formulation for the Light Helicopter Experimental (LHX) Program in 1981-83. In addition, in 1983, I served as the Interim Chief

Scientist for the Army Combined Arms Center (CAC) on a six-month temporary assignment at Fort Leavenworth, KS. These activities will also be addressed in chapters in Book 2.

In addition, from 1974 to 1978 I received two advanced degrees: first, a MA Degree in Business Administration from Webster College (St. Louis) in 1975, and then a DSc in Mechanical Engineering from Washington U. (St. Louis) in 1978.

Also, toward the end of Book 2, I will describe my two Army Reserve positions beginning in 1978 in the Army Reserves: first, as a Mobilization Designee (Mob Des), Assistant Professor, in the Department of Mechanics, USMA; and second, as a Military Academy Liaison Officer (MALO) with the USMA Admissions Directorate for the St. Louis Area.

I went on to serve in these positions until I retired from active reserves in 1996 as a Colonel. I officially retired from the inactive US Army Reserves as a Colonel in 2002. In 1983 AVRADCOM was being brought back together and combined with Aviation Readiness elements to re-form AVSCOM. I was being supported to become the Technical Director, by retiring MG Story Stevens and Mr. Richard B. Lewis II, for this new AVSCOM Technical Director position. However, they were both leaving, and Mr. Charlie Crawford, my mentor for seven years, really wanted the Technical Director position. Also, the new command's two primary mission concerns were readiness (immediate) and research and development (eventual). This was a reversal of priorities from AVRADCOM, and the new AVSCOM would be led by a readiness officer, MG Orlando Gonzales, who had little, if any, research, and development experience. I could see that the Technical Director's role would primarily be to solve field problems.

While this was a significant role for Army Aviation, it was not my desire, as I had established my name and reputation in research and development. I had been approached by the Chair, and the Associate Chair of the School of Aerospace Engineering, Dr. Arnold Ducoffe and Dr. Robin Gray, respectively, in 1982-83, to become the Rotorcraft Design Professor and the Associate Director for the Georgia Tech Army Rotorcraft Center of

Excellence (RCOE), which was initiated and sponsored for long term research in 1982.

I had also been approached by Mr. Bob Lynn, VP Engineering, Bell Helicopter Textron (BHT), to become their Director of Technology, which could lead to VP of Engineering upon Bob Lynn's retirement. However, I thought that being named a full professor in rotorcraft design at Georgia Tech was a wonderful opportunity and the preferred choice for me and my family, who wanted to spend more time together.

In January 1984, I left civil service at AVRADCOM/AVSCOM and became the Rotorcraft Design Professor in the School of Aerospace Engineering at Georgia Tech and the Associate Technical Director for the Army sponsored Center of Excellence in Rotary Wing Aircraft Technology (CERWAT), also called Rotorcraft Center of Excellence (RCOE), and later the Vertical Lift Research Center of Excellence (VLRCOE).

Finally, in Book 2, I try and set the record straight, and answer the question of why it has been so hard for the Army to develop and field new Aviation Systems. It will be seen that it wasn't a high priority with readiness (immediate) and research and development (eventual) for the new AVSCOM. It is still a major issue today where the Future Vertical Lift (FVL) Program didn't start for over thirty years and, hopefully, will complete the development of the FLRAA, FARA, and Future Tactical Unmanned Aerial Systems (FTUAS) by the 2030s. Fortunately, Dr. Bill Lewis, a former colleague, and my PhD student at Georgia Tech, was there to initiate the FVL Program. Hopefully, previous lessons learned in this book and elsewhere can be applied.

AVRADCOM Successes Can't Be Ignored

However, I would be remiss if I didn't document the contributions that AVRADCOM, and particularly the Directorate of Advanced Systems (DAS) made for Army Aviation. A Summary of AVRADCOM was presented by MG Story C. Stevens upon retirement in an Army Transportation Oral History Program interview on 7 June 1985 by CPT David Michael Dobson at MG

Stevens's home on Hilton Head Island, South Carolina. These are unique lessons learned that were not followed after AVRADCOM was dissolved back into AVSCOM in 1984. It is a major reason why no new Army Aviation Systems have been successfully developed over the past forty years. MG Story Stevens, the first and only CG of AVRADCOM presented a candid summary of his view which follows:

"I was fortunate enough to get the command. The first thing we decided was to set up a new directorate. It was called the Directorate of Advanced Systems (DAS), and this was the organization that would essentially pull the programs together over the year, establish the priorities which would interface with the Training and Development Operations Command (TRADOC), and work with the combat developers down there. In fact, we got the whole community pulled together pretty well. I think by about the second year, we were getting our priorities worked up together with TRADOC, and even with the Department of the Army (DA). So, when we went in from the aviation side of the house, we knew exactly what the priority was in our programs; and this gave us an advantage over the other guys, the tankers and everybody else, because they had not done this yet. It worked well for us in regard to getting the funding for our programs. The other thing we did was to sit down and draw up an R&D plan. We spent some time the first year really looking to see what we needed to do. At that time, money was getting short. For a long time, before this, you could have the little pet projects; the laboratories could do their things and pursue them. And perhaps if a real breakthrough came, that would be great, but if it didn't, that was okay too. They were doing their own thing. There were some things they'd been doing for years in the laboratories that they just kept on getting funding for, and they weren't getting anywhere with them. *So, we decided to look at this whole thing and just do the things that were important to the program. The program we were looking at was the development of a new family of air vehicles.*"

Another key mentor who helped me become the Director of Advanced Systems was Mr. Richard B. Lewis II. He became the Technical Director of AVRADCOM in 1977. He and MG Stevens recommended me for SES as

Director of Advanced Systems and Associate Director for S&T. From 1969 to 1985, Mr. Lewis served in a number of civilian technical leadership positions for the US Army and ultimately managed thirty-four Army laboratories, including personnel, facilities, and scientific and technical research and development policy. He joined ITT Corporation in 1985 as its Technical Director, became president of the electro-optical products division, and ultimately served as group executive and corporate vice president. He also then served as the President of Rolls Royce Allison and later as the Director of the Lean Aircraft Initiative (LAI) at MIT, applying Integrated Product and Process Development (IPPD), which I helped pioneer, and used at Georgia Tech to build a graduate program in Aerospace Systems Design. This will be the theme and title of Book 3.

The accomplishments of DAS set the stage for a successful foundation for Army Aviation Research and Development (R&D). Initially, the Advanced Systems and Technology Integration Office (ASTIO) in 1978-79, was a precursor to DAS with Mr. Dean Borgman, as the Chief, ASTIO. It was expanded into DAS with Dean Borgman as its Director in 1979-80, with the responsibility for developing the annual AVRADCOM Research, Development, Test, and Engineering (RDT&E) Plan.

Although an Aviation RDT&E plan had been published previously, it concentrated on technology forecasting whereas the new thrust was to project the available and developing technology into systems. The first results of ASTIO's R&D planning efforts were published as the Army Aviation RDT&E Plan in October 1978. The key elements of the DAS RDTE Planning Process are shown in Figure 5. As DAS was formalized, these elements were transitioned into divisions in DAS, e.g., Concept Design & Analysis (CDA), Advanced Technology Requirements, and User and Threats Requirements, and Technology Base Funding. These divisions were run by very competent engineers and managers, e.g., David Weller, Roger Smith, George Singley III, Lew Feaster, Jim Means, LTC Harry Fraser, etc. The pulling together of the RDT&E Plan was led by George Singley III.

It became the model for other major support command plans in DARCOM/AMC. George went on to hold senior executive positions in AVSCOM, Program Executive Officers (PEO) LHX, and the Department of Defense (DoD), reaching the rank of Deputy Assistant Secretary of the Army (Research and Technology). Dave Weller went on to be the DAS Director and helped transition it to Huntsville, AL, and the Aviation and Missile Command (AMCOM).

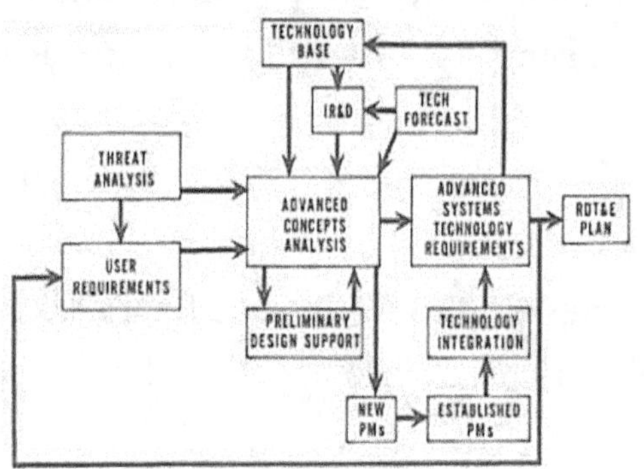

Figure 5. DAS RDT&E Planning Process

The AVRADCOM Six Years of Excellence are identified in the publication cover in Figure 6, flanked by the Army Aviation Advanced Development funding trends before and after AVRADCOM and DAS. You can see that by 1985 the R&D funding had turned the corner.

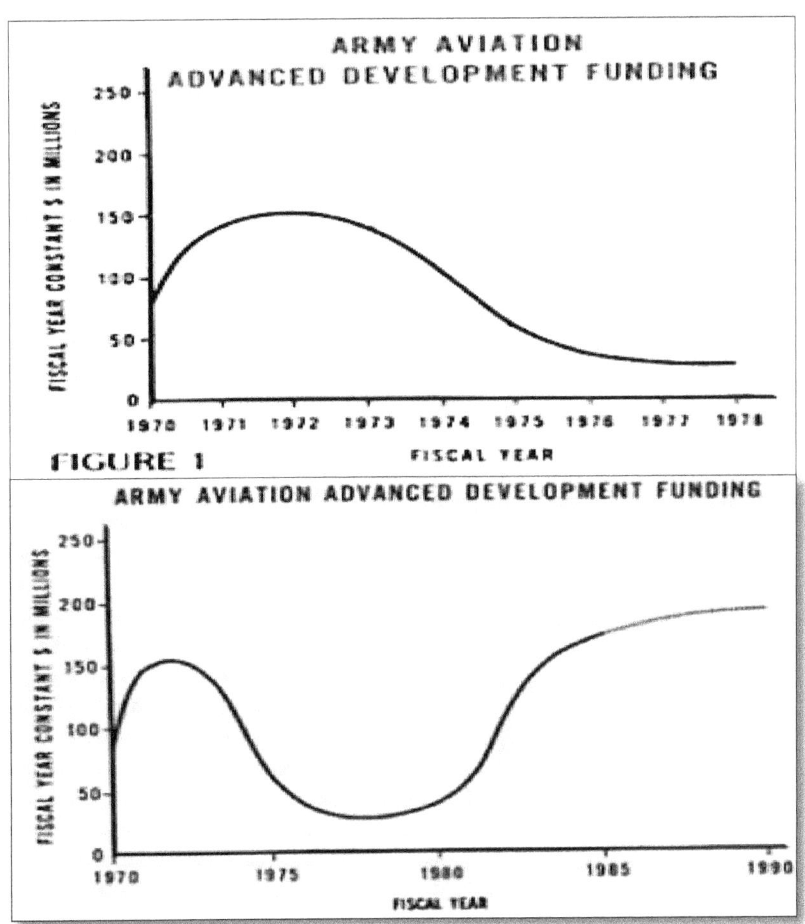

Figure 6. AVRADCOM Six Years of Excellence w/Funding Trends before and after

I led the Light Helicopter Experimental (LHX) Concept Formulation effort, the first major AVRADCOM Initiative, along with Mr. George Singley III, and with the other division chiefs. We accomplished this, along with the Army Aviation Applied Technology Directorate (AATD) support, in achieving the funding for necessary Science & Technology (S&T) programs to mature the critical technologies, as seen in Figure 7, with a peak in 6.3 funding in FY1986.

One S&T integration program example was the Advanced Rotorcraft Technology Integration (ARTI) Program. The objective of the ARTI Program was to determine whether the LHX could be a single-pilot aircraft. While proven technically feasible, the Army user wanted to retain two pilots. One reason was the workload for the Scout Combat Assault Team (SCAT) pilot to look outside the cockpit while handling the team functions in the cockpit. Another potential reason was possibly reducing the number of pilots with one versus two pilots and taking a hit in force structure.

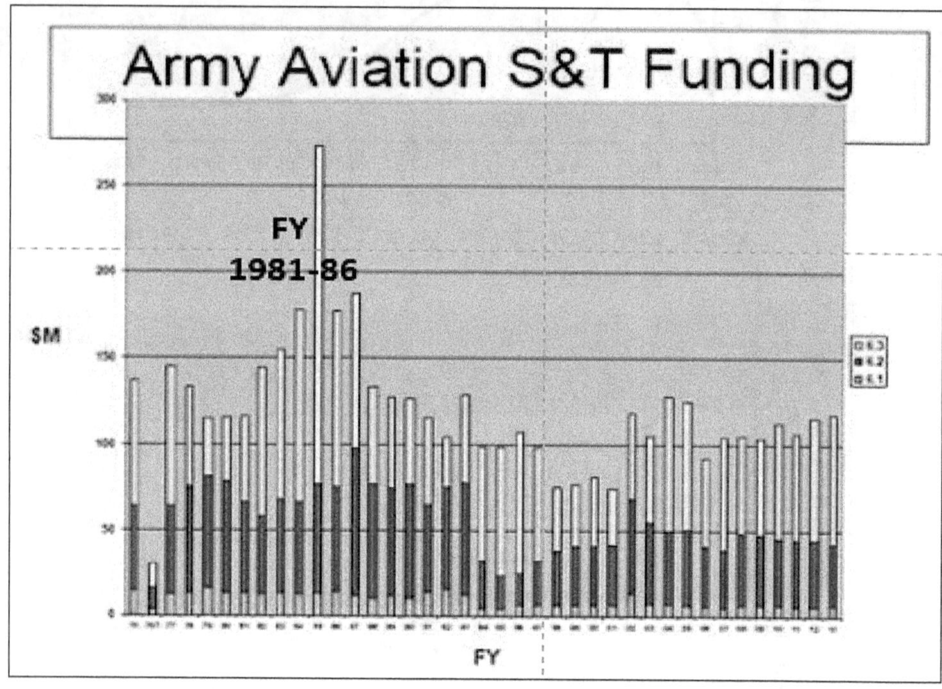

Figure 7. Large AVRADCOM 6.3 Funding Increase, FY 81-FY86

EVOLUTION TO AMCOM: THE RE-ESTABLISHMENT OF AVSCOM, 1984-90

(Ref. 5. From Dr. Kaylene Hughes, US Army Historian)

Effective March 1, 1984, the US Army Materiel Development and Readiness Command (DARCOM) reestablished the US Army Aviation Systems Command (AVSCOM) as a life cycle materiel command. All missions and activities of the US Army Aviation Research and Development Command (AVRADCOM) and the aviation-related missions and activities of the US Army Troop Support and Aviation Materiel Readiness Command (TSARCOM) were transferred to the newly restored organization.

AVSCOM was also assigned the host command mission at the Goodfellow Federal Center joint headquarters complex, which included the Administrative and Installation Support Activity (AISA) and the St. Louis Area Support Center (SLASC). Effective March 30, 1988, the Department of the Army (DA) re-designated the SLASC as the Charles Melvin Price Support Center (CMPSC). The new insignia for AVSCOM is shown in Figure 8.

Figure 8. New AVSCOM Insignia

Literally a coast-to-coast command headquartered in St. Louis, AVSCOM was responsible for the life cycle management of all-Army aircraft and related systems, which included technical and professional guidance for the worldwide support of aviation materiel. It provided complete logistic, technical, and administrative support not only to the Army's fielded fleet of about 9,000 (8,500 by 1990) aircraft but also to another 2,500 aircraft operated by non-Army customers.

On October 1, 1985, the command received its new Distinctive Unit Insignia (DUI) from the Institute of Heraldry. Designed by Bonnie Morris, a visual information specialist in the AVSCOM Public Affairs Office, the device featured a silver Pegasus, the flying horse of Greek mythology, with yellow wings, above a silver bolt of lightning and surrounded by the four silver stars of the Pegasus constellation. The brightest is the star Pegasus shares with Andromeda, represented by a silver disk. The insignia's background is brick red; the color associated with the Transportation Corps. The yellow of Pegasus' wings is one of the colors used to represent Ordnance units. The winged horse is symbolic flight or aviation, the lightning connotes speed and relates to AVSCOM's rapid deployability, as well as its air weapons systems.

Two other physical representatives of the restored command's aviation legacy involved an equipment display at the Smithsonian in Washington, D.C., and the mounting of a UH-1M Iroquois ("Huey") gunship in front of the AVSCOM Headquarters building in St. Louis. In June 1986, a T-700, the first engine specifically developed for use on a rotary wing aircraft, and the 10-tool kit used for its maintenance became part of the Vertical Flight gallery of the National Air and Space Museum. Three years later, AVSCOM held a dedication ceremony on April 14, 1989, for a Huey gunship flown in Vietnam, which was placed on display as a symbol of all aircraft, past and present, fielded by the command. It (was) dedicated to all the men and women of AVSCOM, past and present, who gave their lives and their energies to supporting soldiers and the aircraft they fly.

Assigned Missions

The command's two primary mission concerns were *readiness (immediate) and research and development (eventual). Obviously,* the term eventual meant no new aircraft development for thirty years.

Research and Development (R&D).

The latter aspect of AVSCOM's assigned responsibilities was managed by AVRADCOM from July 1977 to March 1983. After the command was reestablished, its R&D mission and functions continued to rest on the activities of five organizations clustered under the direction of the AVSCOM Research, Development, and Engineering Center (RDEC). Established on June 16, 1985, the AVSCOM RDEC united the command's widely- dispersed life cycle technology, test, and engineering focal points to strengthen ties between the combat developer, the materiel supplier, and the user. Included under the RDEC organizational umbrella were the:

Directorate for Advanced Systems, which served as the chief adviser to the AVSCOM Commander on the research, development, test, and evaluation of Army aviation systems and subsystems through its assessment and evaluation of aircraft and aircraft threat technologies.

AVSCOM's Four Outlying Laboratories

Three of which adjoined NASA complexes, and which were responsible for most of the physical aviation R&D work.

Avionics Research and Development Activity (AVRADA)

Collocated at Fort Monmouth, New Jersey, where it cooperated with the Communications-Electronics Command (CECOM) and the Electronics R&D Command (ERADCOM) on joint programs in communication and navigation research and technology. AVRADA also developed ground systems to provide effective vehicular navigation and air traffic control, and in 1989, it became responsible for the proper integration of all aircraft onboard electronics into the automated cockpit.

Directorate for Engineering

Which provided life cycle engineering support for developmental and operational aircraft systems and related material, served as the focal point for aviation life support equipment (ALSE) and issued airworthiness releases, among other things. By 1989, it also oversaw the value engineering and the Department of Defense (DoD) standardization programs at AVSCOM.

Aviation Engineering Flight Activity (AEFA)

At Edwards Air Force Base, California, which performed all-Army aviation flight tests under monitoring by the AVSCOM Engineering Directorate. In 1988, it became a subordinate activity of the AVSCOM RDEC Management Office, responsible for conducting airworthiness tests of the Army's new and modified aircraft as well as fielded systems.

Readiness

The most significant mission change for the reestablished Aviation Systems Command (AVSCOM)was its resumption of responsibility for the readiness of the Army's aviation systems. Key elements of the readiness mission included the headquarters' service as both a National Inventory Control Point (NICP) and a National Maintenance Point (NMP) for Army aviation, involving functions such as:

Material Management

Initially supported almost 9,000 aircraft and 49,000 line items of supply, along with a depot inventory valued at $2.5 million in 1985. By 1990, the number of supported aircraft had declined to 8,500, but the numbers for stockage and depot inventory value had soared to 122,000 and $4.2 billion, respectively.

Procurement and Production Actions

Involving 29,000 contracts totaling over $3.8 billion, as well as managing more than $20 billion in contracts, monitoring the command's assigned Army plant representative offices (ARPOs), and managing all foreign military sales (FMS) contracts. Although the number of individual contracts declined to about 19,000 by 1990, their value increased significantly to $27.6 billion.

Product assurance oversight of aviation product quality,

Which was guaranteed by establishing reliability, availability, and maintainability (RAM) characteristics in conceptual designs as well as assuming responsibility in 1988 for managing the command's sample data collection, warranty, and flight safety parts programs.

Maintenance Support

By issuing 1,500 to 1,526 technical manuals between 1984 and 1990, managing an annual $400 million overhaul program that escalated by 1989 into a $750 million operation and maintenance program (of which $650 million went to the overhaul or repair of aircraft and components), providing appropriate training for the introduction of new aircraft and associated support items starting in 1988, and furnishing repair parts, tools, facilities, and skilled engineering support to ensure all-Army aviation systems with initial provisions and essential maintenance and modernization. By 1990, AVSCOM was handling about 96,000 provisioning transactions annually.

International Logistics Oversight

About 170 FMS contracts that grew to 300 by 1989, with a program value that varied from more than $1 billion in 1985 to $3.7 billion in 1989 and falling to $2 billion in 1990. The command was also accountable for the dispatch of about 3,500 to 3,700 aircraft to forty-five other nations.

Readiness by Acting as the Command Center

For applying logistical concerns to materiel design supportability and sustainability through the use of logistic support analyses, integrated logistics support (ILS), and the worldwide logistics assistance and readiness analyses programs.

In keeping with DA's Commitment to Greater Cost Effectiveness Through Competition

AVSCOM established a Competition Advocacy and Spares Management Office in 1989 to direct its spare parts program and manage a command-wide effort to open more aviation materiel contracts to competition. One notable example of the command's commitment to savings was the opening of its Depot Engineering and Reliability Centered Maintenance (RCM) Support Office (DERSO) on Jun. 26, 1989. Housed in the AVSCOM Engineering Analysis Facility located at the Depot Systems Command's (DESCOM's) Corpus Christi Army Depot (CCAD) and modeled after a Vietnam-era disposal program, the new office screened aircraft parts slated for disposal for any salvageable items. Within three months, DERSO had an inventory of about 7,000 parts worth over $2.5 million.

Another Noteworthy Indicator of AVSCOM's Resumed Commitment to Aviation Readiness

The extensive activities of the command's Logistics Assistance Representatives (LARs), whose operations "spanned the globe" in support of Army aircraft, the latter characterized in 1987 as the "Third Largest Air Force" in the world. To help maintain a mission-ready fleet, AVSCOM LARs concentrated on preventing problems and quickly identifying corrective actions. Relying on their ability to communicate with soldiers at all levels and making extensive use of their "toolbox" of technical manuals, bulletins, maintenance work orders, and information collected from classroom coursework and exercise training classes, LARs on the flight line worked whenever and wherever needed to meet the ideal of "Warriors' Winged Readiness."

In addition to sustaining Army aviators around the world, AVSCOM LARs also provided emergency logistics support for humanitarian efforts. For example, the command supplied not only necessary parts, but also logistics assistance for the Chinook and Black Hawk helicopters being used to transport firefighters and equipment deployed to fight devastating forest fires in Montana and Wyoming in 1988. LARs from Fort Lewis, Washington, and Fort Carson, Colorado, spent time on-site providing technical assistance for aircraft involved in the fire fighting and serving as the AVSCOM liaison to the crews flying and maintaining the aircraft. Also, in 1988, an AVSCOM LAR represented the command for the first time in a turnover of equipment to a military unit. The fielding was a positive action toward continual improvement of the relationship between LARs in the field and the units they support.

ORGANIZATIONAL CHANGES

On July 15, 1985, AVSCOM's first major reorganization occurred with the establishment of the new Directorate for Readiness, which incorporated elements of the former directorates for system management and force modernization/integrated logistics support (ILS), the former field services activity, as well as the plans division of the former directorate for plans and analysis.

Another important organizational addition that same year was the Office of the Project Manager (PM) for the Light Helicopter Family (LHX), which was tasked with developing about 5,000 LHX aircraft in scout-attack and utility variants that were expected to eventually replace the bulk of the Army's then-current 8,500 helicopters.

Establishment of Program Executive Officers (PEOs) as Life Cycle Managers

The most significant change in the command's organizational structure was the creation of the Program Executive Officers (PEOs) as part of the restructuring of the Army's acquisition process that began in 1987. Previously, from 1984 to 1986, AVSCOM had jurisdiction over several project and product managers (PMs), each of whom exercised full authority for planning, directing, and managing the command's assigned aviation systems or multiple subsystems. In addition to PMs for the OH-58D scout, Black Hawk, modernized Chinook, Cobra, and Apache helicopters as well as the LHX, there were PMs for aviation life support, aircraft survivability, and selected intelligence and surveillance equipment.

One of the major thrusts impacting the reorganization of Army acquisition was the Packard commission, which President Ronald Reagan directed to be formed in July 1985 to study defense management and organization in its entirety. Another important influence was the passage of the Goldwater-Nichols Department of Defense Reorganization Act of 1986.

One of the primary impacts of this act on the Army's organization was the consolidation of the service's research, development, and acquisition efforts. Under this reorganization, the Under Secretary of the Army was appointed as the Army Acquisition Executive (AAE) to supervise the service acquisition process, establish acquisition policy, approve the program baseline, and designate PEOs and the programs under each PEO's oversight. The PEO was to report directly to the AAE on program matters, ensure that program/project managers were properly resourced, and enforce the program baseline.

The project/product managers were to report directly to the PEO and the AAE on all program matters, execute the program within established guidelines and develop the acquisition strategy and the program baseline for PEO/AAE approval. The PEOs/PMs would accomplish their missions through the use of functional personnel and facilities supplied by the major subordinate commands (MSCs).

In essence, the PEOs became life cycle managers, e.g., from the cradle to the grave, without the knowledge or resources for understanding and managing research and development. This often became a barrier to new concept development and the initiation of new rotorcraft development.

The Program Executive Officers (PEOs) instituted in FY 1987 continued to dominate the organizational changes occurring at AVSCOM between 1988 and 1990. Efforts to establish the three initial PEOs--Aviation LHX, Combat Aviation (CA), and Combat Support Aviation (CSA)--were ongoing throughout 1988, with new subordinate offices being created and different PMs transitioning from AVSCOM to the PEOs. On Aug. 4, 1988, the AAE decided to merge the bulk of the CA and CSA PEOs into a single PEO, Aviation, which concentrated all logistics management of all of the Army's major aircraft types under one program executive officer.

On November 15, 1988, the PEO Aviation stood up, with five major Army aircraft systems assigned to it: Apache, the OH-58D, Black Hawk, modernized Chinook, and Cobra. From 1989 to 1990, the PEO LHX exercised authority over the development program of a new rotary wing aircraft to replace the Army's aging light helicopter fleet. Originally expected to produce ~5,000 helicopters in multiple variations, scaled-back projections in 1989 called for the eventual construction of more than 1,000 LHXs in the 1990s. After the LHX PEO reverted to program manager status, the PEO Aviation became the only such organization collocated with and supported by AVSCOM in 1990.

Working through the original PEOs, AVSCOM supported the various aviation projects or product managers that once had been under its jurisdiction. Starting in 1988, however, responsibility for certain programs initially transferred to the PEOs, such as the Light Observation Helicopter Product Manager and the UH-1 PM, returned to the operational control of the AVSCOM Commander.

In 1989, the Aviation Ground Support Equipment (AGSE) and Synthetic Flight Trainer Systems (SFTS) also reverted back to AVSCOM's control. That same year, all of the offices were grouped together under the newly established

Fielded Aviation Systems Management Office (FASMO). By 1990, the AVSCOM Directorate for Fielded Aircraft Systems, as FASMO became known, was responsible for aircraft retirement programs, FMS possibilities, and the modernization of Army aircraft stock. It also managed the development of fixed base and tactical air traffic control equipment and aviation ground support equipment.

Chief among the directorate's aviation assets were:

- More than 2,700 UH-1 Huey "H" and "V" models.
- The AH-1 Modernized Cobra, which employed missiles, rockets, and cannons.
- The three fixed-wing aircraft-the RV-1D, RC-12, and RU-21-equipped with the Army's airborne signal intelligence and electronic warfare gear.
- 4.The light observation combination of OH-58 and OH-6 helicopters remained the "old reliables" for Army aviation battlefield observation and reconnaissance.

In addition, AVSCOM lost four major subordinate elements in 1990. AVSCOM jurisdiction over the US Army Plant Representative Offices (ARPOs) for the McDonnell Douglas Helicopter Company, Boeing Helicopters, and Bell Helicopter Textron (BHT) ended on Jun. 24, 1990, while the Army Engineering Flight Activity (AEFA) at Edwards Air Force Base transferred to the US Army Test and Evaluation Command on Oct. 1, 1992.

The following reorganizations took place in the 1990s.

- 1 October 1992: Army Aviation and Troop Command (ATCOM) was established, consolidating the existing missions of AVSCOM and TROSCOM, less those missions and organizations transferred to other commands.
- 8 September 1995: Congress approves the 1995 Base Realignment and Closure Commission list, disestablishing ATCOM and

transferring its mission and organizations to Redstone Arsenal to merge with the Army Missile Command to form AMCOM.

- 17 July 1997: Army Aviation and Missile Command is provisionally established.
- 1 October 1997: AMCOM was formally established at Redstone Arsenal with the merger of the US Army Missile Command (MICOM) at Redstone Arsenal and the US Army Aviation and Troop Command (ATCOM) in St. Louis, Missouri.
- US Army Aviation and Missile Life Cycle Management Command was an LCMC. Thus, it had an associated contracting center. This LCMC Aviation and Missile Life Cycle Management Command was formerly Aviation and Missile Command (AVSCOM) (1997). This LCMC purchases about $1 billion worth of aircraft and missile parts each year.

THE MICOM STRATEGY WAS TO CONSOLIDATE ALL AMC RDECS IN HUNTSVILLE, AL

It is interesting to note that when I was on the Army Science Board in the 1990s, I saw how Dr. Bill McCorkle, Technical Director Army Missile Research and Development Center (MRDEC), was leveraging powerful Army leaders and Alabama Congressmen and Senators to bring the Army Material Command (AMC) and all of its weapon systems RDECs, to Huntsville, AL. I tried to warn Tom House, then ATCOM RDEC Technical Director, to be aware of this, but to no avail, as he had little control of the situation. Dr. McCorkle argued that there were no laboratories in St. Louis, while Redstone Arsenal had many. Also, since NASA was there, it could collaborate with Army Aviation.

Another hidden reason was that MRDEC was well over its quota of engineers, and with Aviation RDEC moving there, he could solve this problem by training and placing his excess engineers into the Aviation RDEC slots.

Interestingly enough, I taught my award-winning rotorcraft engineering, design, and safety courses from Georgia Tech to the MRDEC engineers through the University of Alabama, Huntsville (UAH), for two years during this transition.

In summary, Dr. McCorkle successfully accomplished his coup d'etat by his retirement in 2009. The exception was the Armament RDEC which remained at Picatinny Arsenal, PA. It is my opinion that even to this date, the missile engineers in Huntsville are given preference over aviation engineers in leadership positions in joint program and technical organizations.

ARMY AVIATION RESEARCH & DEVELOPMENT SHOULD HAVE LED TO NEXT GENERATIONS

As can be seen, Army Aviation Research, Development, and Acquisition (RDA) has been a gypsy for over the past forty years, 1980 to 2020. The only period where Army Aviation RDA was conducted in a successful integrated manner leading to the next generations of rotorcraft was during the AVRADCOM years, 1977-1983.

There should be a common framework for Army Aviation Technology Development coupled with Systematic Operations Analysis for bringing together Army Aviation users and developers. In this regard, I developed and presented a framework, Figure 9, in a paper entitled "Future Rotorcraft Requirements Development Approach for Balancing Technology Push and Pull," at the NATO STO AVT-245/RSM-037 SPECALISTS' MEETING on Future Rotorcraft Requirements, 30 November 2016, in Prague, Czech Republic (Ref.6). It was included in the Final Report from the NATO Specialists' Meeting as a recommended approach.

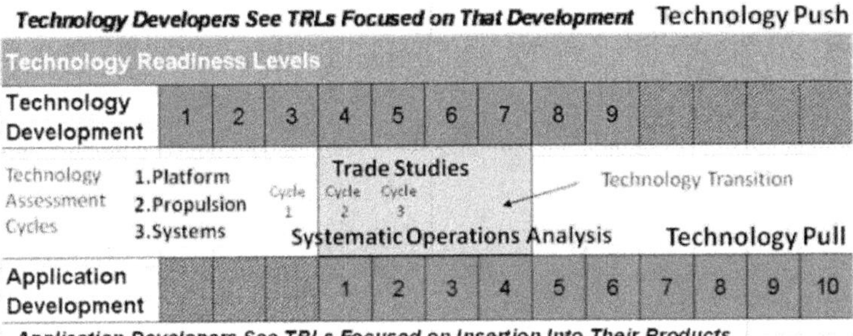

Figure 9. Framework for Execution of Army Aviation System of Systems Operational Effectiveness Analysis with Emerging Rotorcraft Design and Life Cycle Cost Assessment

It was also gratifying to see that in the Army Science Board (ASB) 2015 Study Report, "Army Science and Technology for Army Aviation 2025-2040," their top recommendation was to apply Aviation System of Systems and Operational Effectiveness Analyses for Army Science and Technology for Army Aviation 2025-2040. (Ref. 7)

The next recommendation was to apply it to Affordable Heavy Lift, which is addressed in Chapter 6 of this book. An Affordable Heavy Lift is needed today. A brief explanation of the Figure 8 Framework will be given to provide a basic understanding of how future rotorcraft should be developed. Cycles 1, 2, and 3 conceptual, preliminary, and detailed design iterations.

This is something few people in the science and technology process understand. It will be illustrated as an Epilogue, with an example and addressed in detail in Book 3: Development of a Graduate Program in Aerospace Systems Design. Executing the Framework present in Figure 9 requires a top-level Value Based Acquisition (VBA) Metric and Models, as a ratio of Systems Effectiveness to Life Cycle Cost (LLC), Figure 10. There must be an initial consensus and understanding between the top-level users and developers based on the next generation of rotorcraft Concept of Operations (CONOPS) and Operations Requirements Document (ORD).

Best Value is often quoted when an Army Aviation System is down selected, such as in the recent Future Long Range Assault Aircraft (FLRAA) decision of Systems Operational Effective Analysis with rotorcraft design and life cycle cost analyses.

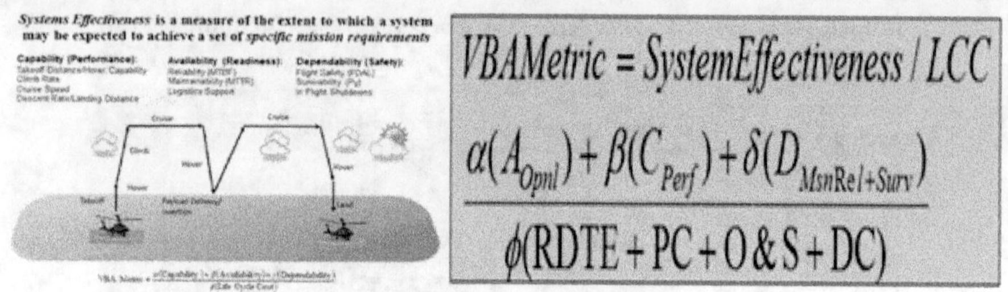

Figure 10. Systems Effectiveness and Specific Metrics related to Mission Requirements

One problem has been that Technology Insertion has been limited by the Army to Concept Design and Analysis (CDA), Cycle 1, by the Army Aviation Design Office at NASA Ames, while Preliminary Design and Analysis (PDA), Cycle 2, is required for relevant tradeoffs. Examples will be given for the LHX/Comanche Program and the Joint Heavy Lift (JHL) Program in later book chapters.

On the Comanche Program, the Bearing-less Main Rotor (BMR) was not designed and analyzed successfully by Sikorsky, Boeing, and the Army for air resonance, while for the JHL Program, the very large two-rotor tiltrotors were not analyzed successfully for scaling, and the Bell-Boeing quad rotors was a more feasible solution. Analytical Models for Operational Availability (A_{Opnl}), Capability (C_{Perf}), and Dependability ($D_{MsnRel\,+Surv}$) can be integrated along with Model-Based Systems Engineering (MBSE) into a System Architectural Model, Figure 11. System Effectiveness is a measure of the extent to which a system may be expected to achieve a set of specific mission requirements. Further model breakdowns can be seen in Figure 11. (Ref.10)

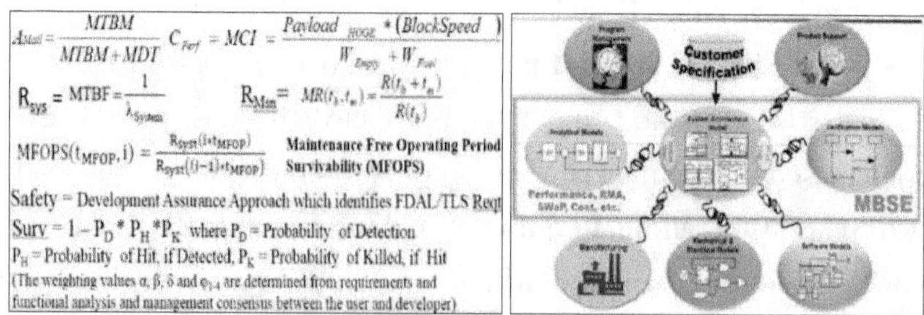

Figure 11. Analytical Models & MBSE Require Integration into System Architectures

Under the Army's Initial FVL Program in 2015, the Georgia Institute of Technology (GIT) and the University of Alabama at Huntsville (UAH) conducted a program for reduced FVL risk using Development Assurance Value-Based Acquisition (DAVBA). Development Assurance (DA) is process used for civil aviation certification throughout the world. The Vee Diagram for DAVBA is illustrated in Figure 12. It also illustrates an Approach for Safety By Design, Airworthiness Qualification and Specification Compliance.

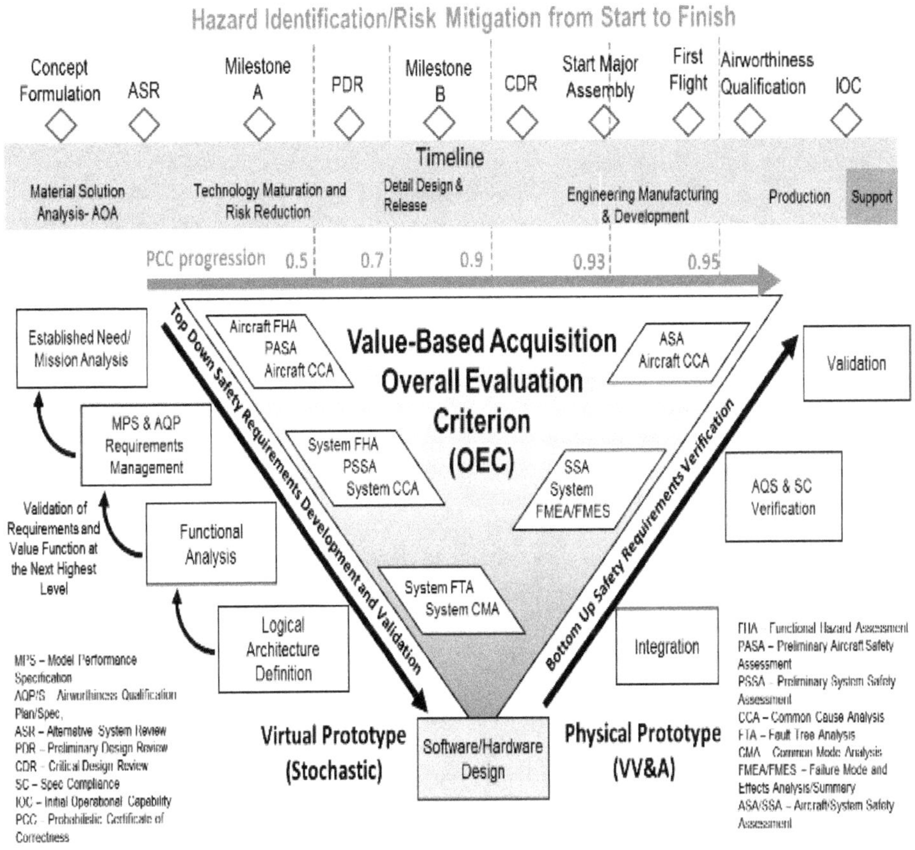

Figure 12. The Georgia Tech-UAH Vee Diagram for DAVBA

A Model-Based Systems Engineering (MBSE) environment was created for development of the DAVBA Stochastic Virtual Prototyping. An objective was to create the FVL Trade Space Definition example for a potential upgrade to the UH-60M into FVL, which is shown in Figure 13. The DAVBA

Approach for Airworthiness Qualification (AWQ) and Specification Compliance (SC) was completed using the UH-60M as a Baseline in Phase I. However, Phase II was not funded due to lack of Army funding. It is interesting that the FARA cancellation and purchase of additional UH-60Ms and AH64s makes them good baselines for both MOSA and expansion of some of them to compound helicopters.

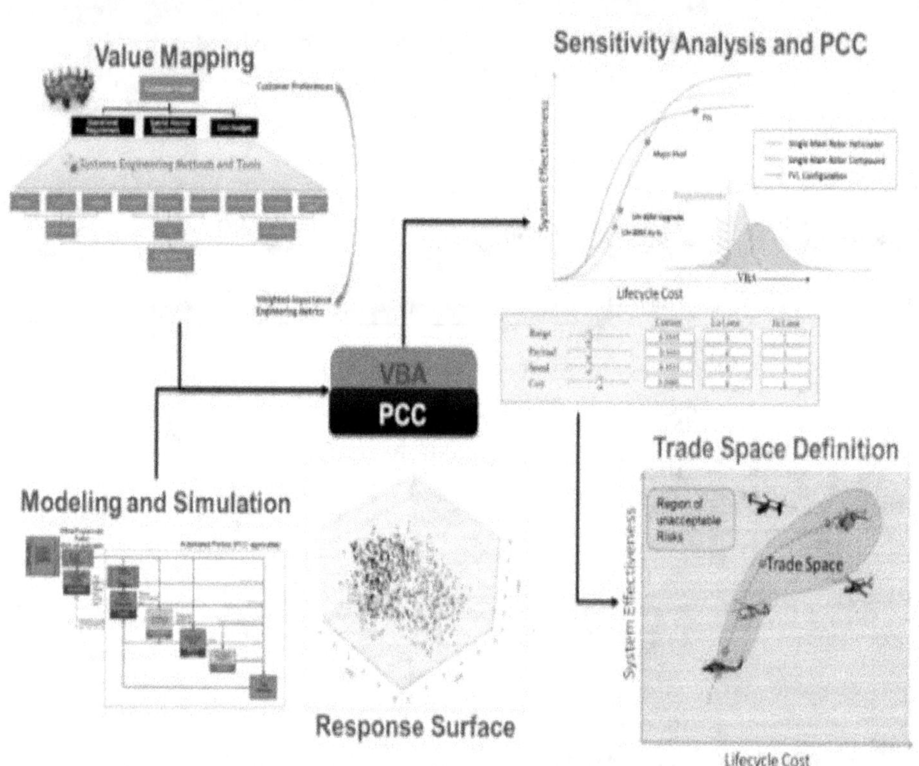

Figure 13. GIT-UAH DAVBA FVL Program (Ref.10)

DAVBA Phase II was proposed for FY16 to apply the approach to FVL with the UH-60M as an "As Is" baseline, as illustrated in Figure 14. Unfortunately, extended funding was not available. The use of a similar United States Air Force (USAF) Cost Capability Analysis (CCA) approach is included in the Defense Acquisition Guide as a Best Practice. (Ref.11)

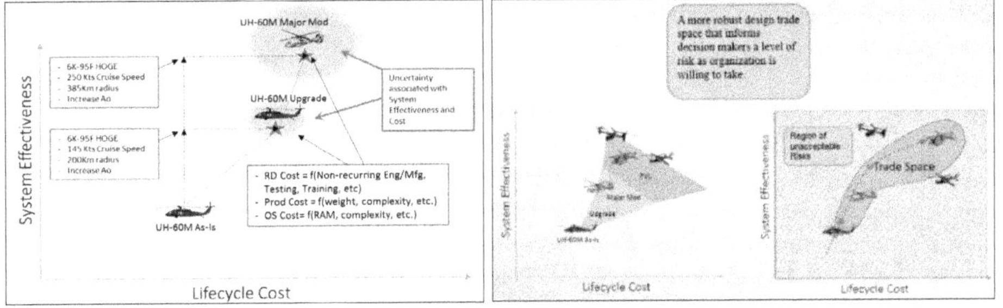

Figure 14. DAVBA Phase II Approach for FVL and Robust Design Trade Space

Below is a summary of the DAVBA Demonstrated Approach, which also includes an innovative approach for Airworthiness Qualification (AWQ) for Military Systems using a combined set of Civil-Military Functional Safety Management Certification Standards.

- Value-Based Acquisition Metric Development to Capture System Requirements.
- Virtual Stochastic Prototype Environment based on Model-Based Systems Engineering (MBSE).
- Identification of Confidence Interval in the Model Prediction through Probabilistic Certificate of Correctness (PCC) as developed by Dassault Systems.
- Develop Draft Aeronautical Design Standard (ADS-X1), Including Example UH- 60L to M as a Recapitalization Case Study.
- This research addressed calls for a new Aircraft Level Aeronautical Design Standard (ADS).

To quote the author and Arterburn, D., "The DoD, FAA and Industry can also use this approach to address multi-core processers airworthiness standards immediately as the consumer market drives suppliers toward the use of multi-core processors over single-core processors" IMPACT Final Report, Airworthiness of Complex Systems, January 15, 2015. (Ref.11)

In the past, e.g., UTTAS, AAH, and LHX, the Concept Formulation Package (CFP) consisted of iterations between the Trade Off Determination (TOD), Trade Off Analysis (TOA), Best Technical Approach (BTA), and

Cost and Operational Effectiveness Analysis (COEA). For later development programs such as the Future Vertical Lift (FVL) Family of Aircraft, e.g., Future Long Range Assault Aircraft (FLRAA), Future Attack and Reconnaissance Aircraft (FARA), and Future Tactical Unmanned Aerial Systems (FTUAS) are required to follow the Joint Capabilities Integration and Development System (JCIDS) Process, Figure 15. (Ref. 12)

The primary objective of the JCIDS process is to ensure the capabilities required by the joint warfighter are identified, along with their associated operational performance criteria (requirements), in order to successfully execute the missions assigned. It includes three processes (acquisition, requirements, and funding), Figure 16A, that support the Defense Acquisition System. The Pre-Systems Acquisition is shown in Figure 16B. The JCIDS process was created to support the statutory responsibility of the Joint Requirements Oversight Council (JROC) to validate joint warfighting requirements. It is, however, a very complex process making it difficult for the primary user and developers to conduct the necessary tradeoffs in an informative and timely manner.

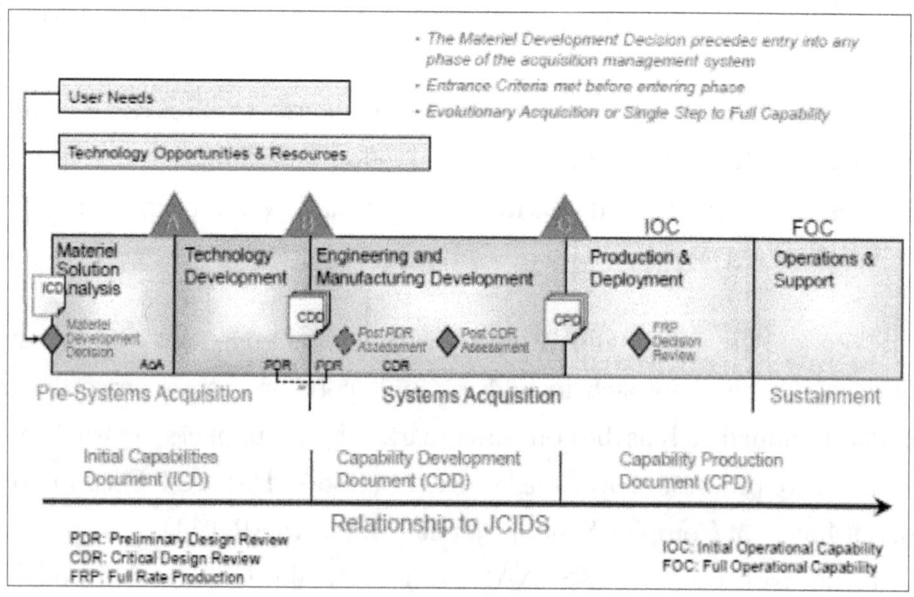

Figure 15. JCIDS Process

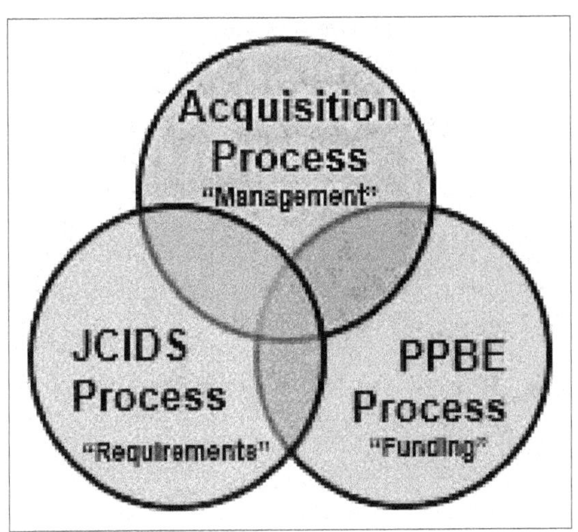

Figure 16A. Three Key Processes: Acquisition, Joint Capabilities Integration and Development Systems (JCIDS) and Planning, Programming, Budgeting, and Execution System

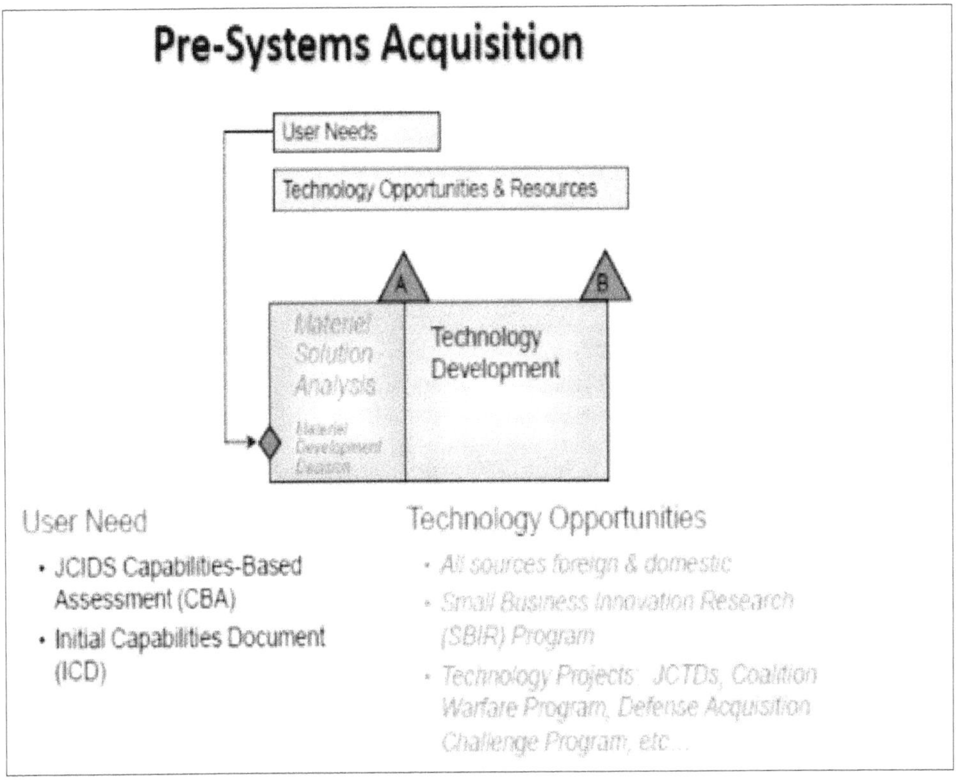

Figure 16B. MSA &Tech Development

In a recent conference (Ref.13), the Army FVL Program Manager, presented a chart, Figure 17, that describes the Good News and the Problem. The entire future material solution process is digital except the Joint Capabilities Integration and Development Systems (JCIDS) process. He goes on to state that this analog process for JCIDS execution occurs repetitively during program execution. While the use of digital and analog tools and environments are useful, they often aren't understood by the key decision-makers. An example of the LHX Concept Formulation Package (CFP) will show how the interaction between the user and developer is far more important than the use of complex tools and environments which aren't understood by most of the participants. This is a lesson I learned from my Army Aviation LHX development, as well as from teaching rotorcraft design and cost analysis for thirty-five years.

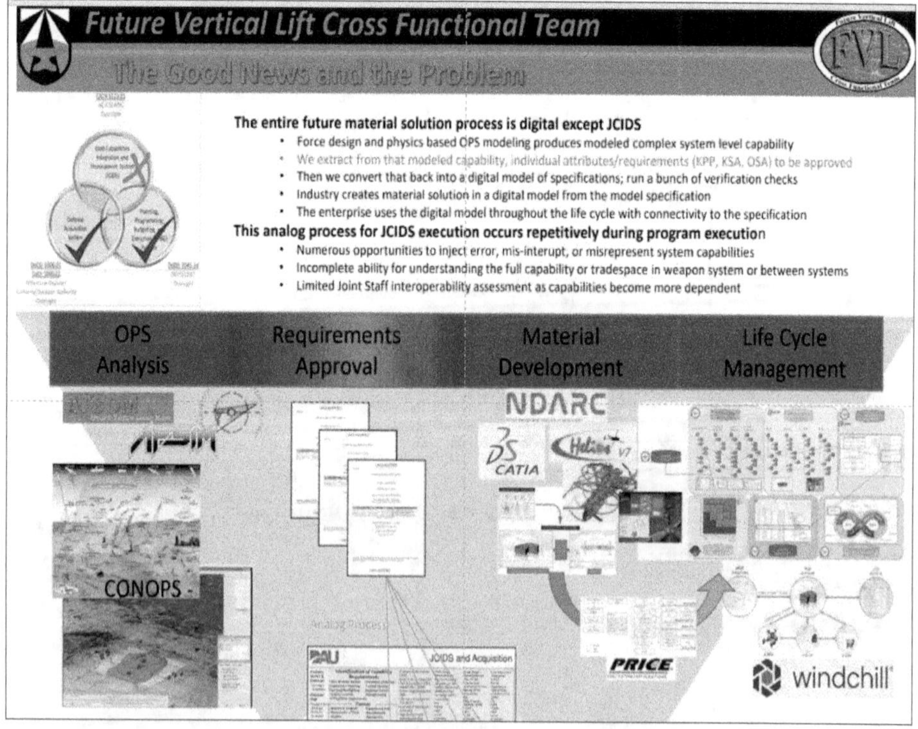

Figure 17. The Good News and the Problem, FVL PMO Chart, VFS Conference, Nov. 2022

Summary of Preface and Background

It was planned for the Preface and Background to give the reader a fundamental understanding of some of the History and Background on Army Aviation Air Mobility and Research, Development, and Acquisition (RDA). It includes reviewing the "shining star AVRADCOM" period along with the politics and barriers that dissolved it. Following are the References outline for the Preface and Background. This will be followed by an Outline of the Chapters for Book 2, which will continue the memoir trilogy from Book 1 and follow on in Book 3.

Endnotes

1. Vertical Flight Society (VFS) Identification of Rotorcraft by Mike Hirschberg, VFS Director, 2018; Modified to Reflect on Next Generations of Army Aviation Systems, by Dr. Daniel P. Schrage, Author, Book 2, 2023.
2. *Army Aviation Foundations of the Modern Fleet* Report, Initiative by the Program Executive Office (PEO) Aviation Public Affairs Office, 1ˢᵗ Addition, Published August 2013.
3. Comanche, The RAH-66 Comanche Helicopter: Technical Accomplishment, Program Frustration, by Arthur W. Linden and the Comanche Team, 2022.
4. Scully, M., *Adventures in Low Disk Loading VTOL Design,* NASA/TP—2018–219981, September 2018.
5. Dr. Kaylene Hughes, Historian: *Evolution from AMC to AMCOM: Parts V- VII: The Reestablishment of AVSCOM,* 1984-1990, By Courtesy January 18, 2018
6. Schrage, D.P. *"Future Rotorcraft Requirements Development Approach for Balancing Technology Push and Pull,"* at the NATO STO AVT-245/RSM-037 SPECALISTS' MEETING on Future Rotorcraft Requirements, 30 November 2016, in Prague, Czech Republic
7. Army Science Board (ASB) 2015 Study Report, Army Science and Technology for Army Aviation 2025-2040, 2016.
8. *Bell's V-280 Valor tiltrotor won the US Army's Future Long-Range Assault Aircraft (FLRAA) competition.* Textron Wins FLRAA Contest, Crushing Boeing And Lockheed Martin, Dec. 06, 2022 11:02 AM ET Textron Inc. (TXT)BA
9. Joint Heavy Lift (JHL) Concept Design and Analysis (CDA) 2005
10. Schrage, D.P., and Van der Velden, A., A Development Assurance Value-Based Acquisition (DAVBA) Approach for Complex VTOL Aircraft Development, 2015 VFS Annual Forum.
11. Schrage, D.P. and Arterburn, D. IMPACT Final Report, Airworthiness of Complex Systems, January 15, 2015.
12. Joint Initial Capabilities Document (JCIDS)
13. Comments, Army Future Vertical Lift (FVL) Program Manager, October 2022.

CHAPTER ONE

RETURN FROM VIETNAM AND ATTENDANCE AT THE FIELD ARTILLERY OFFICERS ADVANCED COURSE

My Returning from South Vietnam

My flight back to the USA from Vietnam stopped first in Oakland, CA with another flight to St. Louis, MO, which arrived on February 20, 1971. My father, Albin, and my young son, Steven, greeted me at the St. Louis Airport, as seen in Figure 18A. Nancy stayed at home with our three-month-old daughter, Susan, as seen in Figure 18B.

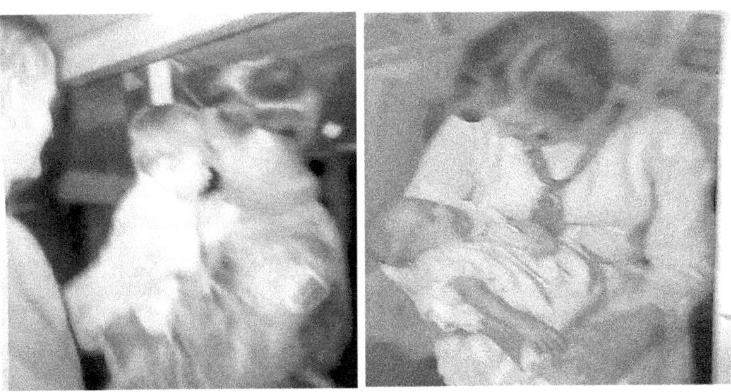

Figure 18A. Dan's Father, Albin, and Steven meet Dan. Figure 18B. Nancy at Home with Susan

Obviously, it was great to be back home again with family for a month before having to report to Fort Sill, OK, for the nine-month Field Artillery Officers Advanced Course. Both during this time Nancy and my parents, Agnes and Marnen, and Mary and Albin, were glad to have some time with us and their grandkids, Figures 19A &B.

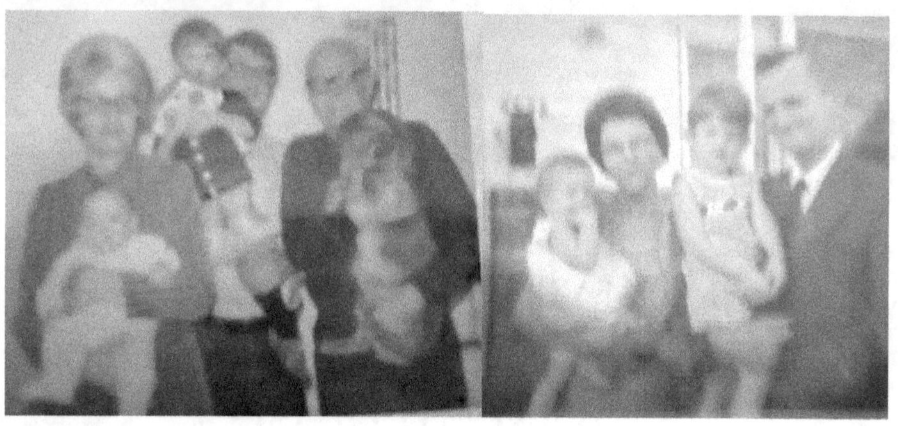

Figure 19A. Agnes, Marnen, and Kurt with kids and Strudel. Figure 19B. Mary and Albin with kids.

I was proud of the experiences and what I had accomplished in Europe and South Vietnam. This was evidenced in my final Officer Efficiency Reports (OERs) from Germany and South Vietnam, which were both outstanding. This was reinforced by a letter I received from the Commanding General of the Field Artillery Branch, which stated that I had performed in an outstanding manner and was being closely monitored for advancement and early promotion.

However, I felt my major interest and where I could make the most contribution was in the rising field of Army Aviation development. I hoped to pursue this opportunity after receiving a Master of Science (MS) Degree in Aerospace Engineering from the Georgia Institute of Technology (GIT /Georgia Tech). Nancy and I were looking forward to being assigned to Fort Sill, OK, and finding a place to live for nine months in Lawton, OK. We also knew we would have lots of friends, as many of my classmates from USMA and their wives were already there and would also be attending the Field Artillery Advanced Course. We found a nice house for rent in Lawton, OK.

Also, my application for advanced civil schooling was accepted to attend Georgia Tech to get an MS Degree in Aerospace Engineering.

My Experiences in the Field Artillery Advance Course

While moving into our rental house, I was approached by a neighbor, the manager of a semi-professional softball team supported by a local Lawton, OK, grocery store chain. He asked me if I had played baseball or softball in high school or college and I told him I had played baseball in high school, at Quincy College, and at USMA, as well as softball in high school and at Fort Wolters during flight school. He then asked me if I wanted to try out for his Lawton, OK, semi-professional softball team. I replied back I would once I settled in and saw how stringent the academic curriculum was in the Advanced Course, as well as the time I needed to spend with my growing family. Shown in Figure 20A, is a picture of me reading a book to Susan and Steven early in the Advanced Course, and in Figure 20B, a picture of Nancy, the kids, and me, at the end of the Advanced Course. There was some concern between Nancy and I, that I didn't spend enough time with our children. However, I did try and also found time to play on the semi-pro softball team.

Figure 20A. Dan reading a Book to Steven & Susan. Figure 20B. Nancy and Dan with kids

The Advanced Course Class I was in was FAOAC 6-71, with over 150 officers. There was good instruction, but centered more on tube artillery, in which I wasn't really interested. My experience in Germany as an Honest John (HJ) missile battery commander and participation in REFORGER I was a wonderful experience, and I learned a lot, but that was enough Field Artillery for me. I was now a seasoned Army Aviator.

I was also excited about being approved for advanced civil schooling and was going to get my MS degree in Aerospace Engineering at Georgia Tech. I then planned to serve the rest of my career with Army Aviation, making the air mobility concept we began learning in Vietnam a bigger reality for the Army. This hopefully included the development of the next generation of Army Aviation Systems, such as an Advanced Attack Helicopter (AAH) and Utility Tactical Transport Aircraft System (UTTAS), which were beginning to be developed by the Army Aviation Systems Command (AVSCOM) in St. Louis, MO.

One thing I was interested in during the Advance Course was the requirement for each student to write monographs/magazine articles on subjects of interest to the Army. The title of my monograph was "Chopper Vietnamization-Will It Work?" As noted in Chapter 2 of Book 1, I was directly involved in Helicopter Vietnamization in the Mekong Delta, including training Viet Nam Air Force (VNAF) pilots and the transfer of equipment and the first Army Airfield, Soc Trang, to the VNAF and their USAF advisers. I had concluded in Chapter 2 and in the monograph that to find the answer one must understand the helicopter, its role in the war, and the Vietnamese people.

I had the winning monograph from the Writing Excellence Award Selection Board and received the Writing Excellence Award, which was presented to me at the FAOAC 6-71 graduation. I also finished in the top 20 percent of the class, despite playing on a semi-pro softball team, coaching the Fort Sill Basketball Team, and finding time to be with my family.

My Experiences in Athletics While Attending the Field Artillery Advanced Course

I got involved in three athletic activities while attending the Advanced Course at Fort Sill, OK.

During the Summer of 1971, I played on the Lawton, OK, Semi-Professional Slow Pitch Softball Team. I was one of two military officers recruited for the team. The other officer was a lieutenant stationed at Fort Sill who had played football as a middle linebacker for the Denver Broncos. He was about 6' 4" tall, weighed about 250 lbs. and could hit balls over most fences.

The team also recruited outstanding players from around the State of Oklahoma, especially from the University of Oklahoma. One player had played shortstop at the University of Oklahoma Baseball Team and was an excellent defender and hitter. Two other Oklahoma football players, offensive linemen, were also on the team for hitting long balls over the fences. While big guys could hit home runs, defensive players were also needed. The centerfielder on the team was a Native American with great speed. He had played Major League Triple A Baseball. Pitching, defense, and players that could hit the ball over the fences were our keys to success.

During the Summer of 1971, we played about sixty games around the State and neighboring states. We played in and won the State of Oklahoma Championship which qualified us for the Four State Regional (Arkansas, Louisiana, Oklahoma, and Texas) which was held in Alexandria, LA.

While we won the Regional Tournament, I broke my wrist when hit by a thrown ball while going into second base to break up a double play. Being in the military, I had to go to Fort Polk, LA, to get my arm set. Shown in Figure 21 is a picture of my son, Steven, and me with my arm in a cast, leaving a motel near Fort Polk, LA, to drive back to Lawton, OK. I went along with the team to the World Championship Tournament in Parma, OH, but I wasn't able to play due to my broken wrist.

The following summer, while I was in graduate school at Georgia Tech, I got a call from our Lawton, OK, softball team manager. He wanted to fly me to Lawton to play in a tournament on July 4, 1973. This was so I could be eligible for the Four State Regional Tournament in Fort Smith, ARK, at the end of August 1973. I did that and played in the tournament. However, we lost to the Louisiana team we beat the year before. They went on to the World Championship (USA) Softball Tournament in Daytona Beach, FL. This sort of ended my semi-professional softball career. It is interesting that a USA National Professional Software League was formed the next year with my Oklahoma team playing in the Championship Finals.

Figure 21. Steven and I leaving the Motel to drive back to Lawton, OK

The second athletic activity I participated in was as the Coach of the Fort Sill Men's Basketball Team. While my broken wrist kept me from playing, I enjoyed being the coach of an excellent team that as a "dark horse" almost won the Fifth Army Basketball Tournament Championship. We lost the Championship Game in overtime to Fort Hood in San Antonio, TX at the FT. Sam Houston, TX Sports Arena, and MFSS Gymnasium, on 24-28 January 1972, Figure 22A. A picture of the Fort Sill team with me shown as coach is in Figure 22B with our runner-up trophy. Figure 22C is a picture of the starting Power Five. San Antonio was and is a beautiful town with the Alamo Fort and its River Walk modernization developed as part of Hemisphere 68.

Fig. 22A. Fifth US Army Tournament. Fig. 22B. Fort Sill team & trophy. Fig22C. Power Five.

With the Army draft still in effect, most of the teams in the Fifth US Army Basketball Tournament had recruited former outstanding college basketball players for their rosters. This was especially true for the Fort Sam Houston Team, which stockpiled players into non-combat medical positions. The top-ranked team, however, was Fort Polk, followed by the Fort Sam Houston Team and then the Fort Hood Team.

Our Fort Sill team was definitely considered an underdog, as I had to revamp the team during the regular season. Several key players on the team were in combat positions and had been transferred out of Fort Sill for the War in Vietnam.

The key power players on the Fort Sill team for the 5th Army Tournament are shown in Figure 22C. The star of our Fort Sill team was Don Griffen, kneeling on the right side in the picture. He was an All-American and a leading scorer for Stanford University, where he played for four years. He was the third draft choice of the Atlanta Hawks in the National Basketball Association (NBA) in 1970 behind Pete Maravich and Butch Beard. However, he ended up playing for a team in the Philippines before being drafted into the U.S. Army. The 6'5" guard was a member of the All Armed Forces team in the fall of 1970. He tried out for the Pan American team and was invited to join the Olympic Basketball Development crew. But he decided to stick with the Armed Forces teams. He and Bill Cotton joined the All Armed Forces team following Fifth Army Tournament, Jan 24-28, 1972.

The other guard kneeling on the left in Figure 22C was Jesse Satterfield, a 5'9" speedster high school player from Richmond, IN. The twenty-year-old averaged seven points per outing and was a good ball handler. Standing on the left side in Figure 22C, is Elbert Giles, who patrolled the lane at the post position. He stood only 6' 4" but weighed in at 220 lb.

The forwards in Figure 7C are LeRoy Outlaw in the center and Bill Cotton on the right side. Leroy Outlaw was a 6'5" West Pointer who ran track but did not play basketball. I added him to the Fort Sill team late in the season to make up for the combat transfers lost to the team. I knew Leroy at West Point as an excellent track athlete. Bill Cotton was the other All American on the team. The 6'5" forward was the NAIA choice for the "dream team" while playing for Hank Iba at Central State in the Oklahoma Collegiate Conference in 1964-65. He was only back from Vietnam for a few months before the tournament but averaged eighteen points per game.

Another key player not shown in Figure 22C was Willy Kelly. Kelly, a forward, stood 6'2" and played small college ball in North Carolina. He averaged thirteen points for the Fort Sill team during the season. He could really jump, was a good scorer, and could play defense. He was also a key to our success. Unfortunately, he was re-assigned before the tournament.

The following assessment after the tournament began was presented in The Fort Sill Cannoneer Newspaper on Friday, January 28, 1972. (On Friday afternoon, we had to win two games versus Ft. Hood—we came close.)

"Coach Dan Schrage's Fort Sill Cannoners have been busy this week practicing their own special brand of hardwood magic on the floor of the Ft. Sam Houston Sports Arena near San Antonio, TX. The action is that of the 5th Army area basketball playoffs, where the visiting Sillmen have already gone on to batter teams who had previously been considered to be "top guns" in the Texas installation tournament. The first was Ft. Polk, the team area hardcourt crystal gazers had gone out on a limb to tab as the real power of the tourney. The Cannoneers toppled their tower of supremacy the first night of the tournament, 86-77. Then, Tuesday night, it was the Ft. Sam Houston Ranger's turn. The defending 5th Army champion Rangers stormed into the hardcourt only to find out that they were disastrously outclassed by the Fort Sill pacers Don Griffin, Bill Cotton, and Leroy Outlaw. By halftime, the Cannoners were running away with the contest, 43-32, in a race for the semifinals that finally ended the Rangers eating Cannoneer dust, 98-68."

However, the Fort Sill Team had to beat a Fort Hood Team twice on Friday, who were also undefeated. Fort Hood was a very talented and deep team that had beaten Fort Sill in three of their four meetings during the season. The Fort Sill Cannoneers got into foul trouble early in the game, losing to Ft. Hood, 93-81, on Wednesday night for their first defeat in the Fifth Army Basketball Tournament. The loss forced Fort Sill to play Fort Polk again on

Thursday in the double-elimination tourney. The winner was Fort Sill, and then we had to face the arduous task, as mentioned above, of defeating Fort Hood twice on Friday, the final day of the tournament if it expected to win the title. Fort Hood had yet to lose in the tournament.

My Story

Having lost to Fort Hood three out of four times during the season and earlier in the tournament, I decided we needed another approach. While Don Griffen was our point guard, leading scorer and handled the ball most of the time I decided to place him at the high post where he could get the ball and pass, shoot, or drive to the basket. This worked very well, and we beat Fort Hood, 93-86, on Friday afternoon in the first game. Don Griffen had thirty-one points (ten field goals and eleven free throws) and more important he was able to feed Leroy Outlaw down low from the high post and Leroy had twelve field goals and two free throws for twenty-six points.

We took the same approach during the second game, for the championship, and led most of the time. With thirteen seconds remaining in regulation play of the championship game, Fort Sill clung to a two-point lead. Don Griffen, the tournament's most valuable player, swiped the ball at that point but missed his ensuing layup. Three Cannoner tip shots followed, and all failed to fall. I believe exhaustion had set in. Fort Hood finally rebounded the ball with less than ten seconds left, and a desperation shot by Fort Hood's Freeman tied it at seventy-eight all and sent the game into overtime. Trailing by a point, Fort Sill stole the ball with ten seconds left in overtime. We threw the ball away before anyone could shoot, however, and Fort Hood added another basket and won the game 89-86.

Following this tournament, Don Griffen and Bill Cotton were named to the All-Army Team. The two players left after the tournament for California, to practice with the team which would represent the Army in the National Amateur Athletic Association Tournament. Two of my former West Point teammates on the 1966-67 USMA Basketball Team were also on the Army Team: Mike Krzyzewski and Jim Oxley. Coach Krzyzewski went on to become the winningest coach in College Basketball History. Jim Oxley went on to become an excellent Army medical doctor.

I received a Letter of Appreciation from Major General R. Wetherill, Commanding General, of Fort Sill, Oklahoma, for my contribution to the Fort Sill Basketball Team and to the Installation Sports Program.

The third athletic activity I got involved in while at Fort Sill was playing on the team handball team. Team handball, also called field ball or handball, is a game played between two teams of seven or eleven players. They try to throw or hit an inflated ball into a goal at either end of a rectangular playing area while preventing their opponents from doing so, as illustrated in Figure 23.

Figure 23. Team handball field and player taking a shot at the goal.

Team handball was being introduced as an Olympics sport, and the military was introducing it to prepare a U.S. Team. It required many of the skills found in basketball and was one reason I enjoyed it. Following these sports activities and graduating from the Field Artillery Advanced Course, it was time for our family to move to Atlanta, GA so I could start graduate school at Georgia Tech.

CHAPTER TWO

CAREER CHANGE TO ARMY AVIATION AEROSPACE ENGINEER

My Experiences in Graduate School at Georgia Tech and a Growing Family

I chose the Georgia Institute of Technology (GIT/Georgia Tech) for my advanced civil schooling for several reasons. First, because it had a long history of involvement with military aviation since 1917, Figure 24A. Also, it was a Public University and had the Guggenheim School of Aeronautics, as one of the original Guggenheim grantees in 1930, and now the School of Aerospace Engineering, as illustrated in Figure 24B.

Second, it was a conservative university that had not had student protests over the Vietnam War. Third, it was one of a few universities that had courses in Rotary Wing Aeronautics and Aeroelasticity under Dr. Robin Gray and Dr. Al Pierce. Finally, it was located in the City of Atlanta, the growing hub of the Southeast with a moderate climate.

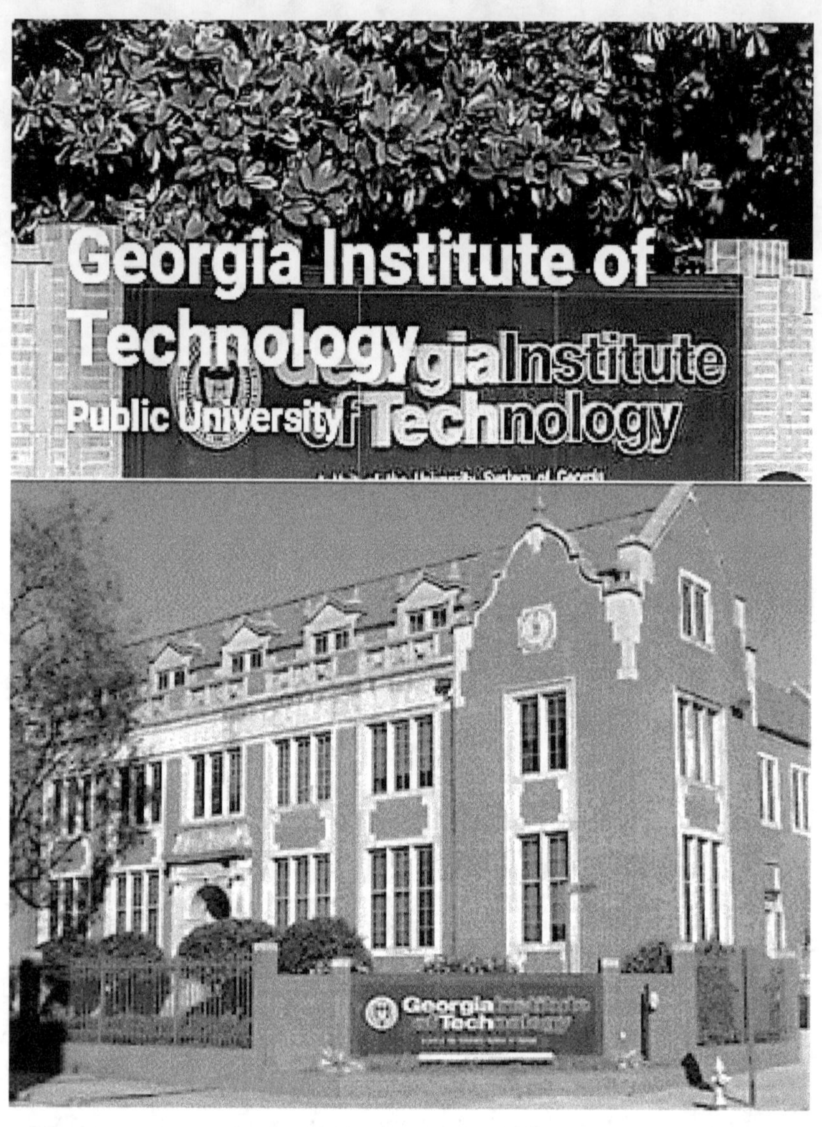

Figure 24A. The Georgia Tech Military School of Aviation Fig. 24B.GIT & Guggenheim School

Nancy, our children Steven, and Susan, and I arrived in Atlanta in early March 1972. We searched to buy our first house. It had to be one we could afford in a neighborhood not too far from Georgia Tech and, hopefully, with a backyard and neighbors who had children close to the age of our children. The house we bought for about $35,000 is shown in Figure 25A. Our daughter, Susan, is shown sitting by our rock garden in Figure 25B.

Fig. 25A. Our Home in Atlanta. Fig. 25B. Susan by our Rock Garden in Backyard

The neighborhood was ideal. We had great neighbors with children close to our children's ages and similar interests. Living next door was a Navy Lieutenant Commander, Gordon Callender, who was working on his PhD in Civil Engineering at Georgia Tech. Across the street was a Delta Airlines Captain and former Navy pilot, Jon Jancuski. Down the street was a manager of an Owens Illinois plant in Atlanta, Vito Pawloski. Also, further down the street was an Atlanta radio disk jockey, Jim Snyder, who was always the voice of a party.

In addition to having children close to our children's ages, our neighbors were also interested in sports. As it turned out, we almost immediately had a tennis foursome that played almost every Saturday morning. In the winter, this included playing on a church basketball team and an Atlanta Lawn Tennis Association (ALTA) tennis team.

The closeness of this set of neighbors has continued for fifty years, and we still try and get together once a year. Another activity that we all participated in was the annual raft race down the Chattahoochee River, something the City of Atlanta initiated in the early 1970s.

When I arrived at Georgia Tech in March 1974, my classes wouldn't begin until June 1972, with the Summer Quarter. Therefore, I was attached to the Army ROTC Detachment on a temporary assignment from 7 March 1972 to 5 June 1972. My principal duty was as an Assistant Professor of Military Science (PMS). I was utilized as an action officer for various projects, such as assisting in the coordination of a two-day conference attended by over 200 college students and faculty members, revising a block of instruction for the Air Defense Artillery (ADA) senior level course, and assisting with the conduct of a Ranger FTX and a Brigade range firing. I received an outstanding Officer Efficiency Report (OER) for this short assignment, which was good for my military career, as I would only get an academic performance report over the next two years.

One outside benefit from this assignment was I was able to get an Army assault rubber raft from the Ranger School at Fort Benning, GA, for the Atlanta Annual Raft Race. My neighbors and friends in the raft race, taking pictures of our families on the bank. Figures 26 shows us approaching the shore to get out of the raft. This was definitely an improvement from my previous raft experiences in Ranger School, such as the night assault across a bay in the Gulf of Mexico during a tropical storm. We planned to come back each year and make the Atlanta Raft Race our annual event; however, the first year back, it rained all weekend. Shortly after that, the City of Atlanta canceled the raft race, as drug abuse became a large part of it.

Figure 26A. Waving to Families on the Bank. Figure 26B. Approaching the Shore

The Georgia Tech Aerospace Engineering program was quite challenging. There were two of my classmates from USMA in the same program that I was enrolled in, as well as an Air Force Officer, a Navy Officer, and several other U.S. graduate students. There were also six graduate students from India under a program Georgia Tech had with India. The adviser for the military students was Dr. Robin Gray. He also taught the Rotary Wing Aerodynamics courses and was the Associate School Chair, Figure 27A, with Dr. Arnold Ducoffe being the School Chair, Figure 27B.

Another key faculty member for me was Dr. Al Pierce, Figure 27C. He taught aeroelasticity, dynamics, and vibrations courses. He was an excellent teacher with practical aeroelastic stability experience gained while working for Rockwell in Dayton, OH, as a flutter and dynamics engineer. Aeroelasticity, Dynamics, and Vibrations (ADV) was the area I decided to concentrate on as my operational experience with helicopters identified these as major problems. Also, Dr. Pierce had three PhD students graduate in Aeroelasticity in the late 1960s, Dr. Eugene Hammond, Dr. Felton Bartlett, and Dr. William White. They worked at NASA Langley in the joint NASA-Army laboratory and were directly involved in conducting research for developing the Army's next generation rotorcraft.

Fig. 27A. Dr. Robin Gray. Fig. 27B. Dr. Arnold Ducoffe. Fig. 27C. Dr. Alvin Pierce

After arriving at Georgia Tech, I realized that evolving Army Aviation leaders, both civilian and military, had come to Georgia Tech's School of Aerospace Engineering in the 1960s to help lead the Army's development of its next generation of aircraft. The key civilian leader was Mr. Charles C. Crawford, Figure 28A, who went on to form and lead the Army Aviation Systems Development and Qualification (D&Q) Directorate in the Aviation Systems Command (AVSCOM) in St. Louis, MO. He got his undergraduate degree in aeronautical engineering at Georgia Tech in 1954 and later obtained a master's degree in Aerospace Engineering under Dr. Gray's guidance in 1969. Mr. Crawford later became a mentor for me and was called the "Father of the UH-60 Black Hawk Helicopter" by both industry and government.

Another key civilian leader was Mr. Robert E. Wolfe, Figure 28B, who also received his undergraduate and master's degrees from the School of AE, Georgia Tech, in the late 1960s. He became the Chief of the Structures and Aeromechanics Division and my first boss in the Aviation Systems Development and Qualification (D&Q) Directorate in AVSCOM.

These two key individuals became the technology development leaders in AVSCOM for the next generation of Army aircraft. I worked for both of them as an ADV engineer from 1974-1978 until I left active duty as the acting Aeromechanics Chief. I became the full Aeromechanics Chief in 1978. In 1979 I became the Director of Structures & Aeromechanics

Figure 28A Mr. Charles C. Crawford. Figure 28B. Mr. Robert E. Wolfe

In addition to these civil servants, there were key Army Aviation military leaders who also received their advanced degrees in Aerospace Engineering at Georgia Tech in the 1960s. The key military leaders, Figures 29A, B, C, and D, were COL Robert L. McDaniel, MS GTAE 1961, MG Carl McNair, MS GTAE 1963, BG Story Stevens, BS and MS GTAE, 1966, and BG Ronald Andreson, MS GTAE 1967.

Fig. 29A.COL R. McDaniel. Fig. 29B. MG C. McNair. Fig.29C.MG S. Stevens.Fig.29 D.MG R. Andreson

Interestingly enough, all had previous operational experience in Vietnam in the same units I served with in the Mekong Delta. They all had key leadership roles in the development of the next generations of Army aircraft, both helicopters, and rotorcraft, e.g., vertical lift aircraft. I would work with them in my next ten years with AVSCOM and then the Aviation Research and Development Command (AVRADCOM).

Some of the key undergraduate and graduate courses I took in my two years at Georgia Tech, 1972-74, were as follows: Dynamics I&II, Fluid Mechanics I-IV, Propeller and Rotor Theory, Mechanical Vibrations, Advanced Engineering Mathematics, Vibration and Flutter, Structural Dynamics, and Aeroelasticity I-III (including Special Problems). I finished with the highest grade point average (GPA) of the U.S. students,3.61, in the program, but not of the students from India. Most of them already had advanced degrees from India.

Several of the Indian students went on to become key technology engineers and leaders in developing the next generation of rotorcraft in U.S. industry, government, and university, e.g., Dr. Ram Janakiram with Boeing and Dr. L. Sankar and Dr. J.V.R. Prasad as faculty members at the School of AE, Georgia Tech.

While these courses were difficult, they put me in a great position to be the ADV engineer in the Aeromechanics Branch in the Structures and Aeromechanics Division of the AVSCOM Development and Qualification (D&Q) Directorate in 1974-78.

Understanding the Rotary Wing Multidisciplinary/Interdisciplinary Problem

It requires an understanding of vertical lift (helicopters/rotorcraft), as illustrated in Fig. 30A&B.

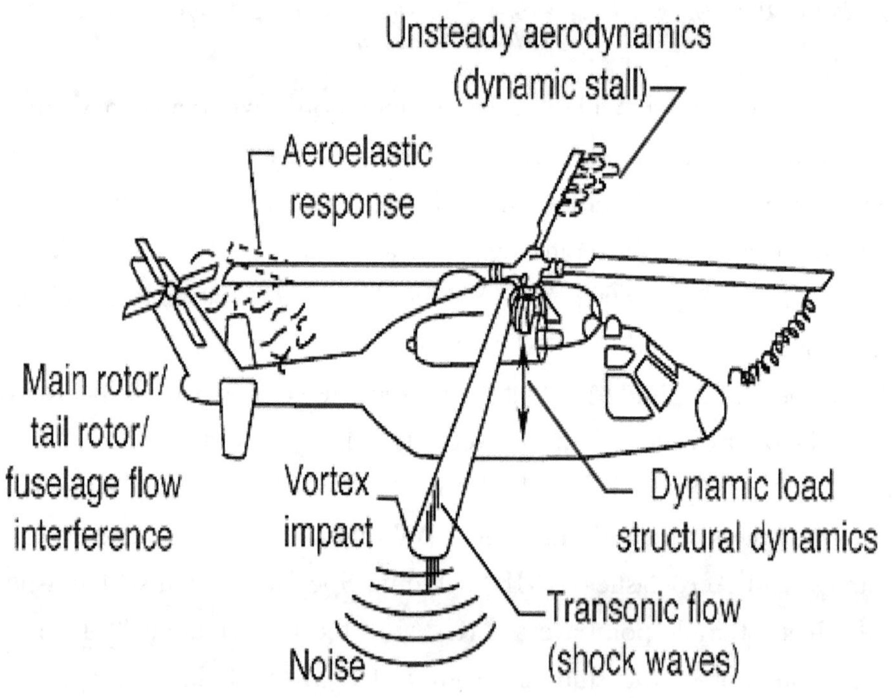

Fig.30. Vertical Lift Interdisciplinary Interactions.

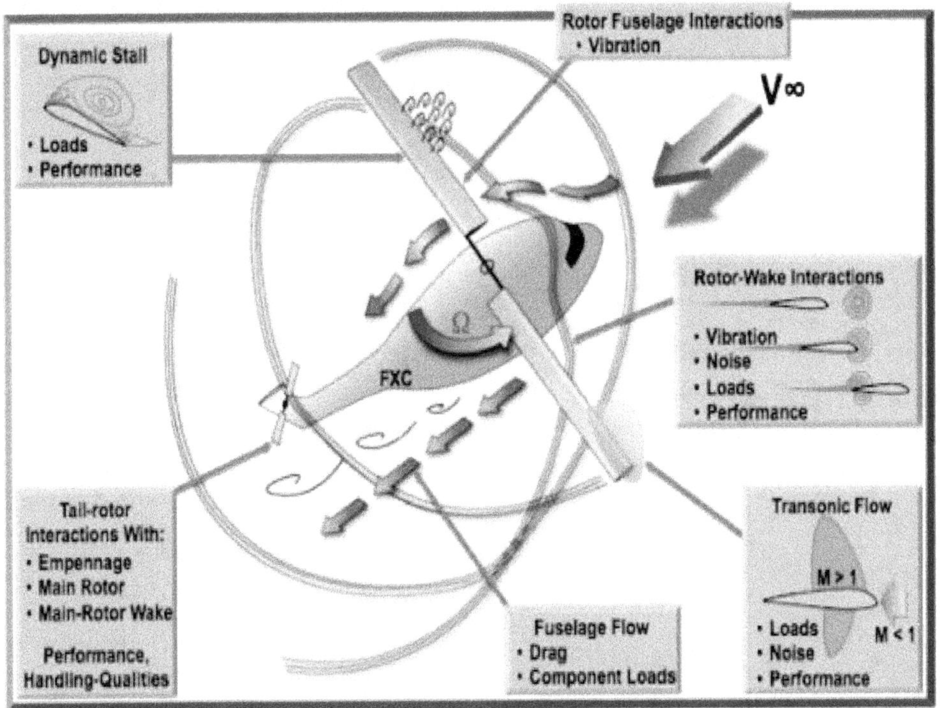

Fig. 31. Vertical Lift Aeromechanics Challenges

The Aerodynamics, Aeroelasticity, Structures and Materials, and Flight Mechanics & Controls disciplines are highly coupled, as illustrated in Figure 31A by Loops 1 to 5. Propulsion is another discipline that could be included. These coupled loops are illustrated in matrix form in Figure 31B. As can be seen, the interactive primary coupled loops (P) are shown in diagonal form, with secondary interactive loops (S) shown in off-diagonal form.

There is still a need for a Multidisciplinary Design, Analysis, and Optimization (MDAO) approach which requires a comprehensive analysis approach. While the Army, industry, and NASA have tried using comprehensive multidisciplinary analysis helicopter programs, such as C-81, 2GCHAS, RCAS, etc., since the early 1970s, MDAO approaches still need to be developed for early preliminary and detailed design assessment, tradeoffs, and analysis in vertical lift development programs. This became evident from my participation in the UTTAS, AAH, and LHX (RAH-66) development efforts.

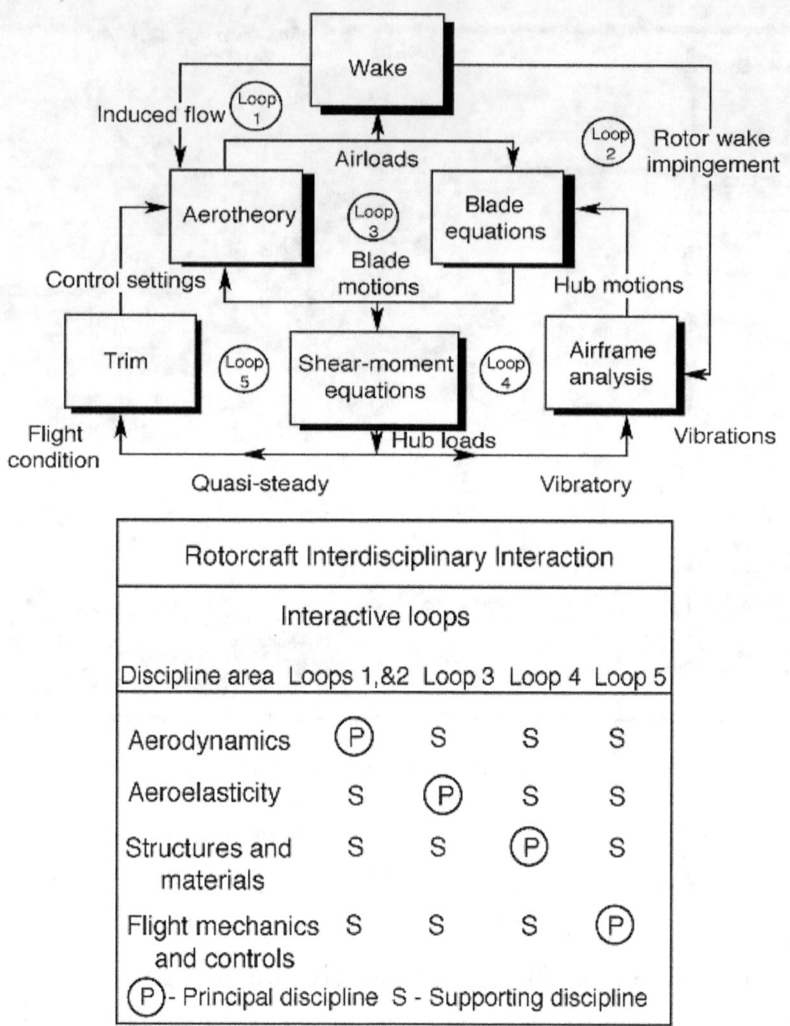

Figure 31A. Coupled Disciplinary Framework. Figure 31B. Matrix Analysis

The two years Nancy and I spent at Georgia Tech in Atlanta, GA, provided stability and the development of our family. The friends we made in our neighborhood became lifelong friendships that still exist today. We sold our house and made enough profit to invest in a larger home in O'Fallon, IL, Figure 32A, which was close to our parents' homes in Carlyle, IL, and St. Rose, IL, as well as close to my assignment at AVSCOM in mid-town St. Louis, MO.

Unfortunately, before our move to Illinois, my father had been diagnosed with a brain tumor that required a major operation. He had retired the year before at age 61 from teaching and principal at St. Rose Grade School from 1956-1973. Figure 32B. He and my mother had made a trip to Brussels, Belgium, to visit my sister, Juanita, her husband Bob, and their family, where Bob was in the Air Force and stationed at North Atlantic Treaty Organization (NATO) Headquarters. Returning home from their overseas trip, my father began feeling the ramifications of his brain tumor. Upon his return home, he had surgery for the tumor. The surgery left him pretty much an invalid for the remaining six months of his life. He lacked coherence, and I spent approximately a month with him in St. Rose, IL. It was during the late winter and early spring of 1974. He could sense that he was dying and wanted me to carry him outside so he could die and be buried. He died and was buried in the St. Rose Cemetery in March 1974. He was a great man and spent his life supporting his family. To me, he was a great father and served as my coach and mentor.

Fig. 32A. O'Fallon Home. Fig. 32B. Albin J. Schrage, St. Rose School Superintendent, 1956-73

CHAPTER THREE

EXPERIENCES AS AN AEROELASTICITY, DYNAMICS & VIBRATIONS (ADV) ENGINEER

Development of Utility Tactical Transport Aircraft System (UTTAS) & the Advanced Attack Helicopter (AAH)

My Introduction to the Aviation Systems Command (AVSCOM)

In June 1974, I arrived at my position as the Aeroelasticity, Dynamics, and Vibrations (ADV) engineer in the Aeromechanics Branch of the Structures & Aeromechanics Division in the Development and Qualification (D&Q) Directorate, AVSCOM. Charlie Crawford had hand-picked me to serve in this position based on recommendations from Drs. Robin Gray and Al Pierce. This included sending me to a Rotor Loads Conference in May 1974 in San Francisco organized by Dr. Bob Ormiston, the Army's top rotary wing aeromechanics engineer before I arrived at AVSCOM. I was the only active-duty military officer serving as an engineer among about ˜100 engineers who were civil servants. Several Army test pilots, active or retired, served in the

D&Q Directorate, providing coordination and direction for the Army Engineering Flight Activity (AEFA) at Edwards AFB, CA.

My top priority was to become familiar with the Utility Tactical Transport Aircraft System (UTTAS) and Advanced Attack Helicopter (AAH) Requirements and Prototypes being designed and developed. This included learning how to run the C-81 Flight Simulation Computer Program to evaluate them for performance and handling qualities. I soon became the troubleshooter for UTTAS/AAH technical and incident problems. This included conducting analysis, witnessing tests, and incident/accident investigations. The following are overviews of the UTTAS and AAH programs, which are followed by my stories.

I couldn't have arrived at AVSCOM in St. Louis, MO, at a more opportune time, as the next generation of Army rotorcraft was beginning development based on lessons learned from the Vietnam War. I felt I could leverage the three relevant experiences I had previously encountered—education, training, and operations.

The first experience came from my training and operational/technical experience as a nuclear weapons battery commander fighting the Cold War in Europe in 1968-69. My participation in REFORGER 1 taught me that repositioning elements of a division, the 24th Infantry Division, along the Rhine River for rapid deployment had to be questioned. This provided me with an understanding and belief that the U.S. needed to shift from the Cold War to Strategic Air Mobility.

The second experience confirmed this air mobility through my flying, commanding combat helicopter operations, and directing their operations in South Vietnam and Cambodia. The third experience followed, obtaining the best technical education and contacts I could receive at Georgia Tech.

Together, these three experiences positioned me with the best knowledge and expertise necessary for contributing to developing the Next Generation of Army Aviation Systems. At AVSCOM, I worked as an aerospace engineer, specializing in Aeroelasticity, Dynamics, and Vibrations (ADV) and supporting the development of the next generation of Army Aviation aircraft.

I became the Army's expert and troubleshooter for ADV on the four Army prototype helicopters, which were being developed by four different contractors: Bell, Boeing, McDonnell Douglas, and Sikorsky.

Additionally, I served on the Source Selection Evaluation Boards (SSEBs) that selected the two winners: Sikorsky for the Utility Tactical Transport Aircraft System (UTTAS) with the YUH-60A Black Hawk; and Hughes Helicopters for the Advanced Attack Helicopter (AAH) with the YAH-64A Apache. I also served four years at AVSCOM in St. Louis on active duty. I supported the development of the winning YUH-60A and YAH-64A prototypes and modifications to the other Army's rotary- and fixed-wing aircraft.

I also took college courses and earned a Master of Arts in Business Administration (MA) from Webster College in 1975 and a Doctor of Science (DSc) in Mechanical Engineering (ME) from Washington University (Washington U) in 1978. Dr. David Peters was my adviser. He returned to Washington U after serving at the Army Aero Flight Dynamics Lab (AAFDL) under Dr. Bob Ormiston and receiving his PhD at Stanford University. At Washington U, he served as a Professor with Dr. Kurt Hohenemser, his former adviser and a world-renowned rotary-wing engineer who had led the Army's technical assessment of the AH-56A Cheyenne half P Hop instability problem. Dr. Peters also served on the first UTTAS and AAH SSEBs in 1972 and 1973. My DSc thesis involved developing a new dynamics and stability analysis tool for evaluating rotor system loads and stability. I used it to assess aeroelastic stability and vibratory loads on the two UTTAS tail rotors on the YUH-60A and YUH-61A helicopters. The next chapter will summarize my PhD research.

PROTOTYPE DEVELOPMENT,

Lessons Learned, and Source Selection experiences for UTTAS &AAH (Leading to the Next Generation of Army Aviation Systems)

The Army Utility Tactical Transport Aircraft System (UTTAS) Program

Beginning around 1967, the Army initiated an aggressive effort to define the requirements for a utility helicopter system that would satisfy projected air-mobile operations in areas like Vietnam and other global points. An intense concept formulation effort between the Aviation Center at Fort Rucker, AL, the Infantry Center at Fort Benning, GA, and the Aviation Systems Command (AVSCOM) and its principal research activity at Fort Eustis, VA, was pursued over two years. Multiple aircraft configurations were considered to include pure helicopters, mixed-wing and rotor-compound designs, and counter-rotating tandem rotors. (*Army Aviation: Foundations of the Modern Fleet*, Prepared_by the Program Executive Office, Aviation, 2013. Ref2).

The Army released its specification to the industry for the UTTAS as part of its Request for Proposals (RFP) in January 1972, as illustrated in Figure 33. Submissions were made in due course by Bell, Boeing-Vertol and Sikorsky. The three companies submitted five proposals. The Army procurement strategy was to select two companies from those who submitted proposals to design and develop prototype helicopters to the same specification and using the same engine models. Boeing-Vertol and Sikorsky were selected to proceed with designing the Boeing YUH-61A and the Sikorsky YUH-60A in the Basic Engineering Development (BED) Phase. Request for Proposals (RFP) in January 1972, as illustrated in Figure 33. Submissions were made in due course by Bell, Boeing-Vertol, and Sikorsky. The three companies submitted five proposals. The Army procurement strategy was to select two companies from those who submitted proposals to design and develop prototype helicopters to the same specification and using the same engine models. Boeing-Vertol and Sikorsky were chosen to proceed in designing the Boeing YUH-61A and the Sikorsky YUH-60A in the Basic Engineering Development (BED) Phase.

Following the BED Phase, the two prototypes would undergo a competitive flight evaluation, Government Competition Test (GCT), by Army personnel. A winner for the sole-source Maturity Phase and Production would be selected based on the results of the competitive fly-off. When the Request for Proposals (RFP) was released, the Army's basis-of-need document established a total requirement of 1,107 aircraft. However, that quantity was later significantly increased during the production period as the UH-60 Black Hawk's versatility began to be understood.

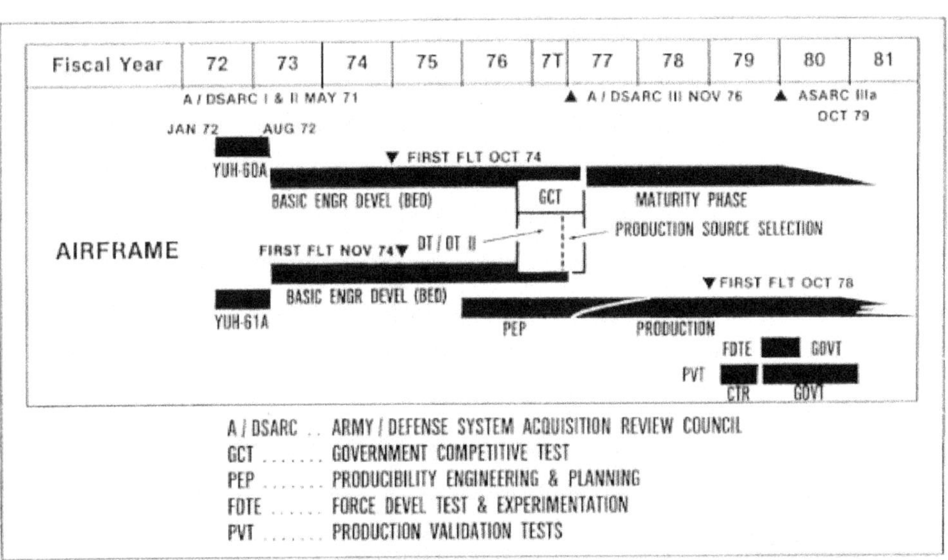

Figure 33. UTTAS Program Structure and Schedule (Ref. 14)

The UTTAS Program Structure is illustrated in Figure 33 with the Basic Engineering Development (BED) phases for the two winning proposals: the Sikorsky YUH-60 and the Boeing YUH-61. The Production Source Selection from the Source Selection Evaluation Board (SSEB) was at the end of 1976 with the selection of Sikorsky YUH-60A.

During the preceding decade, the Army recognized the need for a new utility transport helicopter. It began to draft its specification, basing it firmly on the lessons learned during the war in Vietnam. The YUH-60A was the Sikorsky prototype model, and the YUH-61A was the Boeing prototype model to be designed and developed for the U.S. Army from August 1972 through December 1976.

At the same time, the Army initiated the development of a new turbine engine, Figure 34A, matched to the performance requirements of its planned new helicopter. This new helicopter was to become the prime carrier of the Army's 11-man rifle squad with speed from hover to cruise speed at 145 knots at a 4000'Altitude, 95deg (4K95) day. Other missions included evacuating wounded personnel and resupplying troops in battle, often using low-level and Nap of the earth (NOE) flight. Affordability, supportability, survivability, and transportability were also essential requirements. (Ref.2)

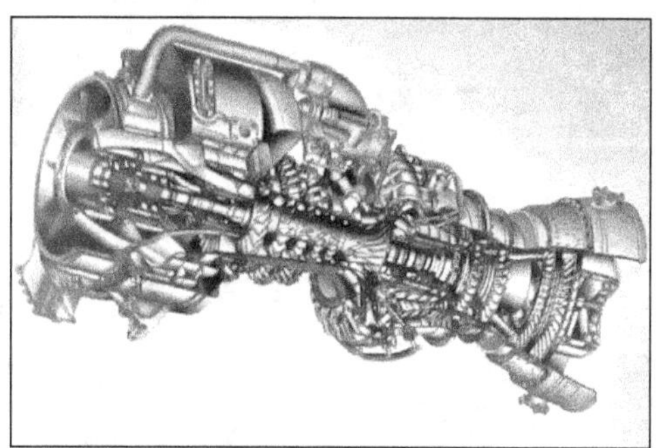

Figure 34A T-700 GE Engine. Figure 34B. 11-man Squad Carrying Helicopter[1]

In addition, the Army required the UTTAS to be easily air transportable within one hour with essentially no disassembly needed to fit inside U.S. Air

Force transports of that period, particularly the C-130 and C-141 models, Figures 35A and B. This air transport requirement turned out to have a significant impact on the UTTAS and the AAH designs, as well as creating significant technical problems that shaped the course of their development.

Figure 35A. YUH-60 Foldup for Air Transportability. Fig. 35B.Insertion into C-141 aircraft

The Sikorsky YUH-6O is shown in Figure 36A, and the Boeing YUH-61 prototype is shown in Figure 36B. The first flight of the YUH-60 was in October 1974, and the first flight of the YUH-61 was in November 1974. Both prototypes are shown in flight for comparison in Figure 36C. While the aircraft looked like each other, they had technical problems that had to be solved during the BED Phase, Figure 33. Sikorsky had more technical issues to solve but made the necessary changes to win.

Fig. 36A. YUH-60 in First Flight. Fig. 36B. YUH-61 in First Flight. Fig.36C. Both in Flight

Major technical and management problems occurred on both aircraft and are summarized in Figure 37.

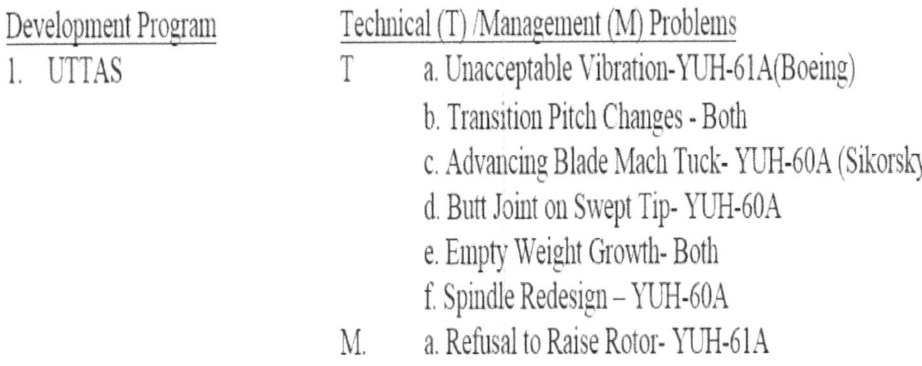

Figure 37. UTTAS Technical and Management Problems

Boeing successfully designed and built a hinge-less main rotor hub design for its YUH-61 prototype, Figure 38A. The main rotor hub is the heart of a helicopter. While the hinge-less rotor can provide simplicity and more maneuverability, it can also have higher vibratory loads. Boeing successfully built the YUH-61 hinge-less rotor with assistance from the German Company MBB. It was a much cleaner, simpler hub design than the Sikorsky YUH-60, Figure 38B, which required extensive use of elastomers, e.g., for bearings and dampers, to keep it from becoming a mechanical and maintainability nightmare.,

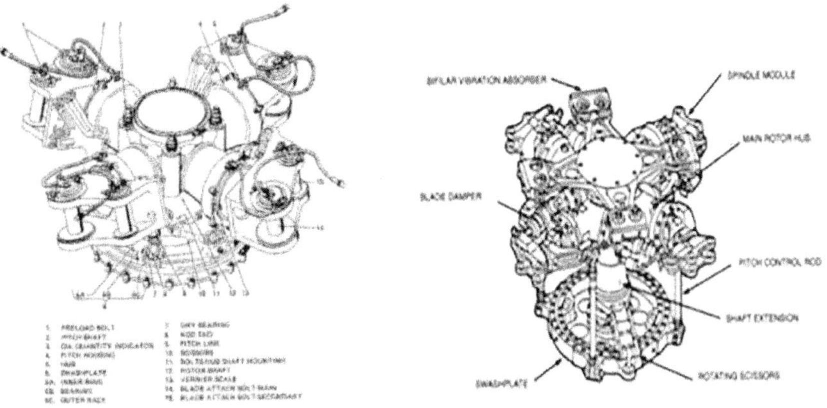

Figure 38A. YUH-61 Hub Design. Figure 38B. YUH-60 Hub Design

Successful changes made by Sikorsky on the YUH-60A are illustrated and described in Figure 39.

> **Sikorsky Changes to YUH-60 during BED Phase Provided Winning UH-60A**
>
> Main Rotor changed from Short Mast to Tall Mast for Vibration Reduction, a Major Design Change
>
> Swept Stabilator changed to Un-swept All Flying Stabilator for Pitch attitude Control & HQ Improvements
>
> - Swept Tips on Main Rotor Blades to prevent Mach Tuck and Reduce High Speed Loads
> - Droop snoot airfoils for > hover performance.
> - Recontoured Forward and Rear Fuselage changed to reduced Drag.
> - Replaced Two Place Window with Single Piece Window
> - Enlarged Exhaust and Recontoured Rear Fuselage for Flow Control and Performance Improvement

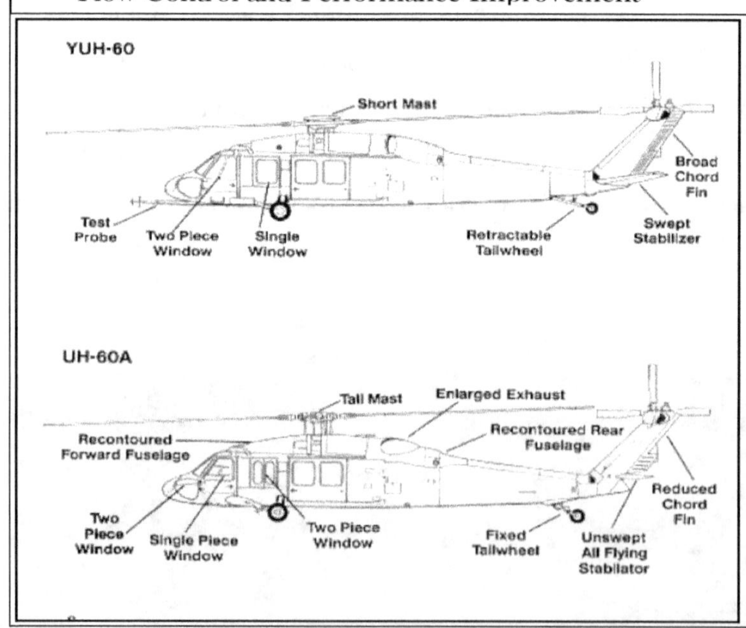

Figure 39. Significant Changes to YUH-60 in the BED Phase Resulted in UH-60A Winning Design

These changes made by Sikorsky during the BED Phase were substantial. Sikorsky solved the air transportability requirement problem with a taller, retractable mast (greater clearance over the tail boom), and they allowed the gear to kneel. This was the primary reason Sikorsky won the Production Contract and propelled the UH-60A Black Hawk Helicopter into the world's predominant transport helicopter.

Why did Sikorsky make the necessary changes to win the UTTAS Production Contract?

When the Army Request for Proposals (RFP) was released to industry in January 1972, the odds of Sikorsky winning one of the two Army contracts for UTTAS were thought to be very low by most industry experts. Sikorsky had not been an Army helicopter supplier for many years after ending its production of S-55, S-56, and S-58 models by the early 1960s. Since its peak output of 470 helicopters in 1958, Sikorsky's output declined for over fifteen years and reached a precipitous level when the UTTAS program began in 1972. A desperate company needs to make aggressive changes. The Sikorsky UTTAS Management Team was aggressive as led by Bill Paul, Bob Zincone, and Ray Leone.

My Stories on the UTTAS Development Program & Source Selection Evaluation Board

My experiences as the Army Aviation ADV engineer and problem solver from the AVSCOM D&Q Director for the UTTAS and AAH Program Management Office (PMO) are described. This position provided an unusual technical oversight. It also helped me form a close relationship with the UTTAS and AAH industry teams' engineers and gave me first-hand knowledge of the problems that needed to be solved. Some examples will now be discussed.

Vibration Problems.

The Army UTTAS and AAH vibration requirements were the most stringent ever applied for helicopters. They are identified in Figure 40, along with the history of helicopter vibration requirements over the years. None of the four competitors in the two fly-off programs could meet the stringent 0.05 Gs at blade passage, n/rev, where n is the number of blades. Because of this, the limit was raised to 0.10 Gs at blade passage but kept at 0.05 Gs below this frequency for the production UH-60A Black Hawk and AH-64A Apache. Thus, the designers were somewhat relieved of this problem, although these new limits were seldom met.

As Ray Prouty, Hughes Helicopters engineer, articulated, "Just because we humans can take high vibration at high frequencies does not mean that equipment and structural components are happy under those conditions. To be really good, the ultimate goal, as stated by some dreamers, is to get the maximum vibration level down to less than 0.03 Gs for all frequencies. That would truly be a 'jet-smooth ride."

A "jet-smooth ride" was beyond state of the art for UTTAS/AAH and is today.

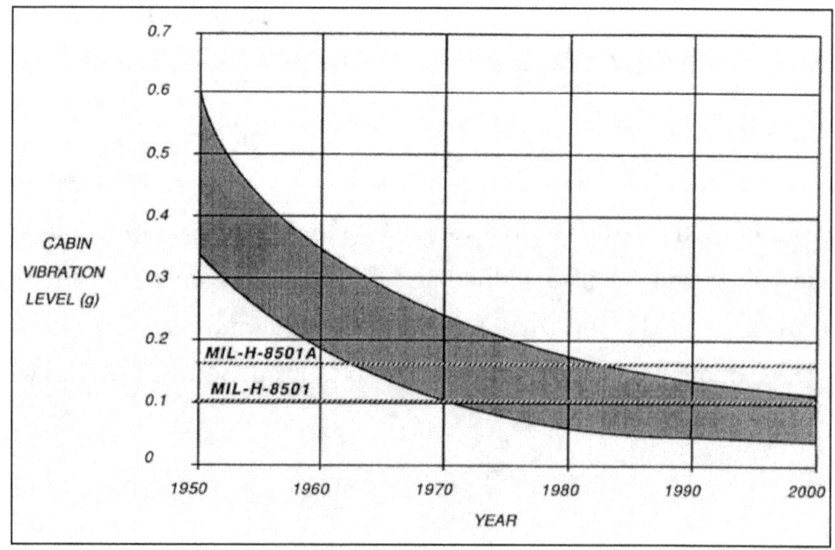

Figure 40. History of Helicopter Vibration

UTTAS/AAH Vibration Requirements

1. <u>Original</u> SpecificationRequirements: Max allowable vibration to be less than .05g's at frequencies at blade passage n per rev (nP) and below at 1P. Above this range, the vibration limit was raised between the one- and four-hour human tolerance limits

2. <u>Relaxed Vibration Requirements</u> the limit was raised to 0.10 Gs at blade passage, but kept at 0.05 below this frequency, for the production

*Requirements (**Helicopter** Aerodynamics Volume II. Ray **Prouty**))*

To make the UTTAS/AAH Vibration Requirements easier to understand and closer to human and equipment intrusion indices, I simplified them to one equation, $G = .004f + .01$. Therefore, if the main rotor turned at ~ 300 RPM, 5 Hz, then the 1P requirement would be .03 Gs. If it was a two-bladed rotor, e.g., Bell YAH-63, 2P would be ~.05 Gs at 10 Hz, and for a four-bladed rotor, e.g., Boeing YUH-61, Hughes YAH-64, Sikorsky YUH-60, 4P would be ~ .10Gs at 20 Hz.

This approach also provided more realistic requirements for higher rotor harmonic resonances, such as at N-1P and N+1P and 2N-1P and 2N+1P. Vibration requirements for second-order aircraft forcing frequencies, e.g., 4P (2-bladed rotors), 6P (3-bladed rotors), and 8P (4-bladed rotors), had often not been specified. However, the 6P on the Boeing Heavy Lift Helicopter (HLH) X62 Prototype was a problem, as illustrated in Figure 41A with the two red circles indicating resonances with 3P (n per rev) and 5P (2n-1 per rev) with second flap-wise frequency. The CH-47D with fiberglass rotorcraft blades also has had a significant vibration problem with 6P due to 5P(2n-1) resonance close with the second flap-wise bending mode. It will be discussed in chapter six.

Illustrated in Figure 41A are blade natural frequencies and two possible resonances for a three-bladed articulated rotor, such as the Boeing-Vertol Army Heavy Lift Helicopter (HLH) prototype or the CH-47D Chinook.

Illustrated in Figure 41B are blade natural frequencies for the Sikorsky X-2 Coaxial Compound Technology Demonstrator, which has good separation from NP, NP-1, NP+1 and 2NP, 2NP-1, and 2NP+1. This separation is very critical for stiff in-plane coaxial rotors, such as the Lockheed/Sikorsky FARA & FLRRA prototypes. The Coaxial Compound would have been a natural for FARA (See Epilogue).

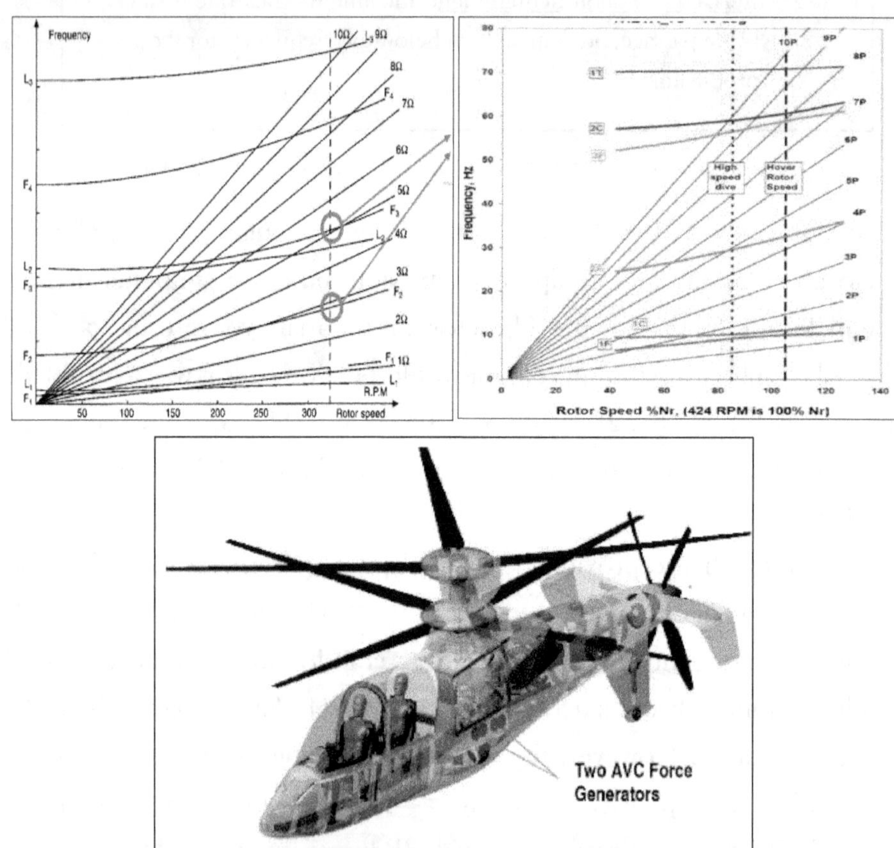

Fig.41A.HLH Blade Resonance Bladed Rotor. Fig.41B. X2 Frequencies & 4Bladed Coaxial Compound

All four of the Utility Tactical Transport Aircraft System (UTTAS) and Advanced Attack Helicopter (AAH) prototypes had rotors too close to their fuselage resulting in vibration problems. They could meet the air transportability requirements and be air transportable inside C-130 and C-141 aircraft with essentially no disassembly and meet the time allowed.

However, both Boeing and Sikorsky had severe vibration problems due to the up flow of air into, through, and out, of the main rotor system, as well as lack of adequate rotor and airframe frequency separation. As a result, Sikorsky carried about 400 lbs. of vibration absorbers in their YUH-60 aircraft to counter the vertical shears and moments transferred into the airframe from the rotor. Their bifilar vibration absorber in their main rotor system only addressed in-plane shear and moment loads.

Also, Boeing had airframe natural frequencies that resonated with the main rotor forcing frequencies. After Sikorsky proposed a successful deviation from the air transportability requirements and raised their rotor, the two AAH contractors, Bell and Hughes, took deviations and raised their rotors too. Boeing leadership, however, decided not to raise the Main Rotor on their YUH-61A, which had a very high-vibration environment. This is listed as the principal management problem in Figure 37 from the UTTAS BED Program. During Basic Engineering Development (BED), I witnessed tests and conducted analyses on the major UTTAS technical problems. I worked with the Boeing UTTAS engineers trying to solve their vibration problems. They knew, and I knew, that raising their rotor would significantly reduce vibration. However, Boeing UTTAS management thought the air transportability requirement was sacrosanct, even though the other UTTAS and AAH contractors took successful deviations.

Possibly a better UTTAS aircraft than the YUH-60 lost out because Boeing would not raise their rotor on the YUH-61. They could not figure out how to reduce the vibration caused by the main rotor exiting the airframe, without raising their main rotor. They tried stiffening the airframe, which increased the empty weight affecting performance without much success.

They tried vibration isolation using the Kaman version of the Improved Rotor Isolation System (IRIS). However, tuning in one direction aggravated vibration in the other directions. Three each of the YUH-60 and YUH-61 prototypes were developed and tested at three different Army test sites, Fort Rucker, AL, Fort Campbell, KY, and at the Army Engineering Flight Activity (AEFA), Edwards Air Force Base, CA.

While serving as the ADV engineer on the UTTAS Production SSEB, I received feedback from all three UTTAS test sites that vibration on the YUH-61A was unacceptable. In addition, the proposal submitted by Boeing recognized they had a vibration problem and that they were continuing to use their company owned UTTAS prototype to find solutions. They proposed to have the Army evaluate their solution at their Boeing Site outside Philadelphia, PA, several months after the Source Selection Evaluation Board (SSEB) had convened for its six-month evaluation.

After three months, Boeing was informed by the UTTAS Production SSEB Chair that they had to demonstrate a proposed successful vibration solution. Thus, the Commander of the Army AEFA, COL Denny Boyle, his senior flight test engineer, John Blaha, the chief test pilot for the AAH Hughes YAH-64 prototype, MAJ Bob Stewart, and I flew up to Boeing Philadelphia to test and evaluate their vibration solution on Boeing's company YUH-61 prototype. Several flights took place with MAJ Stewart flying with the Boeing YUH-61 test pilot. In the evening, John Blaha and I evaluated the recorded test data. This review and MAJ Stewart's and COL Boyle's assessment showed that the vibration problem had not been solved.

MAJ Stewart went on to become the Army's first astronaut. As a result, when I got back to the AAH SSEB in Granite City, IL, I began drafting up the assessment that excessive vibration without a defined solution was still a major deficiency. After review with the UTTAS SSEB Director and other key members, a Red Flag was written, which indicated the vibration was still a significant deficiency for the Boeing YAH-61 Prototype. On a rating scale of 1 to 10, a Red Flag indicated a score of 1. It is interesting to note that the Sikorsky YUH-60 Prototype was given a vibration score of 4 due to moderate vibrations and the approximate 400-500 lbs. of vibration absorbers and isolation it had to carry on the aircraft.

YUH-60A Operational Test Incident at Fort Campbell, Kentucky

During operational testing at Fort Campbell, KY, on the night of August 9, 1976, at 11:15 p.m., the Sikorsky YUH-60 prototype, loaded with 14 Army personnel flying a night mission encountered severe aircraft vibrations. The crew declared an emergency and, thinking they were landing in a cornfield, landed instead in the middle of a pine forest. Ground fog and poor visibility contributed to the error. The aircraft's rotors cut through pine trees up to five inches in diameter, as illustrated and described in Figure 42. Fortunately, the helicopter landed intact with little to no damage to the airframe. No injuries resulted, with the exception of one soldier receiving a minor cut after hitting his head on a tree as he exited the aircraft. The next morning, a path was cleared to the site, and a technical team comprised of Sikorsky and Army personnel inspected the aircraft.

During the controlled descent, the YUH-60A chopped through 40 trees multiple times leaving a forest that appeared to have been cut down by a giant rotary lawn mower. The only damaged parts were the main and tail rotor blades. However, the blade spars sustained no damage. A main blade, the one shown pointing to the right, shed part of its fiberglass skin during flight which caused the strong vibration.

Figure 42. YUH-60A Operational Test Incident at Fort Campbell, Kentucky

When the operational test incident occurred, I was attending a one-week Advanced Aeroelasticity Course at UCLA, taught by Dr. Peretz Friedmann. Monday evening after the incident, I received a call from Charlie Crawford, my boss, as the AVSCOM Director for Development and Qualification. I was told to get to the crash site immediately. I took a "red-eye" flight from Los Angeles, CA, to Nashville, TN. I then drove to Fort Campbell, KY, which is about two hours from Nashville.

When I got to the crash site, I was met by MG Jerry Lauer, the UTTAS Project Manager, and several Army colonels. Although I was an active-duty Army Captain, I was in civilian clothes as I had come from the UCLA Advanced Aeroelasticity Course. MG Lauer told me, "Dan, they (Sikorsky) are going to want to fly the aircraft out." He said, "I want you to go through every inspection you can do on this aircraft." Since there was a clear sudden stoppage of the main and tail rotors, their transmissions would have to be completely inspected.

In addition, shortly after the incident, Sikorsky had an aircraft fly down with key managers, Ray Leone, YUH-60 Chief Engineer, and other engineering personnel, along with new main and tail rotor blades. The new YUH-60 rotors were inspected and installed by Army personnel, Figure 43A. Five days after the crash, all inspections were successful. The aircraft, stripped of all non-essential items such as passenger seats, was brought to a check hover, Figure 43B, and then flown out. Ray Leone, standing by the downed YUH-60A, is shown in Figure 43C.

Figure 43A. Re-installment of Rotors. Figure 43B. Hover Check.,

Figure 43C. Ray Leone comments.

The following quote by Ray Leoni showed how what appeared to be a disaster for Sikorsky turned out to be a successful demonstration of the crashworthiness of the YUH-60. It also inadvertently tested the requirement for its rotor blades to be able to cut through a four-inch tree without failing.

"They took off out of that pine forest to fly back to the base on Fort Campbell from where the testing originated. On the way back to the Sikorsky base, they made a high-speed pass over the Boeing campsite out of the spirit of rivalry. They made a 150-knot pass right over their hangar. Boeing knew that we had a crash. But, whatever (celebratory) feelings they had changed when they saw our airplane zipping by."

The minimal damage sustained by the aircraft earned great praise from the Army, converting what could have been a tragic event into a victory for Sikorsky.

However, a quick solution for the delamination of a main rotor blade had to be made by Sikorsky and the Army. Any plans that I had to return home to St. Louis following my stent at Fort Campbell, KY, were not going to happen. Charlie Crawford instructed me to proceed to Sikorsky and help them solve the main rotor delamination problem illustrated in Figure 44A. It was quickly determined that the swept tip installed temporarily as a butt joint at the end of the main rotor blades was the main source of the delamination. The butt joint was a quick fix after it was deemed that a swept tip was necessary to counter advancing blade Mach Tuck effects in high-speed flight (Figures 44B and C). While Blade Stall had a problem with symmetric airfoils, Mach Tuck is an aeroelastic load producer of highly cambered airfoils, such as those used on the YUH-60 and the AAH YAH-64. The butt joint was eliminated with fibers linking the blade and tip.

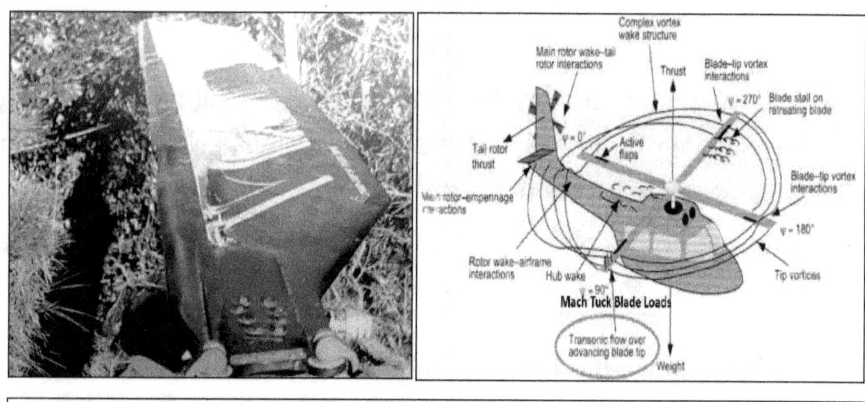

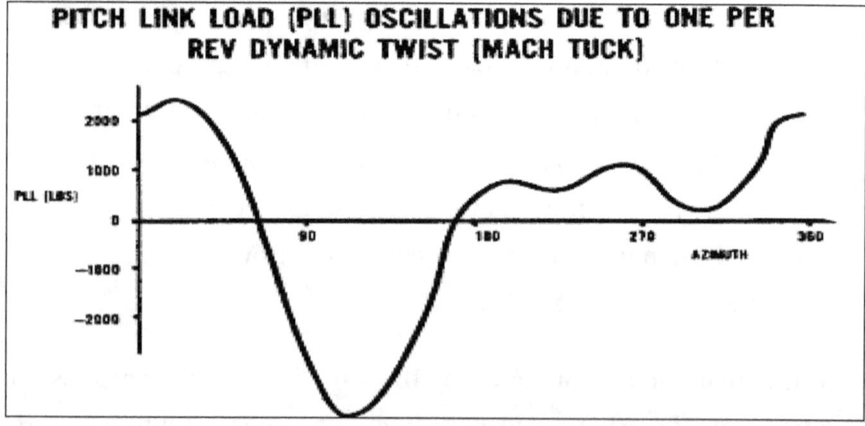

Figure 44A. YUH-60 Delamination. Figure 44B. Aeroelastic Problems. Figure 44 C. Mach Tuck

Shortly after the resolution of the rotor blade delamination problem in the Sikorsky YUH-60 operational incident, and Boeing's failure to demonstrate a vibration solution for the Boeing YUH-61, the down-selection was completed, and Sikorsky was declared the UTTAS competition winner. This was based on an overall lower-risk assessment by the Army. Lower vibration levels compared to Boeing's UTTAS were also a major contributor, but more importantly, Sikorsky had demonstrated a management style deemed impressive to the Army. (*Army Aviation: Foundations of the Modern Fleet,* Prepared by the Program Executive Office, Aviation, 2013, Ref.2)

The Army Advanced Attack Helicopter (AAH) Program

Airmobile operations, as envisioned by the Howze Board, heavily emphasized helicopter air assault. In the Vietnam War, the Huey lift ships and gunships, UH-1B and UH-1C models, Figures 45A and 45B, proved themselves.

This was especially during the 68 Tet Offensive in the Mekong Delta, where they effectively saved three Army Aviation Airfields. In 1969 the AH-1 Cobra attack helicopter was introduced and provided overall better support than the UH-1 gunships. It was able to assume all of the missions presently assigned to older UH-1B/C model gunships. The speed and the increased ordnance load, coupled with the versatility and accuracy of the armament systems, made the aircraft exceed the gunships' capabilities and gave the ability to provide broader tactical capabilities. However, in the Mekong Delta, the UH-1C gunships with four crew; a pilot, co-pilot, crew chief, and door gunner, could provide 360 deg coverage which was more effective against the Viet Cong (VC) in low altitude and low-speed flight. The VC would wait for the AH-1G to hover nearby or pass by and engage the AH-1G from the rear. Therefore, the AH-1G Cobras in Air Cav troops were taught to attack from altitude with the armed light observation helicopters (LOHs or Loaches) providing the necessary low-altitude firepower.

Figure 45A. UH-1B Gunship Figure 45B. UH-1C Gunship.

However, the AH-1G attack helicopter's limitations in payload, flight performance, and vulnerability, fostered the need for a more suitable aircraft with far greater capabilities. In 1965, the Army started the AH-56 Compound Attack Helicopter (CAH), as illustrated in Figure 29A, from bottom to top: the AH-1G Cobra, AH-64 Apache, and AH-56 Cheyenne. Illustrated in Figure 29B is a rear view of the AH-56 CAH with its tail rotor and prop for high speed. The AH-56 Cheyenne was developed by the Lockheed Aircraft Company as an advanced compound attack helicopter and is illustrated in Figure 45C and Figure 47.

Figure 46A. AH-1G Cobra AH-64 Apache & AH-56A Figure 46B. AH-56A Tail Rotor & Prop

Figure 46C. AH-56 Cheyenne in Hovering and in Flight, Figure 31, in 1970

Figure 47. AH-56 Cheyenne Program

The AH-56 project suffered a setback on 12 March 1969, when the rotor on prototype #3 (s/n 66-8828) struck the fuselage and caused the aircraft to crash, killing the pilot, David A. Beil. The accident occurred on a test flight where the pilot was to manipulate the controls to excite 0.5P oscillations (or half-P hop) in the rotor. The 0.5P is a vibration that happens once per two main rotor revolutions, where P is the rotor's rotational speed. The accident investigation noted that safety mechanisms on the controls had apparently been disabled for the flight. The investigation concluded that the pilot-induced oscillations had set up a resonant vibration that exceeded the rotor system's ability to compensate. After the investigation, the rotor and control systems were modified to prevent the same problem from occurring again.

The Cheyenne was a compound helicopter capable of dash speeds in the 220-knot range. Its combination of wing, rotor, propeller, and turboshaft engine, allowed for high-speed maneuverability similar to a fixed-wing aircraft.

Unfortunately for the future of the Cheyenne, two significant issues arose in the early 1970s, which resulted in its arrested development and ultimate termination. Lockheed had encountered difficult technical challenges relating to a rotor aero-servo-elastic instability, designated as half p hop, that took almost two years to resolve. Perhaps more importantly, helicopter-centric operations in Vietnam were countered by the emergence of the enemy's use of shoulder-fired ground-to-air missiles. Cheyenne's capabilities had been based on more traditional fighter plane concepts that involved 'swooping in' from altitude to suppress ground targets with direct fire, but the threat of shoulder-fired missiles made the operational concept impractical, forcing a major change in air assault tactics. To counter the missile threat, aircraft would need to fly slower and closer to the ground to avoid engagement.

While the Cheyenne excelled at achieving higher speeds, it was an impractical aircraft for lower speed, Nap of the Earth (NOE) operations, as illustrated in Figure 48A. UTTAS Required a low-level longitudinal maneuver, e.g., a pull-up followed by a push-over, commonly called the "UTTAS Maneuver." This maneuver pretty much limited UTTAS prototypes to articulated and hinge-less rotors and not semi-rigid rotors, like the AH-1 Cobra; although the insertion of a hub spring, such as on the Navy Cobras, can be more acceptable.

On the other hand, Scout Attack (SCAT) aircraft like the Hughes OH-6A Cayuse "Loach" and the OH-58A/C required longitudinal, lateral, and directional maneuvers, although the OH-58A/C was never accepted as a Nap of the Earth (NOE) aircraft in Vietnam. Figure 48B illustrates how an AAH and SCAT, along with an Air Force A-10 Close Air Support (CAS) aircraft, can be used on the battlefield to defeat Soviet Tanks, and to some extent, Soviet Attack Helicopters such as the HIND, which is also illustrated.

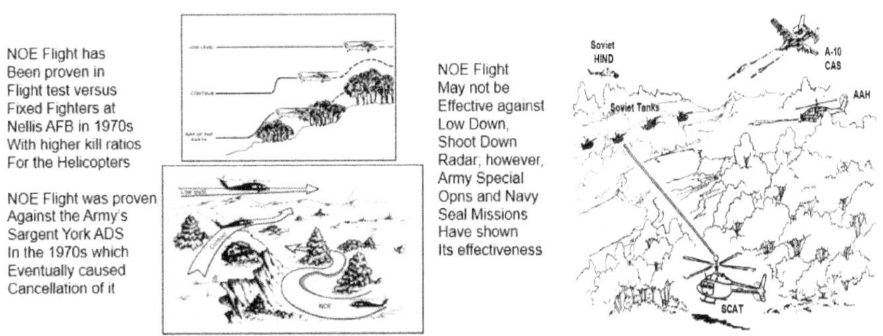

Figure 48A. Required Helicopter M/A on Battlefield. Figure 48B.CAS on Modern Battlefield

The Army went into an accelerated concept formulation effort that resulted in new requirements and specifications calling for a conventional helicopter configured for the attack mission. Emphasis was placed on low vulnerability, crashworthiness, and an integrated target acquisition and weapons system. This led to an accelerated effort between Aviation Systems Command (AVSCOM) and the Aviation and Armor Centers of the Training and Doctrine Command (TRADOC) to develop a revised requirement for an Army Advanced Attack Helicopter (AAH), which was quickly approved by the Department of the Army in 1972. A request for proposal (RFP) was completed and released later in the same year for a competitive fly-off similar to that for UTTAS.

Key requirements for the AAH included the capability of sustained 145 knots cruise speed, a dash speed of 160 knots, extensive agility and maneuverability, the ability to hover out-of-ground-effect (HOGE) at 4k95degs, invulnerability to small-arms fire to include 23-millimeter, high-explosive incendiary munitions, crashworthiness, excellent reliability and maintainability, and the ability to be loaded onto a C-130 in less than an hour, same as the UTTAS.

Five companies responded to the RFP—Lockheed, Boeing, Bell, Sikorsky, and Hughes. All five designs were conventional helicopters. The three losing designs are shown in Figures 49A, B, and C. The winning Bell and Hughes designs are in Figures 50A and B..

Figure 49A. Boeing 235   *Figure 49B. Sikorsky S-71*   *Figure 49C. Lockheed CL-1700*

Figure 50A. Bell YAH-63 1ˢᵗ flight Oct. 1, 1975. Figure 50B. Hughes YAH-64 1ˢᵗ Flight Sept.30, 1975

The air vehicle design was driven by a desire to use the same T700 engine developed for the UTTAS. The Source Selection Evaluation Board (SSEB) convened in 1973 and spent almost six months evaluating proposals and conducting technical and programmatic reviews with the five companies. Lockheed fell short of being fully responsive to the RFP, and Sikorsky and Boeing had major shortcomings in their proposals. The Bell and Hughes proposals were judged to be technically responsive. Both companies were awarded competitive prototype development contracts, as illustrated in Figure 51, with the Bell prototype designated as the YAH-63 and the Hughes prototype as the YAH-64, as illustrated in Figures 50A and B.

Hughes Helicopters YAH-64 was selected and won the Contract Award in December 1976. A picture of the AH-64 Apache with the AH-1 Cobra that it eventually replaced is illustrated in Figure 52.

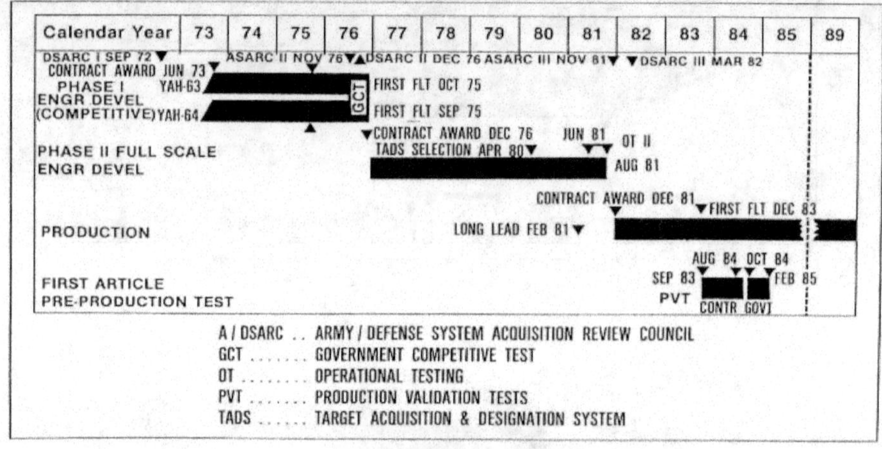

Figure 51. AAH Program Structure and Schedule

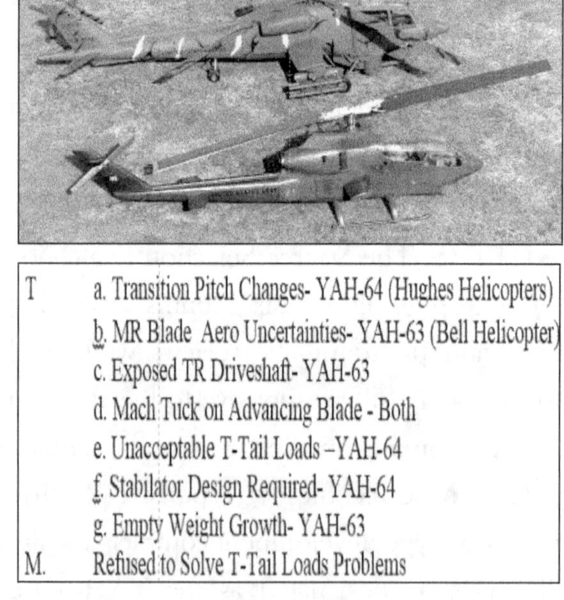

T	a. Transition Pitch Changes- YAH-64 (Hughes Helicopters)
	b. MR Blade Aero Uncertainties- YAH-63 (Bell Helicopter)
	c. Exposed TR Driveshaft- YAH-63
	d. Mach Tuck on Advancing Blade - Both
	e. Unacceptable T-Tail Loads –YAH-64
	f. Stabilator Design Required- YAH-64
	g. Empty Weight Growth- YAH-63
M.	Refused to Solve T-Tail Loads Problems

Fig. 52A. A picture of the YAH-64 Apache & the AH-1 Cobra. Fig. 52B. YAH Tech/Mgt Problems

My Stories on the Advanced Attack Helicopter (AAH) Development Program

Major Technical Problems with the Bell YAH-63 Prototype

The Phase I Engineering Development Program for the AAH was to develop the air vehicle and the necessary interfaces with the Target Acquisition & Detection System (TADS) and Weapon Systems (Hellfire Missiles, Rockets, and Guns). Bell's YAH-63 prototype was a disaster. Major General Ed Browne, then a Colonel, visited Bell shortly after accepting the assignment as the AAH project manager and offered the company some sobering observations. (*Army Aviation: Foundations of the Modern Fleet*, Prepared by the Program Executive Office, Aviation, 2013, Ref. 2)

Bell asked, 'What do you think of our approach?"

I said, "Well I think it is a very poor approach."

Bell said, "What do you mean?"

I replied, "Well first of all, when you've got two rotor blades that weigh over 500 pounds apiece, maintainability issues are going to bite you. Secondly, you've got the pilot up front and the observer in the back, and you have a 13-foot optical relay tube that is going to cause all kinds of problems when you are trying to look through a device like the TADS and have all of that twisting and turning from the natural torques and twists of the airframe. "

A three view of the YAH-63 prototype is illustrated in Figure 53 with some of its major problems.

1. Rotor Blade Weight a Maintainability Problem
2. Pilot in Front Seat Wrong for Army
3. Transition properties between
   Main rotor airfoils not understood.
4. Large rotor deflection contacted open Tail Rotor Drive shaft in Slope Landings
5. Rotor had to be raised to prevent contact and cover for drive shaft
6. Semi-rigid teetering rotor not good for NOE Flight and low g maneuver

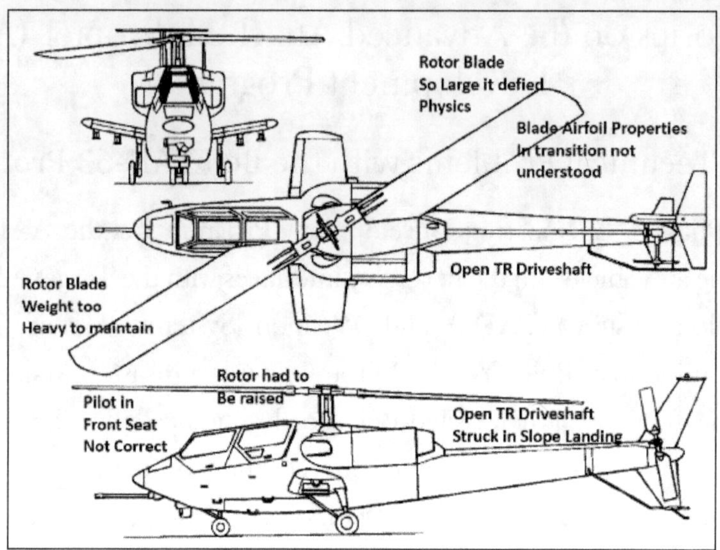

Figure 53. Description and View of Some Major Problems with the YAH-63 Prototype

- **Major Problems 1 and 2** were identified by COL Ed Browne to Bell. My assessment of the Y-63 prototype design from an ADV viewpoint was equally very negative.

- **Major Problem 3** was that the main rotor blade transitioned through several different airfoils, with the aerodynamic and aeroelastic transition airfoils not properly understood. This resulted in the YAH-63 being unable to make its cruise or dash speeds, e.g., 145 KTAS, 160 KTAS.

- **Major Problem 4** was encountered as the first prototype experienced an emergency crash landing which suffered some damage. I served as the technical expert on the accident/incident investigation board, which went to Bell to understand the problem and potential solutions. Bell proposed to the board that the main rotor blade trim tab contacted the open tail rotor drive shaft, causing an initial crack in the drive shaft. Bell believed it occurred during autorotational landings on the previous day of flight. However, my review of flight data did not show a large enough rotor flapping to contact the open shaft. I then reviewed slope landing data from previous flights and saw large flapping, which I believed caused the initial trim tab contact with the drive shaft. I then asked the Bell crew chief to go with me early the next morning to the compass rose at Arlington Airport to check out the sight where the slope landings had been conducted. We then found the broken trim tab from the main rotor blade, which confirmed the contact had been made during the slope landings. Obviously, when this was presented to the Bell investigation management team, they were embarrassed that the Army had identified their problem.

- **Major Problem Number 5** resulted in the main rotor being raised and a cover installed for the tail rotor drive shaft.

- **Major Problem Number 6** was that semi-rigid two-bladed rotors are not adequate for NOE and low G flight. This was confirmed in government flight testing and in Vietnam with the OH-58A/C SCAT, where mast bumping occurred. Bell also had a major management problem with the design and development of its YAH-63 prototype. Its AAH management team was housed in a separate site well removed from the Bell Engineering and Technology Team. While the management team had experienced managers familiar with Bell's proposal experience with two-bladed teetering rotor aircraft and had retired military who thought they knew what the Army needed, they did not adequately use the necessary technical expertise from Bell's strong engineering team. This team was headed by Bob Lynn, VP for Engineering, along with strong key technology managers, such as Rod Balke, Troy Gaffey, Tom Wood, and Jing Yen who should have had more influence on the design.

Aeroelastic, Dynamics and Vibrations Problems Encountered on the YAH-64 Prototype

The Hughes YAH-64 Technical Team was not large, but contained key technology area leads who had successfully developed the OH-6A Cayuse Scout Helicopter, which was the most popular combat helicopter in the Vietnam War. Dr. Kenneth B. Amer was Manager, Technical Department; Sally V. LaForge, Chief, Performance & Computer Engineering; James R. Neff, Chief, Dynamics Analysis; Ray W. Prouty, Chief, Stability & Control; James McDermott, Chief, Structures & Materials and Hank Sawicki, Chief, Transmissions.

However, there were major technical and management problems with the initial YAH-64 prototype that had to be resolved for Hughes to win the AAH Phase II Engineering Development Program, Figure 51. A three-view of the YAH-64 Prototype with identified Major Problem Areas is shown in Figure 54. Lessons Learned from the Sikorsky YUH-60 changes were incorporated into the YAH-64 with Army assistance in Phase I and at beginning of Phase II to move from a T-Tail to a Stabilator.

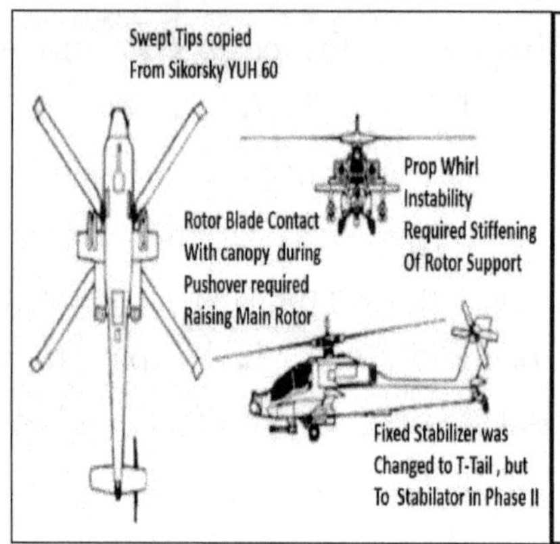

1. Swept Tips were required to control Mach Tuck Loads
2. Prop Whirl Instability shown on Whirl Tower >V required stiffening of rotor support
3. Rotor Blade Contact with Canopy during pullup and pushover required raising rotor
4. Dual Lead-Lag Dampers prevented Ground Resonance with one damper inoperative; however, increased chordwise blade bending loads. Elastomeric dampers cut to reduce loads
5. Low Fixed Stabilizer caused unacceptable aircraft pitch transition; T-Tail Change caused unacceptable loads & vibration in high speed flight

Figure 54. Description and View of Some Major Problems with the YAH-64 Prototype

- **Major Problem 1** was similar to the YUH-60 Mach Tuck Problem, as both used highly cambered airfoils resulting in Mach Tuck pitch link loads in high speed forward flight on the advancing blade, Figure 55A, requiring a swept tip rotor blade as illustrated in Figure 55B.

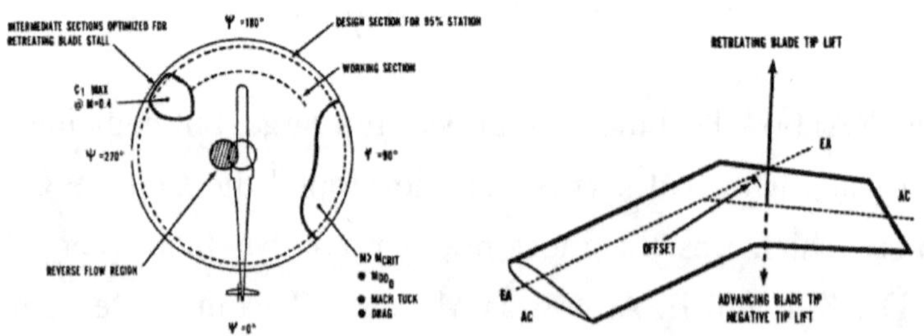

Figure 55A. Critical Regions in Forward Flight. Figure 55B. Offset Load due to Tip Sweep.

- **Major Problem 2** was an unstable main rotor advancing whirl mode predicted near the operating rotor speed, due primarily to the relatively soft rotor support structure summarized in Figure 56, and the coupling of cyclic pitch changes with hub rolling and pitching motion relative to the fuselage. Subsequent tests confirmed the instability on the whirl tower at Rye Canyon, CA, and the Ground Test Vehicle (GTV) at Palomar Airfield, Carlsbad, CA. I was the government test witness for both tests and documented them to the AAH Program Office (PMO).

Summary of YAH-64 Whirl Mode Instability

Soft Rotor Support to reduce & isolate 4P Vibrations from Rotor to Airframe. Whirl Mode symmetric at ~ 3P

Military Requirement for Freedom from instabilities is 1.15 x Max Rotor Speed, N_R = 1.15 X 110% = 126.5 N -

Whirl Tower Instability occurred at ~ 125 N_R with stiffer support

GTV Instability occurred at ~ 108 N_R with YAH-64 softer rotor support.

Therefore, Rotor Support Stiffness required   -

Design Changes in Phase 1 Development for Phase 2 eliminated the whirl mode instability with a flight envelope to 126% N_R and 3.5g load factor.

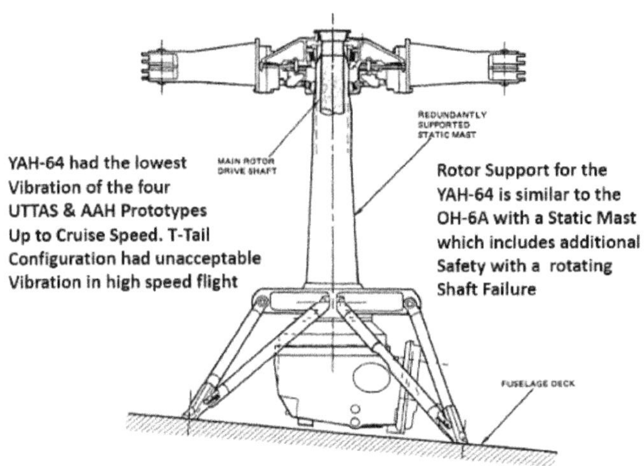

Figure 56. Summary of YAH-64 Whirl Mode Instability and How It was resolved in Phase 2

I received a Letter of Appreciation on 14 July 1975 from BG Samuel G. Cockerham, AAH Program Manager, for my contributions when I was assigned the task of reviewing, witnessing, and reporting, all the contractor actions relating to the early whirl tower and GTV operations. He stated that in all respects, Captain Schrage performed these functions in an exemplary manner. BG Cockerham was also impressed by the depth of Captain Schrage's technical knowledge. His ability to translate these technical aspects into an overview status, together with the ability to put his reports into their proper context, was truly outstanding. "His obvious leadership and management potential with his demonstrated technical excellence result in a well-rounded individual who is a valuable asset to the Army. I would certainly recommend this fine young officer for further military schooling and command opportunities at the earliest possible time."

- **Major Problem 3.** There was a requirement to raise the main rotor as a result of a pull-up, push-over maneuver in flight, similar to the UTTAS Maneuver. Slight contact with the cockpit canopy during a pull-up, push-over maneuver required this change to a raised main rotor system.

- **Major Problem 4.** Dual Lead-Lag Dampers prevented Ground Resonance with one damper inoperative as required in the AAH System Requirements. However, increased chord loads resulted in Hughes reducing the Chordwise stiffness of the dampers to get acceptable loads.

- **Major Problem 5.** This was by far the most significant air vehicle problem throughout Phase 1 and early Phase 2. Satisfying the aircraft pitch attitude changes during transition flight required a major decision by Hughes Helicopters. It was determined shortly after early Phase 1 flight testing that a fixed low-mounted stabilizer wasn't going to be acceptable due to rotor wake interaction causing large pitch attitude changes during transition and low-speed flight. Hughes conducted an empennage tradeoff study, as illustrated in Figure 57. (Ref.2) Their development of the empennage shows the original proposal, the interim T-tail, the final T-tail, the interim stabilator, and the final stabilator production.

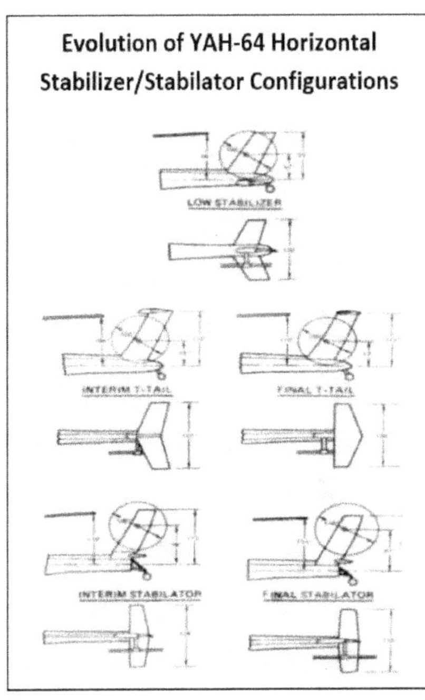

YAH-64 Horizontal Stabilizer/Stabilator Development

1. **Initial Low Stabilizer Development**
- **Unacceptable Airframe Pitch Change during Transition from Hover to Forward Flight**
- **Vertical Fin Blockage too great**
- **Tail Rotor too small**
2. **Interim and Final T-Tails**
- **Unacceptable Loads on Tail in High Speed Flight**
- **Vertical Fin Load Blockage even Greater**
- **Both Symmetric and Anti-symmetric Modes Unacceptable**
- **Tail Rotor still too small**
3. **Interim and Final Stabilators**
- **Solved Airframe Pitch Changes throughout flight**
- **Allowed complete redesign of the Tail**

Figure 57. YAH-64 Horizontal Stabilizer/Stabilator Tradeoffs Evolution (Ref.2)

Going to a stabilator design, with automatic flight control implications, such as Sikorsky had on the YUH-60, was something Hughes felt they did not have the expertise to implement in a timely manner. Instead, they decided to go to a T-Tail configuration with which they had successful implementation on their light commercial helicopter, e.g., Hughes 500 series. However, the weight and size of the YAH-64 tail rotor on the vertical tail were highly excited by the main rotor wake at 4P on the stabilizer in high-speed forward flight. A number of vibratory modes, both symmetric and anti-symmetric, were excited, which Hughes tried to control in Phase 1 with tip weights and planned stiffening for Phase 2, Figure 57A.

Two primary Modes, 10&11, Horizontal Stabilizer Anti-symmetric and Symmetric were excited as illustrated in Figure 57B. (Ref. 1982 AHS Forum Technical Paper, THE YAH-64 EMPENNAGE AND TAIL ROTOR – A TECHNICAL HISTORY by R. W. Prouty and K. B. Amer).

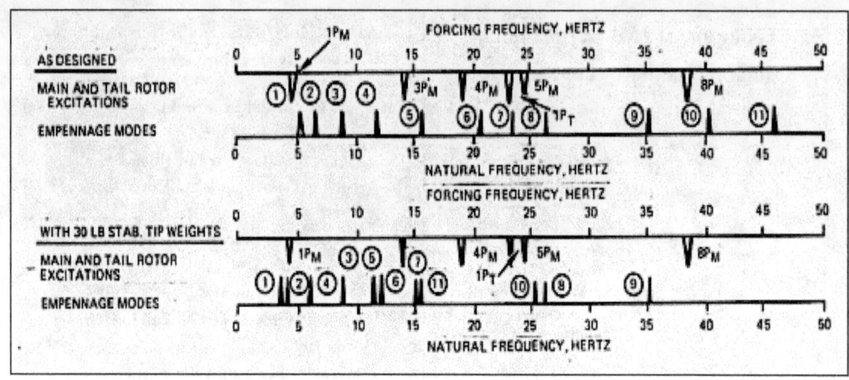
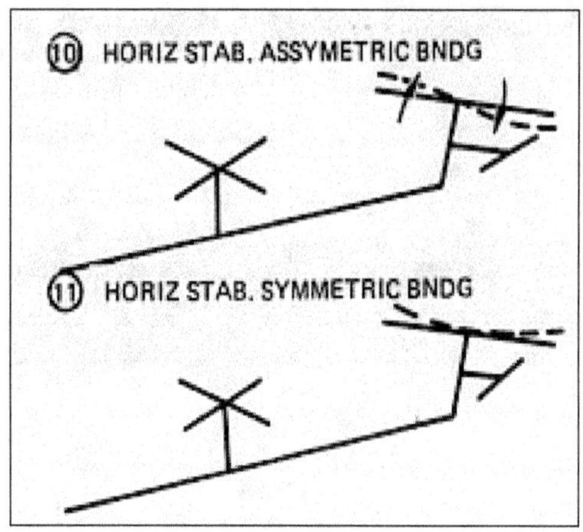

Fig. 57A. YAH-64 Empennage Vibratory Modes w&w/o Tip Weights. Fig 57B. Key Modes

Sikorsky also had key vibratory modes during transition flight, but not as severe on their low-mounted YUH-60 Stabilator. They temporarily solved it with an elastomeric bushing to decouple and reduce Horizontal Stabilator Anti-symmetric Bending mode effects, which stayed on the UH-60A Black Hawk in Production. The YAH-64 Empennage vibrations were much worse as it involved the entire T-Tail and Tail Rotor Blade Loads. As the ADV engineer/trouble shooter I was very much aware of the severity of these vibratory loads and worked closely with Jim Neff, Hughes Dynamics Chief, and Ray Prouty, Stability and Control Chief, to solve. These YAH-64 T-Tail loads even became the primary design fatigue load in the tail rotor as a main rotor 4P vortex excitation.

While I briefed the severity of the situation to the AAH PMO, they were informed by Hughes that they had the problem under control. However, the problem became so large it resulted in a change in the Hughes Helicopter Presidency. The scenario leading to the firing is discussed in the next section, as documented in the *Army Aviation: Foundations of the Modern Fleet,* Prepared by the Program Executive Office, Aviation, 2013. (Ref.2)

The severity of the YAH-64 T-Tail vibration problems was beginning to be understood by the AAH PMO when LTC Bill Leach, then Apache Project Manager, went for an orientation and evaluation flight with the Hughes YAH-64 Chief Test Pilot, Bob Ferry.

The following is the dialogue that took place between the AAH PMO and the Top Management at Hughes Helicopters regarding Problems with the YAH-64 T-Tail, which I helped to prepare at the request of PEO Aviation. A quote on this flight between LTC Bill Leach, Bob Ferry, and me was:

"One of the aircraft started up, and I looked up and saw that the whole tail section was just dancing and twisting and turning," I said to Bob Ferry, who was my instructor pilot and was going to get me checked out.

"Man, that thing looks weird. Does it always shake like that?" Bob replied. "We were up to 63 pounds tip-weight, trying to balance that out," and I said, "Oh man, that looks kind of scary.

LTC(R) Bill Leach Former Apache Project Manager

LTC Leach later accompanied AAH PM BG Ed Browne to a meeting with Hughes Helicopter President Tom Stuelpnagel to express the Army's dissatisfaction with the T-Tail design. The T-tail, Browne said, had to go. Stuelpnagel resisted and insisted that the design, integral to the Hughes Company logo, would not be changed. Stuelpnagel went on to say:

"This is our logo. We'll fix it (and) we'll make it work."

Ed Browne said, "I don't think so. I think maybe you need to think about taking that weight and stuff off of the top."

"No, it's our LOGO. We'll stay with that," Stueplnagel said, adding, "we will make it work."-LTC(R) Leach

BG Browne then directed Bob Hubbard, AAH PMO Technical Manager, to assemble a group of outside experts to assess the problem.

"I put together a group of guys that were made up from NASA Laboratories, the Naval Air Systems Command, Army Laboratories, and we even went to academia."-Bob Hubbard

I, CPT Schrage, briefed this Panel of Experts on the magnitude of the YAH-64 T Tail Problem and potential solutions, including the installation of a stabilator, similar to what Sikorsky had made. The Panel concluded that the stabilator appeared to be the required solution to save the AAH Program. The issue came to a head later during a meeting at Hughes that was attended by the Chairman of Summa Corporation, Will Lummis, which included Hughes Helicopters in its holdings.

"We had this big meeting out there at Culver City, and Stuelpnagel, Perry, and their gang had invited their chairman, Will Lummis, to this meeting. They were really going to put on a high-powered show. They got up and gave their presentation about how this and this, and it was more time, and so forth, and we can make this work. So then I got up, and I said, 'I would like to introduce my blue ribbon team, and Hughes did not know about it. Each one of (the team) went through the area that he had examined and summed it up, and said, 'It is our conclusion that it will never really do the job.' So about then, the meeting broke up. It was a Friday afternoon, and Will Lummis asked Bob Baer if he would mind joining him in the office for a little discussion. He did, and Stuelpnagel started to get up, and he said, 'no, I just want to talk to General Baer.' So, he went in. They came out and left, and Monday it was announced that Stuelpnagel was no longer President'." – MG(R) Ed Browne

A new tail design based on the stabilator concept previously developed by Sikorsky for their UTTAS quickly needed to be developed and tested. Once the decision was made, I recruited several other key engineers from the Army Aviation Development and Qualification Directorate, AVSCOM, who were familiar with the Sikorsky YUH-60 Stabilator design. We worked closely with Dr. Ken Amer and his engineering team to redesign the YAH-64 Empennage.

A summary of these changes is included in the 1982 AHS Forum Technical Paper, "The YAH-64 Empennage and Tail Rotor — A Technical History" by R. W. Prouty and K. B. Amer. Below is a summary from this paper of these changes that solved a number of problems that needed to be made on the YAH-64 for Phase 2 and Production.

"The empennage change to the stabilator configuration required raising the tail rotor gearbox. This change permitted the diameter of the tail rotor to be increased to provide better performance in right sideward flight (including high altitude operations) where unexpectedly high tail boom side loads were encountered. The maximum pitch of the tail rotor was chosen to satisfy the sideward flight requirement without encountering excessive drive system loads during reversals of hover turns. The vertical stabilizer was cambered to unload the tail rotor at high speed to prevent high stresses."

The schedule for the incorporation of these changes is illustrated in Figure 59. As can be seen, it spread into Phase 2 of the AAH Program, and even into Phase 3 production. An illustration of the two designs, YAH-64 T-Tail and AH-64 Production aircraft are shown in Figures 60A and 60B.

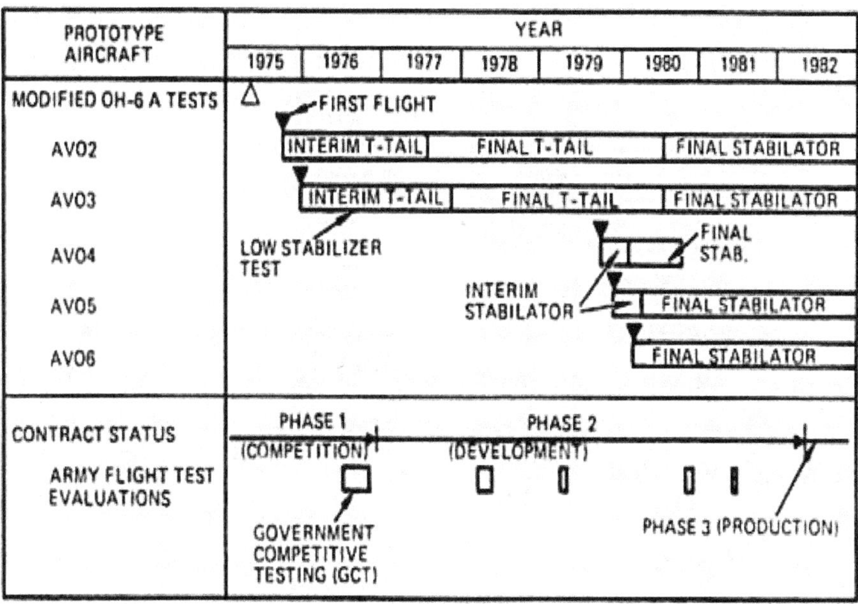

Figure 59. AAH Program Schedule Leading to Production

Figure 60A YAH-64 T-Tail Configuration. Figure 60B. AH-64A Production Configuration

The AH-64 has gone on to become a World Class Attack Helicopter. While initially developed by a very competent but small Hughes Helicopters technical team, its success was achieved with the help of a strong Army Aviation technical team. This was followed by the purchase of Hughes Helicopters by McDonnell Douglas Aircraft and then the purchase of McDonnell Douglas Aircraft by The Boeing Company.

There are writings by former Hughes AAH Management Team members, such as John Dendy, the design manager during development of the prototype Apache, that the T-Tail was not a problem that had to be solved. He was completely wrong. (*Howard's Whirlybirds, Howard Hughes' Amazing Pioneering Helicopter Exploits,* by Donald J. Porter, 2013. Other conflicting comments were made by Dick Simmons, Former Apache Program Manager."

He stated the following in *Army Aviation: Foundations of the Modern Fleet,* Prepared by the Program Executive Office, Aviation, 2013, Ref.2. "In Phase 1, which was really a mechanical system program, we did not have a lot of problems. I mean, we hit the first flight date, we hit all of the marks on speed, handling qualities, vertical rate of climb-we hit all of those numbers and everything that we needed to do with that aircraft. It was only when we got into the Phase 2 type thing that Hughes really went out and said we need help" - Dick Simmons, Former Apache Program Manager for Boeing

However, previously as was stated by Dick Simmons, an engineer involved with the design of the YAH-64 tail surfaces, in the book *Howard's Whirlybirds,* "At about 100 knots, the T-tail caused all kinds of problems in forward flight. The airflow over the T-Tail would cause problems with pitch. I can remember a comment by Morrie Larsen [a company test pilot] who said that at 100 knots, you could move the cyclic stick fore and aft an inch or so and get no pitch attitude change."

At the 75th Forum of the Vertical Flight Society (VFS) in Philadelphia, PA, on May 25, 2019, I gave a paper in the History Session entitled, "Setting the Record Straight on Failures and Successes in Army Aviation Development Programs over the Past Forty-Five Years." It summarized that Army Aviation Development Programs have had more failures than successes since beginning with the successful Utility Tactical Transport Aircraft System (UTTAS) and Advanced Attack Helicopter (AAH) development programs in the early 1970s.

While these follow-on failures have often been blamed on either the Army user, Army developer, or industry, they often come with biases from sideline participants without firsthand knowledge. The author of this book has had mostly this firsthand involvement and knowledge of these Army Aviation Development Programs over the past 50 years, and he has sought to set the record straight. I was told that this was the largest audience to attend an AHS/VFS Forum History Session in recent memory. As a result, the VFS Director sent out after the Forum a free copy of the paper to members of the VFS Community.

Some Major Problems/Solutions in UTTAS Maturity Engineering Development and AAH Phase II Full-Scale Engineering Development

YUH-60A Housatonic River Crash.

The first prototype Sikorsky YUH-60A Black Hawk, 73-21650, crashed on May 19, 1978, into the Housatonic River, during testing at the Sikorsky plant in Stratford, Connecticut, killing three. All three fatalities were civilian employees of Sikorsky. The subsequent Army investigation revealed that during routine maintenance the night before the fatal flight, the airspeed sensor for the tail-plane actuating system was inadvertently left unconnected. As the aircraft transitioned from hover to forward flight, the tail-plane did not automatically change its angle. as speed built up, it forced the helicopter's nose down until an attitude was reached, from which recovery was impossible. A manual backup system was available and functioning. It could have been used to correct the tail-plane angle, but for some unexplained reason, it was not used, possibly due to failure to analyze the nature of the problem in time. Minor modifications were introduced as a result of this accident.

A major feature of the YUH-60 prototype was a fixed-incidence horizontal stabilizer having an area large enough to provide adequate longitudinal stability to compensate for the aft CG placement described, and to meet the stability and control requirements in the appendix in the UTTAS System Specification.

However, its unusually large area, ~45 ft² resulted in undesirable flying qualities during landing approaches, and quick stop maneuvers caused by main rotor downwash impinging on the horizontal tail surface. Therefore, the YUH-60A Horizontal Stabilizer was converted to a stabilator with connection with the Automatic Flight Control Systems (AFCS), as illustrated in Figure 61. Its size and high pitch attitude in hover, ~42 deg, is shown in Figure 62, along with the UH-60A Stabilator characteristics.

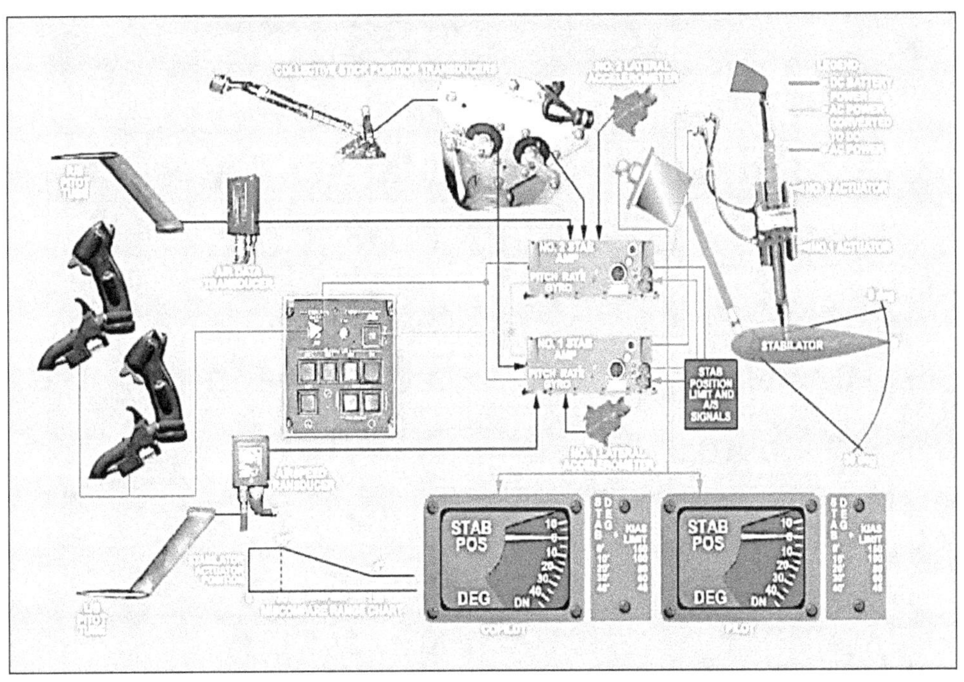

Figure 61. UH-60A Stabilator Interface with AFCS (UH-60A Flight Manual)

UH-60A/L Horizontal Stabilator	
Span	14.38 ft (172.5 in.)
Area	45 ft^2
Root Chord	3.67 ft (44 in)
Tip Chord	2.54 ft (30.5 in)
Sweep (1/4C)	0 deg
Aspect Ratio	4.6
Airfoil	NACA 0014
Incidence/Dihedral	0 deg

Figure 62. UH-60A in Hover and Horizontal Stabilator Characteristics.

Summary of Successful Solutions for the YAH-64 Empennage and Tail Rotor Design

(Reference. *The YAH-64 Empennage and Tail Rotor-A Technical History*, R. W. Prouty and K. B. Amer Hughes Helicopters, Inc., 'American Helicopter Society, May 1982).

Since the original design of the YAH-64 in 1973, several substantial configuration changes were made to the horizontal and vertical stabilizers and to the tail rotor of the YAH-64. These changes have been made in partnership between Hughes Helicopters and Army Aviation development engineers to solve various technical problems, which were not evident until the aircraft entered its flight test program.

By incorporating these changes, the YAH-64 and AH-64 met all the requirements of the Stability and Control appendix to the AAH System Specification. The horizontal stabilizer progressed from a low, fixed position which produced a large, nose-up trim shift at low speeds, to a T-tail which produced a large nose-up fuselage attitude in climb to a movable stabilator which solved both problems at the price of some increased complexity.

I led the Army Aviation development engineers to solve these problems and work with Hughes to determine if a smaller area stabilator, e.g., 33 ft^2 versus 45 ft^2, would work for the AH-64. It was verified through simulation and validated with a flight test that the smaller area was acceptable. The key from the flight test proved the smallest area to be satisfactory when used with a modest amount of pitch rate input, and so it was adopted for the production design.

The low-speed sensor above the YAH-64 rotor for weapons firing was also key to this success. The empennage changes to the stabilator configuration required raising the tail rotor gearbox. This change permitted the diameter of the tail rotor to be increased to provide better performance in right sideward flight (including high altitude operations) where unexpectedly high tail boom side loads were encountered. The maximum pitch of the tail rotor was chosen to satisfy the sideward flight requirement without encountering excessive drive system loads during reversals of hover turns. The vertical stabilizer was cambered to unload the tail rotor at high speed to prevent high stresses.

The analysis of the various primary and secondary problems and their successful solutions represents a body of experience that should be of value to those developing the next generation of helicopters. With the passing of Ken Amer and Ray Prouty, the Hughes technical designers, it is essential that these lessons learned, be passed on to the next generation of rotorcraft designers. Reliance should not be left to YAH-64 and AH-64 government and industry managers who have incomplete and often conflicting understandings of the lessons learned.

Other AAH Phase II and AH-64A Production Lessons Learned

With the tail issue resolved, Hughes and Army engineers focused their attention on other developmental hurdles. The next major challenge centered on the integration of avionics, fire control, and weapons systems. The AAH benefitted from what was probably the first true Army-integrated product team (IPT). Hughes and Martin-Marietta joined Army engineers to work together to complete the integration of the Target Acquisition and Designation System (TADS)/Pilot Night Vision System (PNVS). Shown in Figure 47A is the AH-64 Weapon Systems, and in Figure 47B, the AH-64 in Flight with the TADS/PNVS mounted on the nose of the aircraft.

Figure 63A. AH-64 Weapon Systems. Figure 63B. AH-64 Armed in Flight

The Armament and Mission Configurations are illustrated in Figure 64.

Mission[42]	Hellfire	30 mm rounds	Hydra 70	Maximum speed (knots)	Rate of climb (feet/min)	Endurance (hours)
Anti-Armor	16	1,200	0	148	990	2.5
Covering Force	8	1,200	38	150	860	2.5
Escort	0	1,200	76	153	800	2.5

Figure 64. The Armament and Mission Configurations with Hellfire missiles, M231 30 mm Chain Gun, and Hydra 70 Rockets

My Story on the Lightweight Rocket Launcher System Vibration Design Environment

The launcher system for the Hydra 70 Rocket System was designed to be lightweight and disposable after a certain number of firings. However, during a design review of it at General Dynamics (GD) in the San Fernando Valley, CA, the GD Chief Engineer explained how it was impossible to design the lightweight launcher to meet the requirements in MIL-STD-810C. They learned this lesson when their Hellfire missile launcher failed during the Vibration Test in accordance with the MIL-STD-810C.

Vibration testing in a laboratory for external stores mounted on a helicopter in MIL-STD-810C include rigid attachment for sinusoidal frequency input for sweep testing, dwell at critical frequencies, and testing as summarized in Table 1, e.g., Table 514.2-IVA below. The frequency sweeps and dwell time input are illustrated in Figures 65A and 65B. The ridged attachment is a cantilevered boundary condition compared to a free-free condition boundary condition in flight. Test results on the GD Hellfire Launcher for the controlled inputs of 1.0g and 2.0g resulted in responses of up to 50 g, which resulted in unrealistic failure conditions. The Lightweight Rocket Launcher on AH-64, Fig.65C.

Table 514.2-IVA. Cycling Period and Rate, and Time Schedule Chart for Externally Carried Stores for Helicopters, Category d.3

Excitation Axis	Sinusoidal cycling (See Test Procedure 4.6.3.3.3)			(See Test Procedure 4.6.3.3.4)				
	Time Schedule		Curve (See figure 514.2-4B)	Test time per dwell	Test Level Curves			
	Sweep time 5-500-5 Hz	Total test time			11 Hz	22 Hz	33 Hz	44 Hz
Vertical	15 Minutes	See table B	A	See figure 514.2-4C	See figure 514.2-4D			
					VA	VB	VC	VD
Transverse	15 Minutes	See table B	B	See figure 514.2-4C	See figure 514.2-4E			
					TA	TB	TC	TD
Longitudinal	15 Minutes	See table B	C	See figure 514.2-4C	See figure 514.2-4F			
					LA	LB	LC	LD

Table 514.2-IVB. Sinusoidal Cycling Test Time

Number of Missions	Cycling Time per Axis (Minutes)
0-50	30
51-100	60
101-∞	90

Table 1. Summary of MIL-STD-810C Vibration Testing for External Stores on Helicopter

114

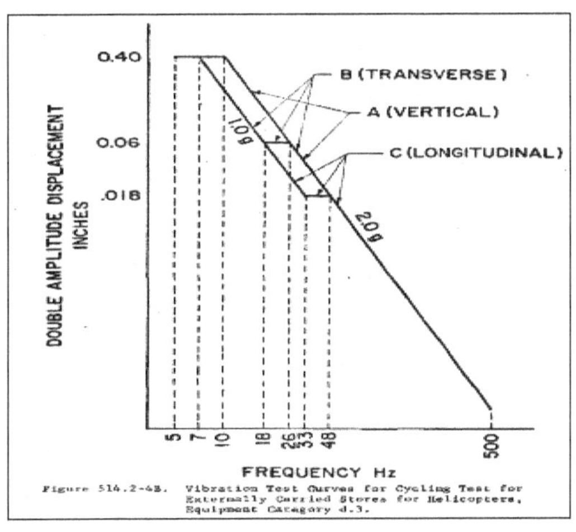

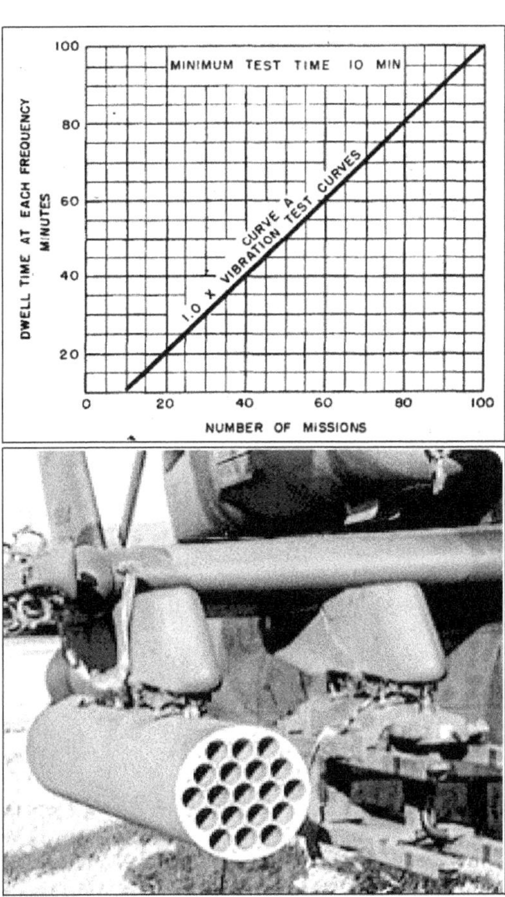

Fig. 65A. Frequency Sweeps with 1.0 and 2.0 Controlled Inputs. Fig. 65B. Dwell Time Frequencies versus # of Missions. Fig. 65C. Lightweight Rocket Launcher

MIL-STD-810, from *the Environmental Engineering Considerations and Laboratory Tests,* is a United States Military Standard that emphasizes tailoring an equipment's environmental design and test limits to the conditions that it will experience throughout its service life and establishing chamber test methods that replicate the effects of environments on the equipment rather than imitating the environments themselves.

These environmental management and engineering processes can be of enormous value in generating confidence in the environmental worthiness and overall durability of equipment and materiel.

Still, the user must recognize that there are limitations inherent in laboratory testing that make it imperative to use engineering judgment when extrapolating from laboratory results to results that may be obtained under actual service conditions.

Based on my experience with vibration analysis and flight testing, I knew that 50g's response was completely unrealistic. I knew that the Army Engineering Flight Activity (AEFA) at Edwards AFB, CA had an AH-1 Cobra Testbed and that GD had an available lightweight rocket launcher for instrumentation. I made a call to AEFA and arranged a flight test evaluation of an instrumented GD lightweight launcher on the AH-1 Cobra Testbed.

In a short period of time, the AH-1 Cobra flew a flight loads survey with the instrumented GD lightweight launcher. Results showed that the vibratory response never exceeded 5g's in any maneuver. Therefore, we used engineering judgment and obtained results under actual service conditions. The results ended up changing the controlled vibration inputs in MIL-STD-810C to a controlled response in the MIL-STD-810D revision in 1983. I believe this is another of my contributions to the AAH Development Program.

Chapter Four

COMPLETION OF GRADUATE DEGREES

Webster College, MA, and Washington University (St. Louis), DSc Mechanical Engineering

Advanced Degrees from Webster College and Washington University (St. Louis)

I took advantage of the GI Bill to further my education in business, management, and engineering. I enjoyed continuing my advanced education and felt it would be advantageous for my military career or in the civil sector.

Webster College, now University, was offering a Master of Arts (MA) in Business Administration at the end of each day in the AVSCOM Headquarters Building, the Mart Building, in mid-town St. Louis, MO. The classes lasted from 4-6 p.m. four days a week. I was thus able to complete the MA in one year, Spring 1975.

This experience would come in handy later, when I became the Director for Advanced System(DAS), the Associate Director for Science and Technology (S&T), and took responsibility for overseeing the Planning, Programming, and Budgeting System (PPBS) for AVRADCOM. When I finished this program, the opportunity came about for me to start working on a Doctor of Science (DSc) degree in Mechanical Engineering at Washington University in St. Louis, MO.

Dr. David A. Peters, who had served on the initial UTTAS and AAH SSEBs in 1972 and 1973, had recently joined the Washington University faculty as a professor. He received his PhD in Aeronautics from Stanford University while working as a civil service research engineer with the Army Aviation Aeromechanics Laboratory at NASA Ames Research Center, Moffitt Field, CA. He had received his BS and MS degrees in Mechanical Engineering at Washington University (St. Louis) while under the tutelage of Dr. Kurt Hohenemser, a recognized rotorcraft academic expert, who had led an Army panel that reviewed the aeroelastic problems of the AH-56 Cheyenne compound helicopter in the early 1970s. This provided me with an excellent opportunity to further my expertise as an ADV engineer.

I simultaneously worked as a DSc student and as an ADV engineer in AVSCOM from 1975 to 1978 while helping my wife, Nancy, raise a family. I was also helping solve UTTAS and AAH technical problems and serving on their SSEBs, leading to major decisions which were very challenging. One of my major challenges was taking required courses and preparing for my DSc qualification exams while often traveling. Fortunately, MAJ Grady Wilson, another Army Aviator at AVSCOM, was also taking similar courses at Washington University and would share his class notes for those classes I missed.

There were four DSc topic areas for the DSc qualification exams, from which I had to choose three. The four areas were Applied Mathematics, Fluid Mechanics, Solid Mechanics, and Thermodynamics and Heat Transfer. I chose the first three areas and avoided the fourth one, Thermodynamics and Heat Transfer since I had little exposure to it while taking courses and earning my master's degree in aerospace engineering (AE) at Georgia Tech. Initially, I was required to take written exams in the three areas chosen and had to pass two. I then had to take an oral exam in the area I did not pass, which in my case was Solid Mechanics, which I then passed. Having studied for the DSc qualification exams for 3-6 months, I felt that this was the most knowledgeable I would ever be about engineering.

For each of the written exams, I could have an 8.5 by 11-inch piece of paper with formulae and notes with me in each exam. In retrospect, I believe this was the most appropriate and fair way of taking DSc or PhD qualification exams.

Having served on the PhD exam committees during most of my 35 years as a professor in the School of Aerospace Engineering (AE), where all PhD qualification exams are verbal only, I felt that only taking oral exams is not fair to the students. Often the oral exams given by professors in the School of AE at Georgia Tech were thought up by the professors the night before exams. While a specific set of courses was identified for each exam, the professors often did not use material from these courses. Occasionally, some professors never even taught these courses. There were instances where students were successful in these courses, e.g., received an A or B grade, while professors claimed in the open discussion faculty meeting following the PhD exams that this was the worst student they had ever evaluated. The pass rate for first time PhD exams by students is usually less than 50%. Junior faculty without tenure are also under pressure from senior faculty to flunk PhD candidates to maintain this failure rate. While given a chance to retake the exam, if failed in one of the three areas evaluated by four faculty, failure in two areas is not allowed to retake the exam. Also, the retake failure rate is still close to 50%. I believe DSc and PhD exams should be a combination of written and oral exams; however, opportunities to change to this approach in the Georgia Tech School of AE have not been accepted. More and more students in the School of AE at Georgia Tech are opting to take their PhD qualification exams at other schools where a fairer evaluation occurs.

Achieving the DSc Degree in Mechanical Engineering and Relevant Research for Army Aviation

I was fortunate for my DSc research at Washington University in that I could develop a new aeroelastic methodology approach that could be applied to real Army Aviation aeroelasticity problems, both loads and stability, identified in the UTTAS development program. The aeroelasticity problem was the design and development of the bearing-less flex-strap tail rotors for the Boeing and Sikorsky prototype tail rotors, as illustrated in Figures 66 and 67

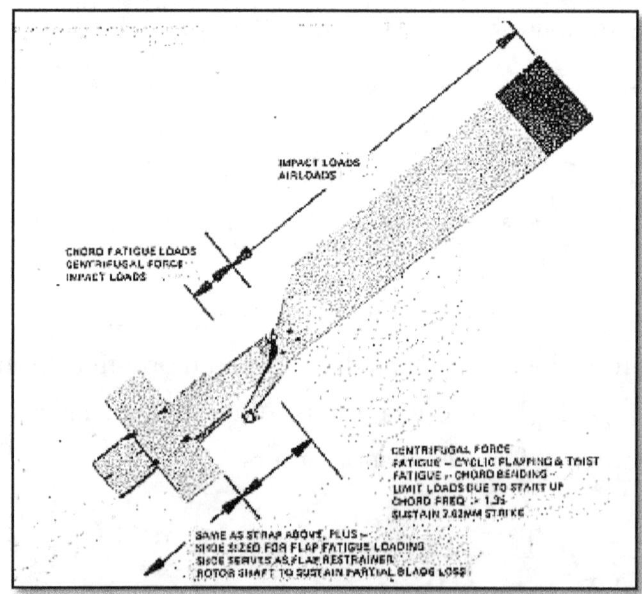

Figure 66. Boeing & Sikorsky Stiff In-Plane Bearing-less Flex-strap Rotors, Loads & Wind Tunnel Test.

The Boeing Vertol YUH-61A (UTTAS) tail rotor was a new, completely bearing-less flex-strap design which used fiberglass composite blades and straps to achieve composite blades and straps to achieve significant gains in simplicity and survivability. Predominant loadings at the blade root section result from blade flapping. Flapping deflections produce the flap-wise moment

of the flex-strap and that portion of the chordwise moment due to Coriolis forces. These moments are readily calculated once flapping is known. For flight conditions that create large aerodynamic excitations, such as arrested yaw conditions, the theory is inadequate. An empirical equivalent flapping was derived from wind tunnel and flight test data. The wind tunnel 8/10 scale model setup is illustrated in Figure 66.

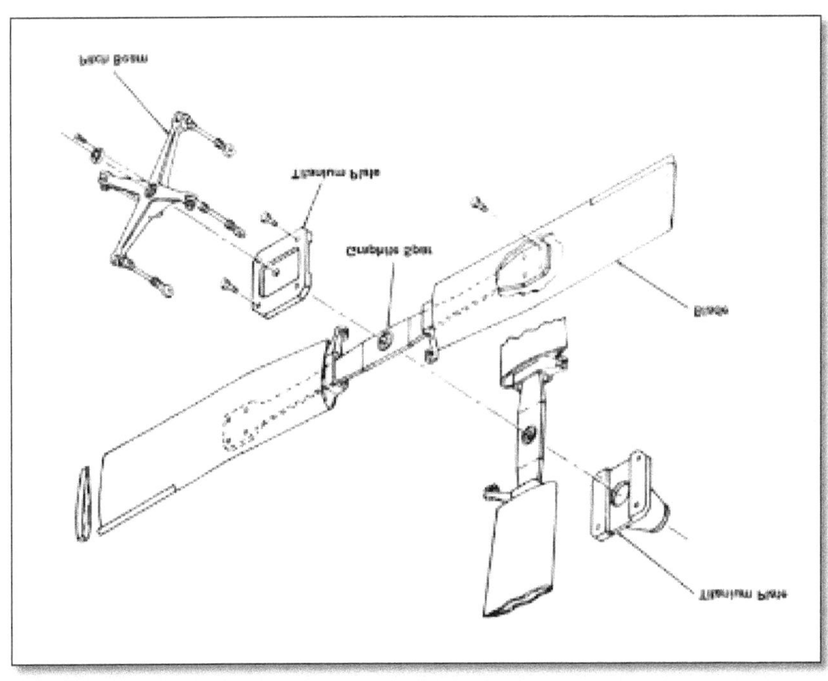

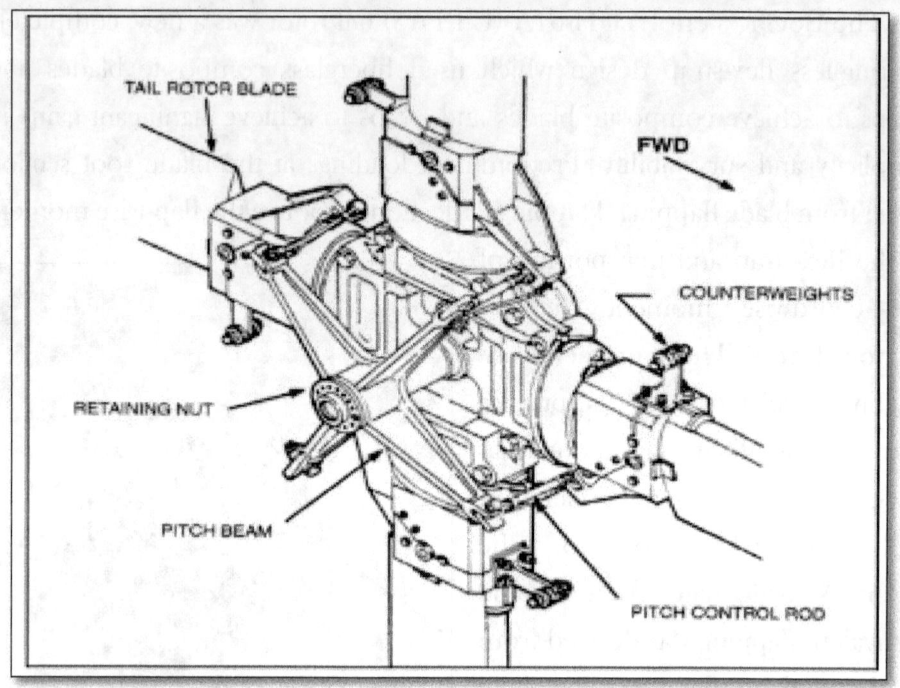

Figure 67.. Sikorsky Stiff In-Plane Bearing-less Flex-strap Tail Rotor Elements and Installation

The Sikorsky UTTAS was designed with its 4-blade tail rotor canted upwards twenty degrees to provide approximately 400 pounds of extra lift, Figure 67. This lift at the tail permitted the aircraft's center of gravity to be placed aft of the main rotor. This allowed the nose and cockpit to be moved aft, thereby shortening the fuselage length to better fit inside the C-130 cabin during air transport. The added lift contributed by the canted tail rotor was provided at very low power that further helped improve the overall Figure of Merit of the aircraft. The canted tail rotor thrust resulted in yaw-to-pitch coupling that was resolved by incorporating appropriate counter coupling in the flight controls mixing unit. The crossbeam tail rotor was designed with composite materials such that no bearings were needed to provide pitch blade change. It consisted of two continuous paddles, each eleven feet long, mounted at ninety degrees to each other with a central hole through which passed the pitch control members.

My DSc Thesis Research:

Effect of Structural Parameters on the Flap-Lag Forced Response of a Rotor Blade in Forward Flight, May 1978

This thesis reported on research conducted to obtain a better understanding of the structural coupling parameters contributing to rotor blade forced response and stability in forward flight. A new analytical method, named eigenvalue and modal decoupling analysis, was developed to solve the flap-lag equations of motion with periodic coefficients. An extension of the Floquet Transition Matrix Method, the eigenvalue and modal decoupling analysis is used to obtain the forced response of the flap-lag equations. Application of this analysis to study four different rotor configurations provided new insight into the stability and response problem. The results provide the rotor design engineer with valuable guidelines to study the tradeoffs necessary during the design phase of new rotor systems. The analysis was performed on an IBM 360/65 computer. The model used with a rigid blade and appropriate root end couplings; R is Flap-Lag structural coupling, θ_β is Pitch-flap coupling, and θ_ζ is Pitch-lag coupling in Figure 68.

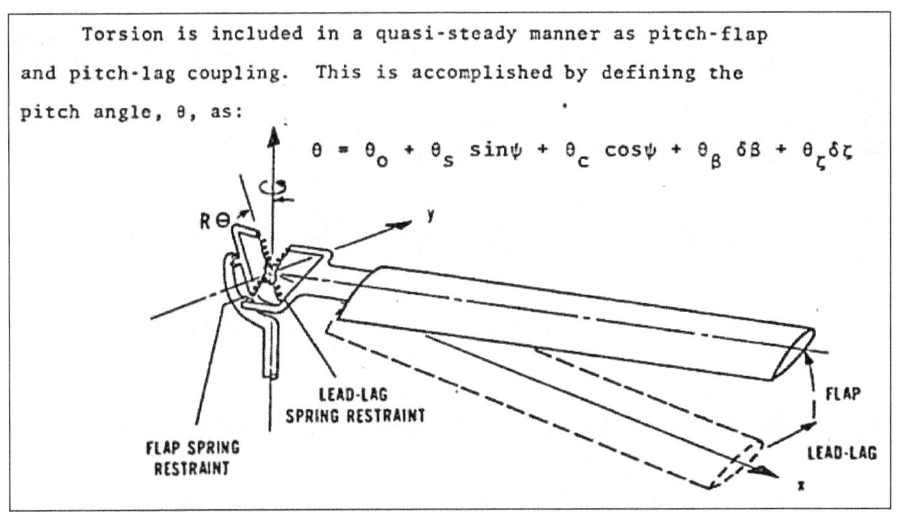

Figure 68. Rotor Blade Model for My DSc Research

The Equations of Motion in the Eigenvalue and Modal Decoupling Analysis are shown follows:

$$\left\{ \begin{matrix} \delta\ddot{\beta} \\ \delta\ddot{\xi} \end{matrix} \right\} + [C(\psi)]\left\{ \begin{matrix} \delta\dot{\beta} \\ \delta\dot{\xi} \end{matrix} \right\} + [K(\psi)]\left\{ \begin{matrix} \delta\beta \\ \delta\xi \end{matrix} \right\} = [L(\psi)]\left\{ \begin{matrix} \bar{\dot{\beta}} \\ \bar{\beta} \end{matrix} \right\} + [M(\psi)]\left\{ \begin{matrix} \bar{\theta} \\ \bar{\phi} \end{matrix} \right\} + [N(\psi)]\left\{ \begin{matrix} \ddot{\bar{\beta}} \\ C_d \end{matrix} \right\}_{o/a}$$

The nine-step Eigenvalue and Modal Decoupling Approach follows:

Eigenvalue and Modal Decoupling Analysis

Step   **Computation**

1. Put Flap-Lag Equations into State Variable Format:
$$\{\dot{x}(t)\} + [D(t)]\{x(t)\} = \{f(t)\}$$

2. Determine the State Transition Matrix, $[\phi(t)]$, and FTM, $[Q]$, by Numerical Integration Over One Period:
$$\{x(t)\} = [\phi(t)]\{x(0)\}, \quad [Q] = [\phi(T)]$$

3. Find Eigenvalues and Eigenvectors of FTM:
$$[A(0)]^{-1}[Q][A(0)] = [\,^{\backslash}e^{\Lambda T}_{\backslash}\,], \quad [\,^{\backslash}\Lambda_{\backslash}\,] = [\,^{\backslash}\eta_{\backslash}\,] + i[\,^{\backslash}\omega_{\backslash}\,]$$

4. Obtain the Characteristic Functions, $[A(t)]$, for Variable Substitution, $\{x(t)\} = [A(t)]\{y(t)\}$, for Each Time Increment:
$$[A(t)] = [\phi(t)][A(0)][\,^{\backslash}e^{-\Lambda t}_{\backslash}\,]$$

5. Compute Inverse Matrix, $[A(t)]^{-1}$, to Obtain:
$$[A(t)]^{-1}\{f(t)\} = \{g(t)\}$$

---

6. Obtain the Uncoupled Equations by Variable Substitution:
$$[A(t)]\{\dot{y}(t)\} + \left([\dot{A}(t)] + [D(t)][A(t)]\right)\{y(t)\} = \{f(t)\}$$
$$\{\dot{y}(t)\} + \left([A(t)]^{-1}[\dot{A}(t)] + [A(t)]^{-1}[D(t)][A(t)]\right)\{y(t)\} = \{g(t)\}$$
$$\{\dot{y}(t)\} - [\,^{\backslash}\Lambda_{\backslash}\,]\{y(t)\} = \{g(t)\} = \sum_{n=-\infty}^{+\infty}\{a\}_n e^{int}$$

7. Obtain $y(t)$ From Complex Fourier Analysis:
$$\{y(t)\} = \sum_{n=-\infty}^{+\infty}\left[\,^{\backslash}\tfrac{1}{in-\Lambda}_{\backslash}\,\right]\{a\}_n e^{int}$$

8. Obtain $\{x(t)\}$ From Matrix Multiplication, $\{x(t)\} = [A(t)]\{y(t)\}$, and Convert $\{x(t)\}$ Information in Time Domain to Vibratory Shears and Moments

9. Fourier Analyze Shears and Moments to Obtain Strength of Harmonic Components

Two soft in-plane rotors and two stiff in-plane rotors were analyzed. The two stiff in-plane rotors were representative of the two UTTAS bearing-less prototype tail rotors, e.g., the Boeing and Sikorsky Flex-strap tail rotors. By being able to analyze both rotors for vibratory loads and aeroelastic stability from the same analysis showed that Sikorsky had designed a more stable rotor but with higher vibratory loads, while Boeing had designed for lower loads but with less stability.

A major conclusion from the research was that the eigenvalue and modal decoupling methodology allowed for aeroelastic stability and vibratory loads to be obtained from the identical set of equations and allowed for a direct comparison of the effects of coupling parameters. Thus, the internal consistency ensured that stability-forced response correlations were sound.

Accomplishments and Recognitions that Continue the "Why" for this Book

I was given the 1976 AIAA Young Professional Engineer Award by the St. Louis Section in Recognition of my work on solving the aeroelastic instability characteristics of helicopters. Given by William H. Gillispie, Chairman, St. Louis Section, AIAA.

Endorsed by MG Eivind A. Johansen, Commanding General, AVSCOM

The Letter of Commendation from COL Charles A. Bullock, Director of Research, Development and Engineering:

"I am pleased to add my congratulations to you for your performance of duty and technical achievements that resulted in your selection for this fine award. This is the first time anyone from this Command has been so honored and also the first time a military officer has won the award. Recognition by your professional peers is certainly one of the highest tributes attainable. I understand the St. Louis Section has over 600 members and has been ranked as the outstanding Section in AIAA for the past two years. Therefore, you have been honored by some of the best aerospace engineers in the free world."

Promotion to Army Major from MG Story C. Stevens and for CPT Schrage's Contributions to the UTTAS and AAH Programs

In January 1978, I was promoted to Major by MG Story C. Stevens, Commanding General, AVRADCOM. A picture of MG Stevens pinning on my Major oak leaf cluster with assistance from my wife, Nancy, and my two children, Steven and Susan, Figure 69.

Figure 69. CPT Daniel P. Schrage's Promotion to MAJOR

Army Commendation Award w/First Oak Leaf Cluster, Major Schrage, FA, USA

On 30 June 1978, MG Story C. Stevens presented MAJOR Schrage with this Award based on the following:

"Outstanding service from 6 July 1974 to 30 June 1978 while assigned as Aircraft Dynamics Engineer and, subsequently, as Acting Aeromechanics Team Leader for the Directorate for Development and Engineering, United States Army Aviation Research and Development Command (AVRADCOM), St. Louis, Missouri. Throughout this period, CPT Schrage demonstrated the highest level of leadership and technical competence in complicated and demanding assignments. His extreme dedication to a heavy workload and his technical expertise enabled him to make major contributions to the development of new and modified Army Aircraft. Of particular note was his technical contribution to the development of the UH-60 Black Hawk and the AH-64 Apache Helicopters. Major Schrage's professional attitude and ability reflect great credit upon himself and the United States Army."

CHAPTER FIVE

TRANSFER TO THE U.S. ARMY RESERVES

Hiring as the Aeromechanics Branch Chief,

Directorate for Development and Engineering, U.S. Army AVRADCOM, July 1, 1978.

I had decided in 1978 to leave active duty as a Major and accept a civil service GS-14 engineering position (equivalent to a Lieutenant Colonel) as the Aeromechanics Branch Chief, overseeing ~ 30 engineers. I transferred to the US Army Reserve (USAR) as a Major and served two weeks a year as a Mobilization Designee (MOBDES) faculty member in the Department of Mechanics and later the Department of Civil and Mechanical Engineering at USMA. I also served as a Military Academy Officer (MALO) for the St. Louis Area with the USMA Directorate of Admissions until 1996.

During my assignment as a MOBDES faculty member, I helped in transitioning the USMA Curriculum from a General Engineering Program into specialized engineering programs into Civil and Mechanical Engineering

(CME), Electrical Engineering and Computer Science (EE&CS), and Systems Engineering (Sys Eng).

This included developing design courses for CME and Sys Eng. I stayed in the US Army Inactive Reserves until 2004 when I officially retired as a Colonel. Based on my contributions and near-perfect Officer Efficient Reports (OERs) at USMA over the years, I was informed by the Pentagon that I was eligible for consideration for promotion to Brigadier General (BG). However, I felt that I received enough recognition in my other careers, so I didn't pursue this promotion.

In 1978 AVSCOM was converted into the Aviation Research and Development Command (AVRADCOM) with Aviation Readiness combined with the Army Troop Support Command (TROSCOM). This provided an entry into the "Golden Years of Science & Technology (S&T) for Army Aviation Research and I had decided in 1978 to leave active duty as a Major and accept a civil service GS-14 engineering position (equivalent to a Lieutenant Colonel) as the Aeromechanics Branch Chief, overseeing ~ 30 engineers. I transferred to the US Army Reserve (USAR) as a Major and served two weeks a year as a Mobilization Designee (MOBDES) faculty member in the Department of Mechanics and later the Department of Civil and Mechanical Engineering at USMA. I also served as a Military Academy Officer (MALO) for the St. Louis Area with the USMA Directorate of Admissions until 1996.

During my assignment as a MOBDES faculty member, I helped in transitioning the USMA Curriculum from a General Engineering Program into specialized engineering programs into Civil and Mechanical Engineering Development". It also provided me with a rapid promotion growth path to become the youngest Senior Executive Servant (SES) Level Three (equivalent to a Major General) at the age of 37 in the U.S. Army Development and Readiness Command (DARCOM).

With this SES selection, I became the Director of Advanced Systems (DAS) and Associate Tech Director for S&T in AVRADCOM, including the joint Army-NASA Labs at NASA Ames, Langley and Lewis, the Army Aviation Applied Technology Directorate (AATD) at Fort Eustis, VA and the Army Avionics Research and Development Activity (AVRADA) at Fort Monmouth, NJ.

I served under the guidance of Mr. Richard B. Lewis, AVRADCOM Technical Director, and MG Story Stevens, Commanding General (CG) AVRADCOM, who served as my mentors. Also, in 1983 I spent six months on a temporary assignment at Fort Leavenworth, KS, as the Acting Chief Scientist for the Army's Combined Arms Center (CAC) under LTG Jack Merritt, CG, CAC. I worked in an office next to BG Colin Powell, who was serving as the Commander of the Army Combined Arms Development Activity (CACDA). I got to know him and his family quite well.

An interesting side story was that when BG Powell and I would travel together to the Army Training and Doctrine Command (TRADOC) Headquarters at, Fort Monroe, VA, I would get better quarters than he would as my position was equivalent to a two-star general, as an SES Level Three, and he was a BG. I offered to exchange quarters with him as I was only an LTC in the Army Reserves. We both got a laugh over this.

Experiences as the Aeromechanics Chief in AVRADCOM and Technical Chief for the Army Helicopter Improvement Program (AHIP)

Aeromechanics Branch, Structures & Aeromechanics Division, Development & Qualification Directorate, AVRADCOM

Chapter Five addresses my transition into Civil Service and promotion to increased levels of responsibility with Army Aviation. When I resigned from active duty as a Major at the end of June 1978, I was the acting Chief of the Aeromechanics Branch in the Structures and Aeromechanics Division in the

AVRADCOM Engineering Directorate. I became Chief of Aeromechanics as a GS-14 Engineer, which was equivalent to an LTC, in July 1978.

Aeromechanics engineering disciplines included Performance and Weights, Flight Dynamics and Controls, and Vibrations & Dynamics. There were approximately 30 engineers in the Aeromechanics Branch responsible for aircraft development and airworthiness qualification by providing technical support to the Army Aviation Program offices.

As a senior engineering manager and later as a senior executive in Army Aviation, I was directly involved with the initial design and development of almost all of today's Army Aviation aircraft systems. This included serving as the Technical Director for the AHIP SSEB and successful development support for the OH-58D Kiowa.

In 1979 I was promoted to Chief of the Structures and Aeromechanics Division. This included overseeing the airworthiness qualification and technical modernization of the CH-47D Chinook Helicopter. This will be addressed in Chapter Six.

In 1981, I was selected as a Senior Executive Service (SES) Level 3 and became the Director of Advanced Systems (DAS) and the AVRADCOM Associate Technical Director for Science and Technology (S&T). This led to the technical effort in 1982-84 for the Concept Formulation for the Light Helicopter Experimental (LHX), which is addressed in Chapter Seven.

Chapter Eight will also address my reasons for leaving AVRADCOM and accepting a full professor faculty position in the School of Aerospace Engineering at Georgia Tech, where. I would then lead their Army sponsored Center of Excellence in Rotary Wing Aircraft Technology (CERWAT) and serve as the Rotorcraft Design Professor. I will describe in Book 3 my role as a professor, science and technology director, aerospace leader, and consultant to government and industry from a leading technical university.

Background Leading to the Army Helicopter Improvement Program (AHIP) Source Selection Evaluation Board (SSEB)

In 1960, the United States Army issued Technical Specification 153 for a Light Observation Helicopter (LOH) capable of fulfilling various roles: personnel transport, escort and attack missions, casualty evacuation and observation. Twelve companies participated in the competition, and Hughes Tool Company's Aircraft Division submitted the Model 369.

Two designs, submitted by Fairfield-Hiller and Bell, were selected as finalists by the Army-Navy design competition board. The U.S. Army later included the helicopter from Hughes as well. Bell went on to develop the D-250 design into the Model 206 aircraft, with the HO-4 designation being changed to YOH-4A in 1962. Bell produced five prototype aircraft for the Army's test and evaluation phase.

Following a fly-off of Bell, Hughes, and Fairchild, the Hughes Helicopters were selected in May 1965. The Army awarded Hughes a contract for production in May 1965, with an initial order for 714 that was later increased to 1,300 with an option for another 114. (Ref. 1)

Hughes's price was $19,860 per airframe, less engine, while Hiller's price was $29,415 per airframe, less engine. In addition to the Bell visual LOH image problem, their helicopter lacked cargo space and only provided cramped quarters for the planned three passengers in the back. Their solution was a fuselage redesigned to be sleeker and aesthetic, adding sixteen cubic feet of cargo space in the process. The redesigned aircraft was designated as the Model 206A, and Bell President Edwin J. Ducayet named it the Jet Ranger denoting an evolution from the popular Model 47J Ranger.

In 1967, the Army reopened the LOH competition for a second buy of LOH aircraft, Bell resubmitted for the program using the Bell 206A and its improved airframe, which had substantial growth capability. Fairchild-Hiller failed to resubmit their bid with the YOH-5A, which they had successfully marketed as the Fairchild Hiller FH-1100. In the end, Bell underbid Hughes to win the contract and the Bell 206A was designated as the OH-58A. Following the U.S. Army's naming convention for helicopters, the OH-58A was named Kiowa in honor of the Native American tribe.

It was reported that Howard Hughes had directed his company to submit a bid at a price well below the actual production cost of the helicopter during the first bid to secure this order, resulting in substantial losses on the U.S. Army deal, with the anticipation that an extended production cycle would eventually prove financially viable.

In 1967, Hughes submitted a bid to build an additional 2,700 airframes. Stanley Hiller complained to the U.S. Army that Hughes had used unethical procedures. Therefore, the Army opened the contract for rebidding by all parties. Hiller did not participate in the rebidding, but Bell did. After a competitive fly-off, the Army asked for sealed bids. Hughes bid $56,550 per airframe, while Bell bid $54,200. Reportedly, Howard Hughes had consulted at the last moment with his confidant Jack Real, who recommended a bid of $53,550. Howard Hughes, without telling him, added $3,000 to the bid and thus lost the contract. (Ref. 1)

However, the OH-6A Cayuse had become a very popular scout helicopter in Vietnam, especially when used in air cavalry units and as part of a hunter killer team with attack helicopters, especially the AH-1 Cobra. The pilots dubbed the new helicopter "*Loach*," a word created by pronunciation of the acronym of the program that spawned the aircraft, LOH (light observation helicopter). In fact, when the air cavalry squadrons in Vietnam were told to turn in their OH6As for OH-58As, they all refused except for one unit. (7). Both the OH-6A and OH-58A went on through modernization programs to become bigger and better for other evolving Army Aviation military operations which became the Army Helicopter Improvement Program (AHIP) in 1978.

MY STORY

I was serving as the S-3, 13th Combat Aviation Battalion (CAB), in Soc Trang, South Vietnam in Summer and Fall 1970. LTC Sauers, the 13th CAB, had been involved in the source selection of the second buy of scout helicopters and felt selecting the OH-58A over the OH-6A was a big mistake. As a result, he tasked me to do an in-country evaluation of the two scout helicopters, the OH-58A Kiowa which I flew as the S-3 and an OH-6A Cayuse which was flown as a scout helicopter in our C Troop, 16th Air Cav, Figure 70.

Scout-Attack Team

Figure 70. My OH-58A and C Troop Air Cav Aircraft. OH-6A "Loach" Outcasts, and AH-1 Cobra Mustangs

I, therefore, spent a week at Vung Tau, South Vietnam, getting transitioned into the OH-6A helicopter, Figure 71.

Figure 71. OH-6A Training Helicopter at Vung Tau and My Approach Flying into Vung Tau

Vung Tau was a resort city, some said for both the South Vietnamese and the Viet Cong. The military mayor of Vung Tau City was Dean Hansen, an Army Captain given this assignment, who had been my roommate at West Point and one of my best friends. He was an outstanding linebacker on the Army Football Team.

During the Spring of our first class "senior" year, we both played defense for our Company E-4 Lacrosse Team. I couldn't play baseball, as I had my nose fixed from a deviated septum so that I would be eligible for going to flight school following a combat arms assignment. I did this in Germany, 1968-69, as an Honest John missile battery commander.

Following my evaluation of the OH-6A Cayuse versus the OH-58A with our Cav Troop I wrote up an evaluation report for LTC Bob Sauers. It pointed out the superiority of the OH-6A "*Loach*" versus the OH-58A Kiowa, especially in NOE flight. Later, in 1978, while serving as the Technical Chief on the Army Helicopter Improvement Program (AHIP) Source Selection Evaluation Board (SSEB), my evaluation was different, as Bell Helicopter had made major changes in the OH-58D, such as a four-bladed hingeless rotor with a larger engine and transmission and the growth 206 airframe. This made it far superior to the baseline OH-6A proposed by Hughes Helicopters. They had proposed minor changes such as new main rotor airfoils which couldn't accommodate the growth necessary for the AHIP requirements.

Interestingly, at the first meeting of the AHIP SSEB Directors, the Source Selection Authority (SSA), the CG of the 18th Airborne Corps at Fort Bragg, NC, made the statement that "all of the aviators at Fort Bragg had told him not to let them pick another OH-58". I knew our work would be cut out for us, but also knew that the OH-58D proposed by Bell Helicopter would be a far superior aircraft.

The author of this book has had substantial involvement in subsequent Army Aviation modernization programs, including both the AHIP/OH-58D Kiowa Warrior and the CH-47D Chinook Modernization. This will be followed in Chapter Seven addressing my leadership in the Concept Exploration for the LHX/RAH-66 Comanche.

This chapter will first introduce the modernization of the OH-58A/C to the OH-58D through the Army Helicopter Improvement Program (AHIP) and its arming into the OH-58D Kiowa Warrior. It will address how the AHIP accomplished a number of first-time integrations, e.g., the first integrated digital cockpit, the first digital fuel control, the first completely composite rotor blades, and hinge less rotor hub, and the first integration of a multi-sensor mast mounted sight (MMS). (Ref. 2)

The AHIP Program was followed with an integrated weapon system program, converting it from a LOH to a light attack and reconnaissance helicopter, the OH58D Kiowa Warrior. (Reg.3) The retirement of the OH-58D Kiowa Warrior, which had proven itself in battle, has led to the current need for a Future Attack and Reconnaissance Aircraft (FARA) program. The FARA Program was initiated by the U.S. Army in 2018 to develop a successor to the Bell OH-58 Kiowa Warrior by 2030 as part of the larger Future Vertical Lift (FVL) program. (Ref. 4)

However, the FARA Program has currently slipped nine months, as both the Bell and Sikorsky prototypes have difficulty in meeting the government requirements with a single ITEP Engine. This chapter will identify lessons learned from the AHIP/OH58D Kiowa Warrior Program and, in a later chapter, the LHX/RAH-66 Comanche program, which should be applicable to the FARA Program.

However, it doesn't appear that the FARA and FLRRA programs have the technical experience or interest in understanding the lessons learned presented in these chapters. This is particularly relevant to empty weight to gross weight ratio and the use of a single engine.

The AHIP Source Selection Board and Selection of the Bell OH-58D Kiowa

The background leading to the AHIP Source Selection Evaluation Board (SSEB) is as follows:

- In 1977 Army designs were formulated for an Advanced Scout Helicopter (ASH) to accompany the new YAH-64 Apache attack helicopter.

- Within a year, the ASH concept was terminated in 1978 at Congressional direction, and work began on a Near Term Scout Helicopter (NTSH), which resulted in the US Army Helicopter Improvement Program (AHIP).

- The AHIP competition was won by Bell Helicopter in competition with Hughes Helicopters and an upgraded OH-6A Cayuse in September 1981.

- Bell's Design included in extensive modification and modernization of the OH-58A Kiowa, to fulfill the Army's desired requirements for an advanced scout helicopter. I was told by Bell it would have been easier to build a new OH-58D from scratch, rather than modify OH-58A/Cs.

The schedule for the AHIP Development is shown in Figure 72A. The Bell OH-58D Development Program Delivery was less than four years as a fixed-price contract. This was remarkable in that a complete modification of an OH-58A/C to the OH-58D AHIP took place. (Ref. 2) As the AHIP Program Manager, COL Bud Forester stated the following.

"AHIP was an aircraft of 'firsts'- the first digital cockpit, the first digital fuel control, the first completely composite rotor blades, composite rotor head and it just went on and on from there. (It also had) a digital (Stability Augmentation System) – we found out that digits did not work too well in that Yuma sun unless they were properly grounded – and just meticulously put together. So, (development testing) was not a fun experience, but we learned one heck of a lot from it and were able to implement those corrections." (Ref. 5)

The Aggressive AHIP/OH-58D Development is shown in Figure 72A.

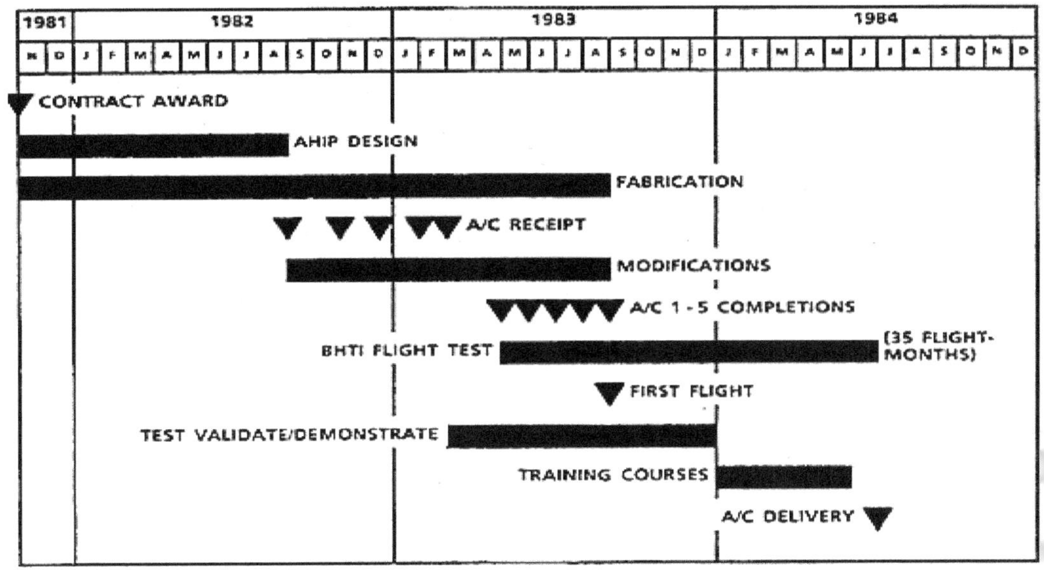

Figure 72A. Aggressive AHIP Development to OH-58D Program

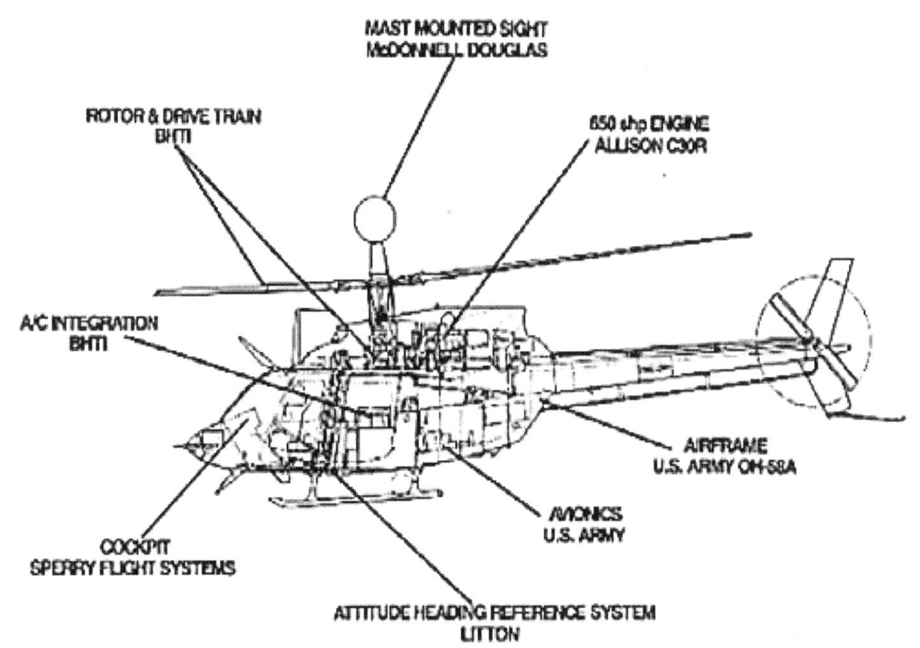

Figure 72B. Major new systems in OH-58D

The AHIP SSEB Organization is illustrated in Figure 73A. Ron Gormont, AVRADCOM Structures & Aeromechanics Chief, was a very competent

139

AHIP SSEB Chair. I served as the Director of the AHIP SSEB Technical Area and had approximately 100 engineers working for me, with more than half being electronics/avionics engineers and human factor engineers, while the others were aerospace and mechanical engineers. LTC Skip Neuwein, Deputy, Combat Development Directorate, Fort Rucker, AL, served as the Director of the AHIP Operational Suitability, SSEB. AHIP candidates are in Figures 73B and C.

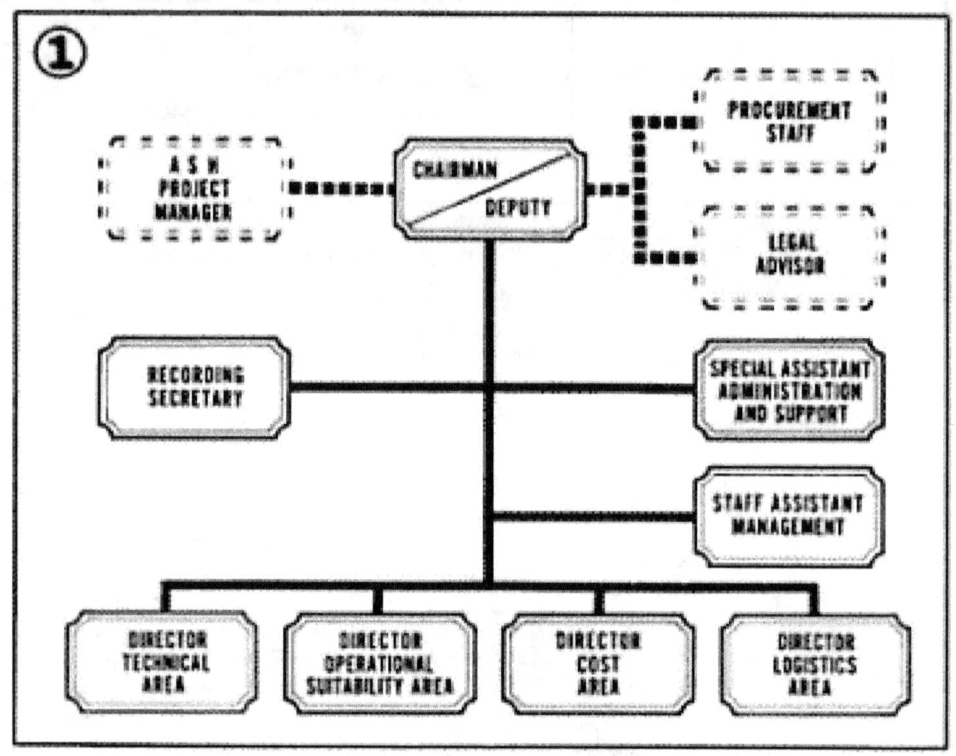

Figure 73A. AHIP SSEB Organization.

Figure 73B. OH-6A Cayuse. Figure 73C. OH-58A Kiowa

A major approach taken by Bell was to understand the required and desired tradeoffs that had to be made. Bell's design included extensive modification and modernization of the OH-58A Kiowa, to fulfill the Army's desired requirements for an advanced scout helicopter. The Army's need, given the competition, appeared to range from "fundamentally required" characteristics to "really desired" capabilities. With this possible range in mind, the Bell team summarized the homework to date with the configuration logic tree shown below in Figure 74A and Components in Figure 74B.

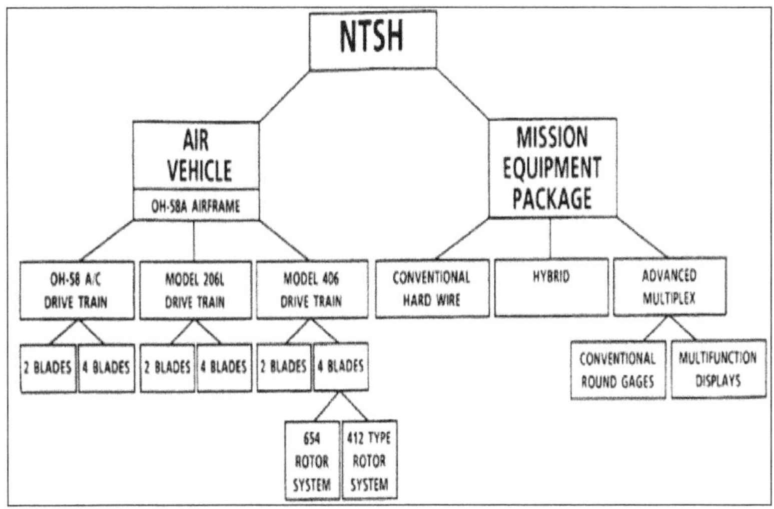

Figure 74A. NTSH Broken Down into Air Vehicle &MEP.

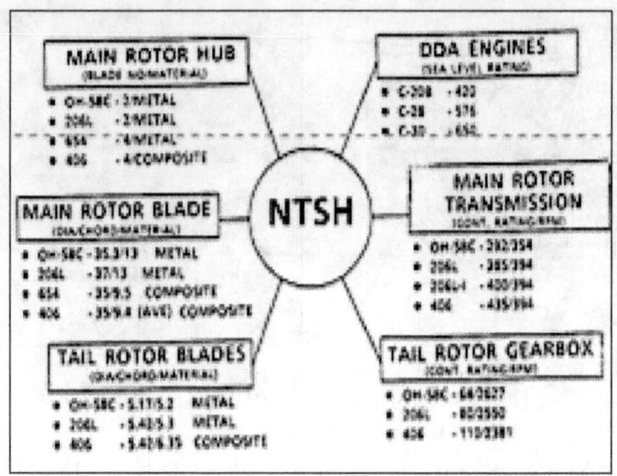

Figure 74B. Enough Components to Draw Upon.

Evolution of the Army's AHIP Precedence of Requirements are in Figures 75A&B.

3.8.1 PRECEDENCE OF REQUIREMENTS

Ranking Of Major Characteristics In Order Of Relative Priority Is:

A. TARGET ACQUISITION, RANGEFINDING, DESIGNATION, AND ACCURACY
B. TARGET HANDOFF CAPABILITY / DATA LINK
C. IMPROVED NAVIGATION CAPABILITY
D. OPERATIONAL SUITABILITY
E. SYSTEMS INTEGRATION AND EFFECTIVENESS
F. MISSION EQUIPMENT GROWTH POTENTIAL
G. SIMPLIFIED INTERFACE
H. IMPROVED COMMUNICATION EFFECTIVENESS AT NOE
I. AIRCRAFT FLIGHT PERFORMANCE / PAYLOAD / ENDURANCE
J. WEIGHT
K. LOST
L. TRAINING AND SUPPORT
M. SURVIVABILITY / VULNERABILITY
N. RELIABILITY, AVAILABILITY, MAINTAINABILITY
O. SAFETY AND HEALTH
P. HUMAN FACTORS
Q. DEPLOYMENT
R. TRANSPORTABILITY

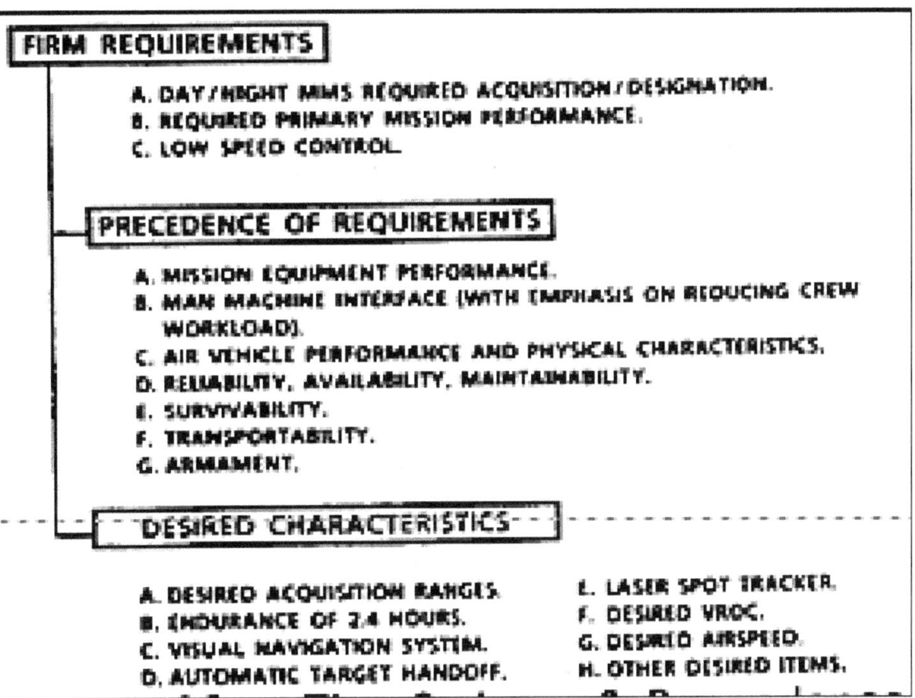

Figure 75A. Initial Order of Precedence of Requirements 9/1/80   Figure 75B. Final Order, 1/7/81

- From the Mission Equipment Package (MEP) Team's point of view, the desired acquisition range to ensure even greater standoff distance became a number 1 goal.

- The MEP Team noted the Army's desire to have laser spot tracking capability (although no reference specification paragraph was given).

- The Air Vehicle Team saw the elevation of desired endurance to higher priority, and they saw a priority introduced for desired vertical rate of climb (VROC).

- In regard to the three firm requirements, A and B, were addressed and understood; however, C, Low Speed Control and the understanding of Agility was not.

Alternate AHIP/OH-58D Configuration elements were selected from the Logic Tree and are Illustrated in Figure 76 under MINI, MIDI and MAXI.

143

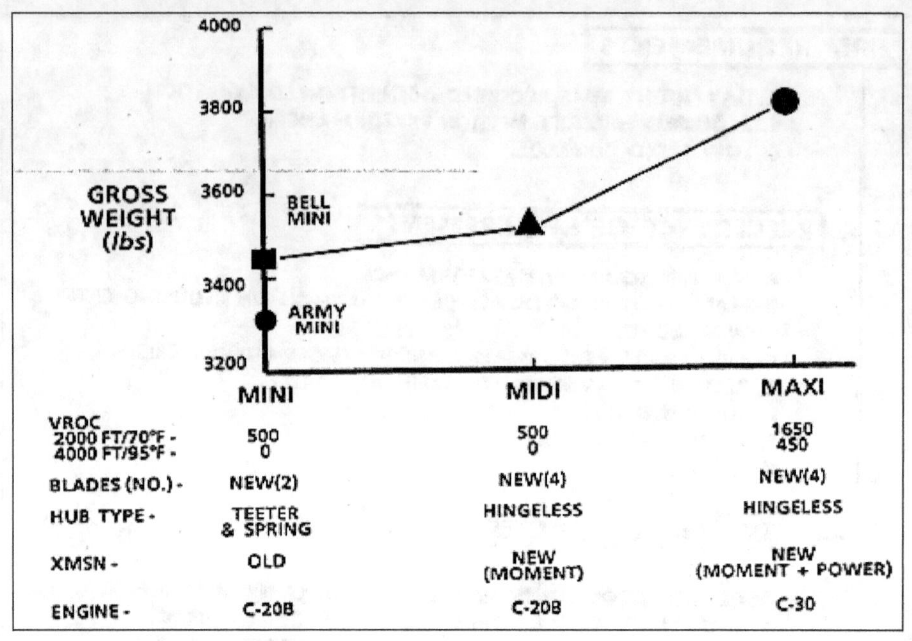

Figure 76. Alternate AHIP/OH-58D Configurations Selected from the Logic Tree

- The MINI was the OH-58C modified to meet requirements.

- The MIDI was thought to specifically address agility with the 654-rotor system.

- The MAXI reached further consideration because it proposed using the Allison C-30 Engine.

Bell went with the MAXI in their proposal, and it made the Bell OH-58D the winner.

AHIP Draft System Spec for Low-Speed Agility & Mobility included desired characteristics, Figures 77 and 78. They were included based on my experience in flying and comparing scout helicopters in the Vietnam War.

"It is desirable to have an average lateral acceleration to 35 knots true airspeed (KTAS), left and right, of 0.40 g's at 2000 ft pressure altitude and 70°F at the primary mission gross weight. The maneuver shall be initiated from a zero wind out of ground effect (OGE) hover and executed without loss of altitude and the handling qualities during this maneuver shall satisfy the requirements of MIL-H-8501."

"The operational/organizational concept of this helicopter includes nap-of-the-earth (NOE) and contour flight and hence requires a very agile and highly maneuverable vehicle. Accordingly the vehicle control and response characteristics shall be tailored to the extent possible to achieve the optimum configuration. Figures 1 and 2 provide some design guidance relative to this requirement."

Figure 77. AHIP Draft System Spec Words for Low-Speed Agility & Mobility

Figure 78. Nap of the Earth (NOE) and Contour Flight Requires Low-Speed Agility and Maneuverability

The desired intent was to convey to industry that the Army needed an AHIP aircraft with excellent low-speed agility and maneuverability for Nap of the Earth (NOE) and Contour Flight. Since the AHIP was a fixed-price contract, the shaded regions in Figure 79 were desired but not required. However, Bell was determined to meet the Desired.

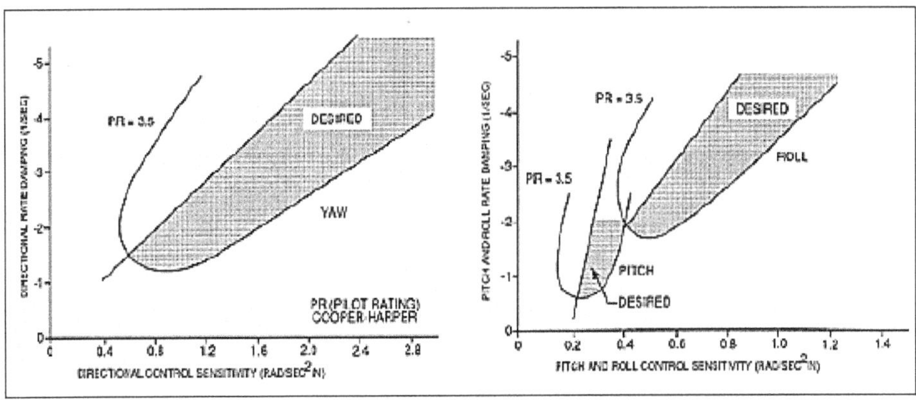

Figure 79. AHIP System Specification Illustrates Desired Pitch, Roll Yaw Control Power vs. Damping

My Story

Understanding Agility was a Key to Bell's Successful AHIP/OH-58D Design. Perhaps the most productive discussion about power margins (as well as main rotor type and tail rotor thrust margins) required for agility occurred in early December 1980.

The Bell AHIP Program Director, Ted Hoffman, with the Bell AHIP Chief Engineer, Frank Harris, went to Fort Rucker, AL to talk "agility" directly with as many scout crews as they could. They asked them to describe the one maneuver they needed to do that.

- Virtually to a man, they described a "duck and run" situation that—from start to finish – took something like 15 secs. This "bob down and dash off maneuver" was carried back to Bell test pilots who tried it with the Model 206LM.

- The "re-masking" maneuver, as flown with this C-20B powered and 4-bladed hinge-less rotor helicopter, yielded some real eye-opening engineering data. The maneuver began in an Out of Ground Effect (OGE) hover. Power was then reduced to produce a rapid vertical descent. Recovery was initiated by increasing power to arrest the descent rate and to accelerate the helicopter in sideward flight with transition into forward flight.

- Based on the Model 206LM tests a peak power of 420 shp was required to stop the descent rate, but engine output exceeded hover power required for only about 6 sec. A 29% power margin over hover and a max vertical load factor of 1.20 were required to successfully complete this maneuver successfully.

- This Maneuver required the MAXI Solution, Figures 80-85.

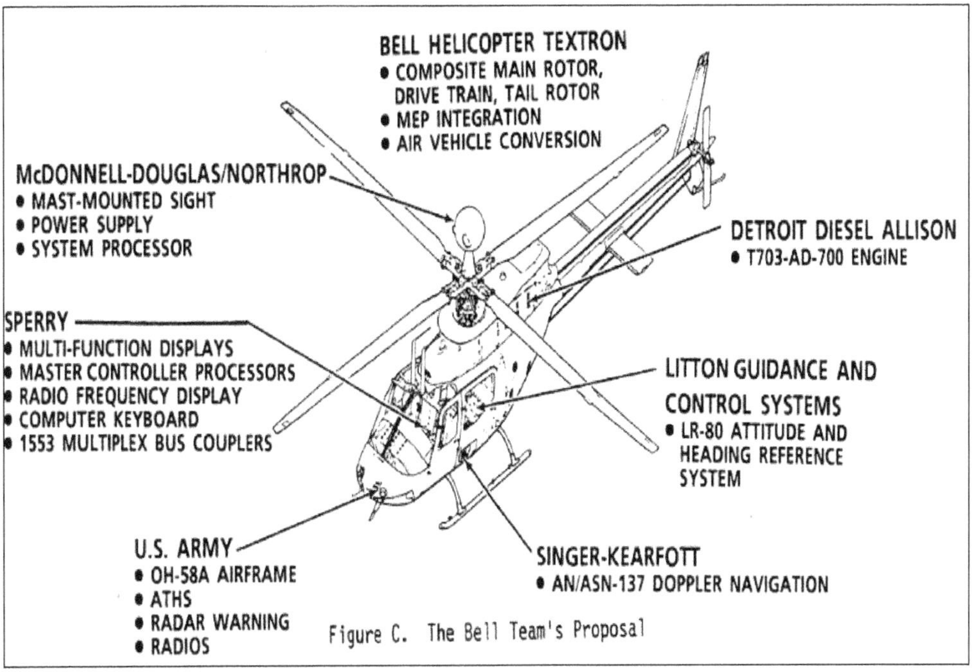

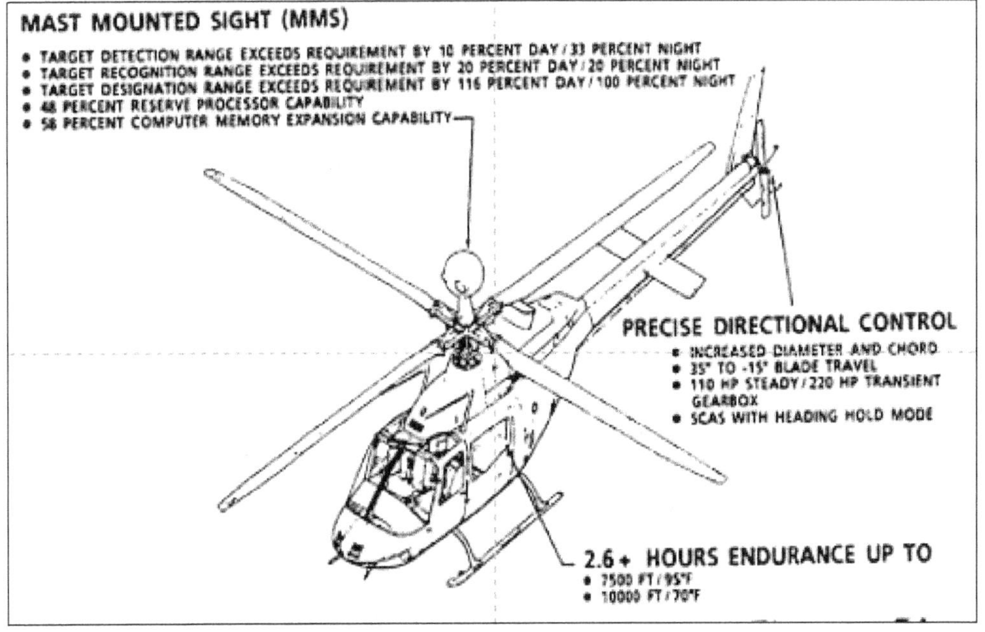

Figure 80. Bell AHIP Maxi Proposal

Figure 81. How the Bell AHIP Exceeded the Three Firm Requirements

AIR VEHICLE ⟶ 1900

ROTOR 320
TAIL 42
BODY 367
GEAR 48
ENGINE SECTION 54
AIR INDUCTION 22
PROPULSION 703
FLIGHT CONTROLS 159
HYDRAULICS 17
ARMAMENT 40
FURNISHINGS & EQUIPMENT 85
A/C 28
LOAD HANDLING/CONTINGENCY 15

MEP ⟶ 746

MAST MOUNTED SIGHT 202
CONTROLS & DISPLAY 89
COMMUNICATION 88
NAVIGATION 59
ELECTRONICS 84
INSTRUMENTS 40
ELECTRICAL 184

WEIGHT EMPTY (WE) ⟶ 2646

WE 2646
CREW 470
ARMAMENT 113
CREW ARMOR 115
OTHER 29
FUEL 574

PMGW 3947

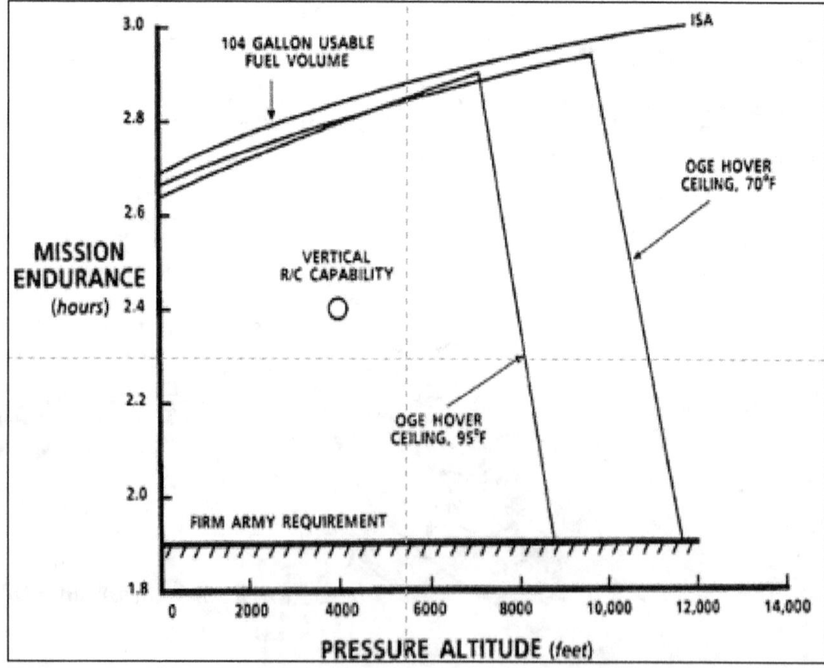

Figure 82. Bell Proposed Weight Statement Figure 83. Mission Endurance Achieved

148

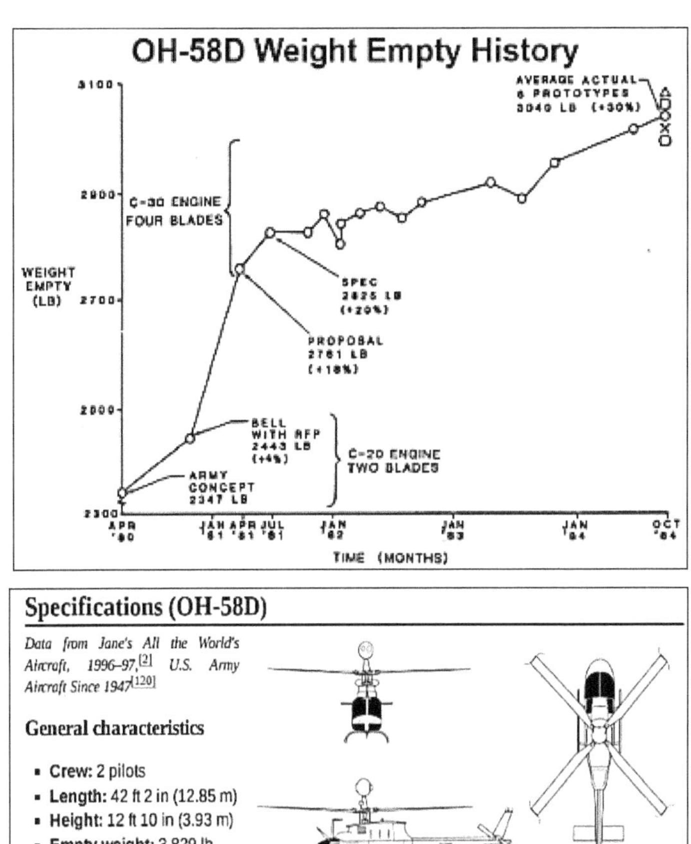

Specifications (OH-58D)

Data from Jane's All the World's Aircraft, 1996–97,[2] *U.S. Army Aircraft Since 1947*[120]

General characteristics

- **Crew:** 2 pilots
- **Length:** 42 ft 2 in (12.85 m)
- **Height:** 12 ft 10 in (3.93 m)
- **Empty weight:** 3,829 lb (1,737 kg)
- **Gross weight:** 5,500 lb (2,495 kg)
- **Powerplant:** 1 × Rolls-Royce T703-AD-700A turboshaft, 650 hp (485 kW)

Figure 84A. OH-58D Weight Empty History   Figure 84B. OH-58D Performance Specifications

Early Bell Design Trade Studies led to the MAXI OH-58D Configuration to meet Desired Requirements, and Flattened the OH-58D Weight Empty History, Figure 84A. The Model 206A Airframe Structure modified by Bell in the 1960s for the OH-58A had sufficient Growth Margin to accommodate the larger engine needed for the OH-58D. The Bell OH-58D Human-Machine-Integration (HMI) was state-of-the-art, as illustrated in Figure 85A. The cockpit visibility outside of the OH-58D was greatly enhanced as viewed in Figure 85B.

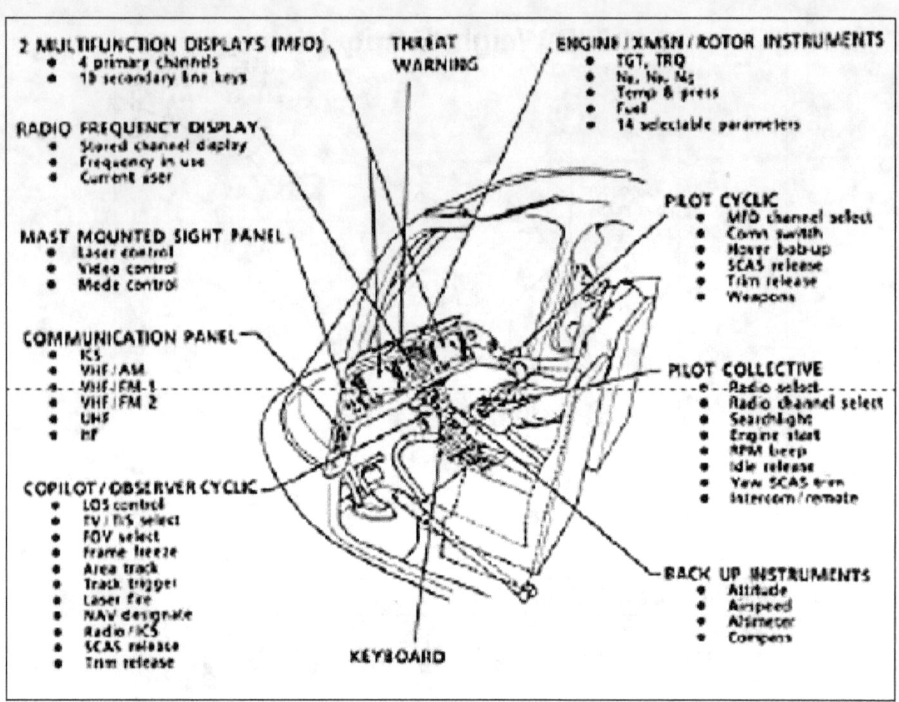

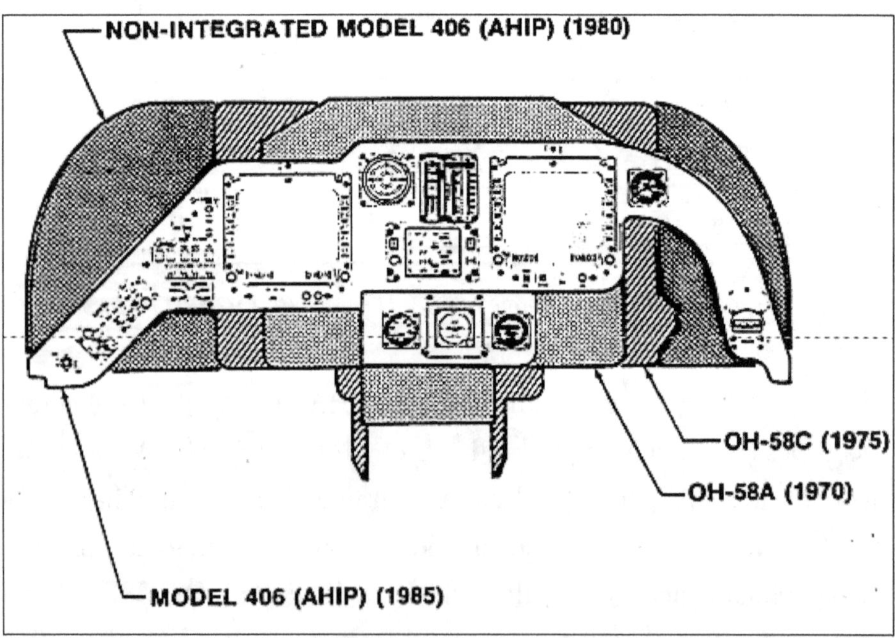

Fig. 85A. OH-58D HMI was State of the Art. Fig. 85B. Outside Visibility from the Cockpit was Enhanced.

150

The integration of the MMS into the Aircraft was a tremendous success, as illustrated in Figures 86 A, B, & C and Figure 87.

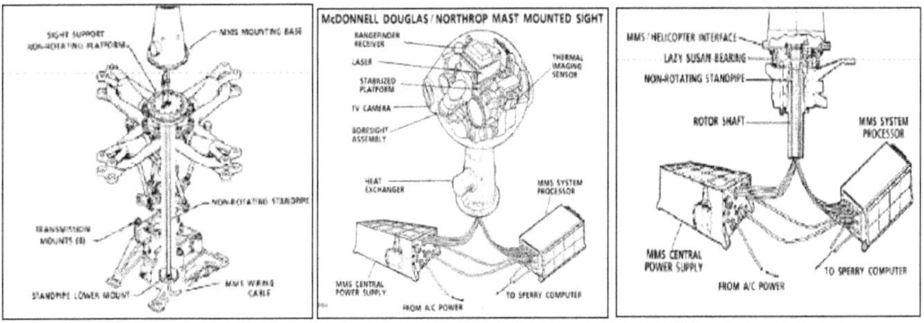

Fig. 86A. MMS-Based Integrated in Rotor Shaft. Fig. 86B. Key Elements in MAA. Fig. 86C.MMS/Helicopter Interface

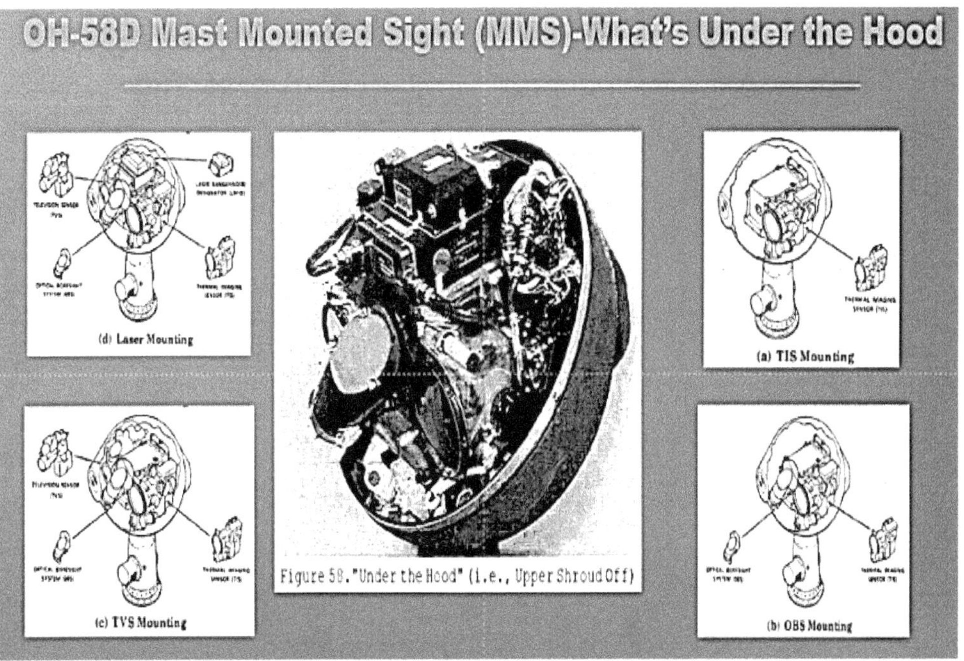

Fig.87. The four major mounting systems for the four functions: Thermal Imaging System (TIS), Observation System (OBS), TV System and Laser Mounting System are illustrated.

1. The success of these mounting systems was based on two things:
2. The Army McDonnell-Douglas developed an Internal Bearing Stabilized Sighting Unit (IBSSU) for MMS which isolated the functional systems from the rotor vibrations.

3. The reduced vibratory environment was with the four-bladed soft in-plane hinge-less rotor on the OH-58D vice the OH-58A/C two-bladed teetering stiff in-plane rotor systems. With the mass of the MMS above the teetering rotor the chordwise bending mode approached resonance with one/rev and very high vibratory loads. Bell had measured the vibratory loads with a mast mounted sight on a Bell two-bladed teetering stiff in-plane helicopter with the Armament Research and Technology Lab at Picatinny Arsenal. They tried to pass on these loads to McDonnell Douglas. However, I stepped in and made them provide the lower four bladed soft in-plane rotor loads to McDonnell-Douglas through LTC(Ret) Skip Neuwein who had retired and then worked for McDonnell-Douglas.

The Army McDonnell-Douglas Developed Internal Bearing Stabilized Sighting Unit (IBSSU) for MMS is illustrated in Figures 88 A, B, and C. The IBSSU and the lower loads from four bladed soft-in-plane hingeless rotor gave longer TV and FLIR ranges.

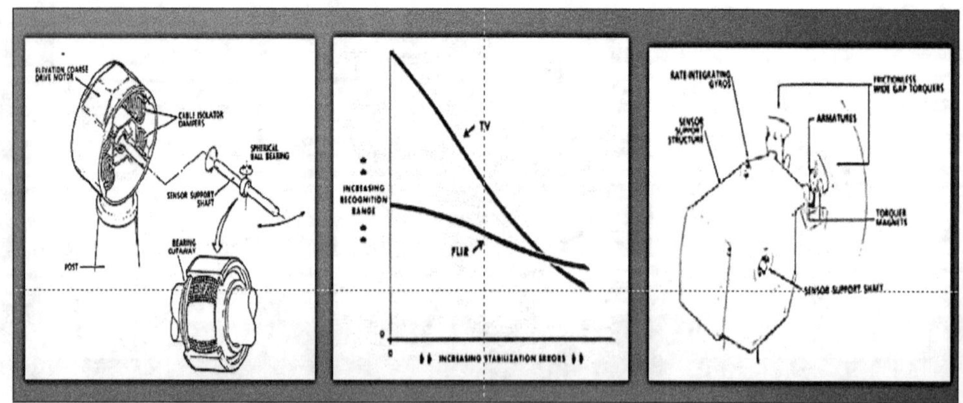

Fig.88A.IBBSU for Vibration Isolation. Fig.88B.Impact of Stab Errors on Recognition Ranges. Fig.88C. IBBSU for Sight Stab.

The concept of parallel mounting of the isolator and spherical bearing with the torquing actuator had been developed for the Army by McDonnell-Douglas to introduce passive isolation into stabilized gimbal platforms. The concept was given the name IBSSU which stands for Internal Bearing Stabilized Sighting Unit. During the AHIP SSEB MMS modeling and simulation illustrated below was used to evaluate the impact of vibratory loads and MMS base motion on the required SSEB required ranges, as in Fig. 89 (Ref. 6)

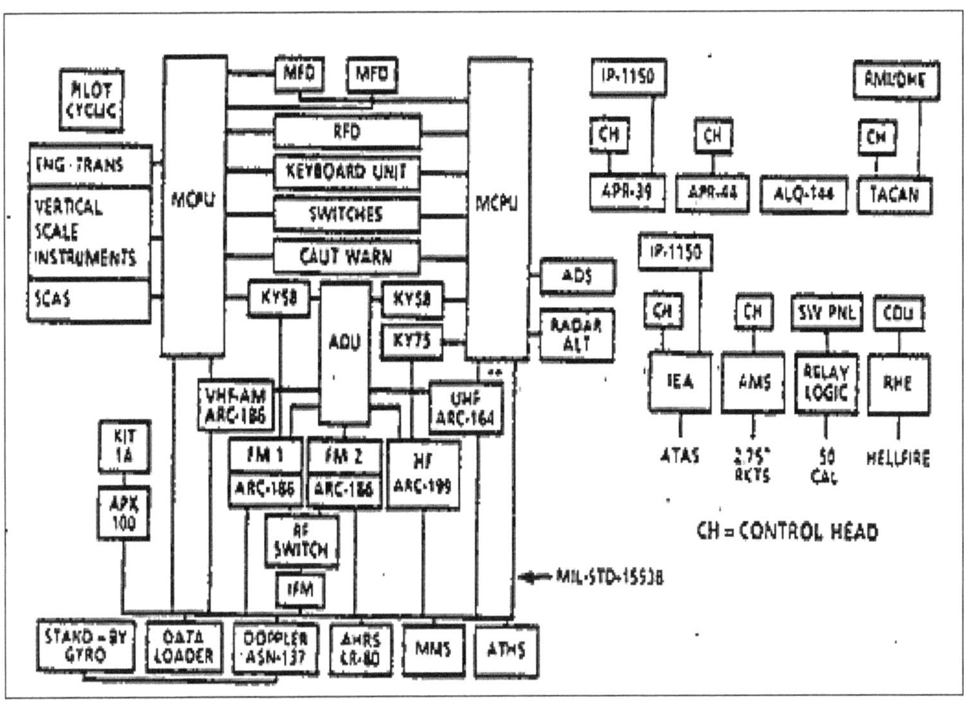

Fig. 89. OH-58D Basic Production MEP Avionics Components

Lessons Learned from the AHIP and Application to the OH-58D Kiowa Warrior from both Industry and Government

Evolution from OH-58D Kiowa to OH-58D Kiowa Warrior

Initially intended for attack, cavalry, and artillery roles, the Army only approved a low initial production level and confined the OH-58D's role to field artillery observation. The Army also directed that a follow-on test be conducted to further evaluate it due to perceived deficiencies.

153

On 1 April 1986, the Army formed a task force at Fort Rucker, Alabama, to remedy deficiencies in the AHIP. [Ref. 3] During 1988, the Army had planned to discontinue the OH-58D and focus on the LHX. However, Congress approved $138 million to expand the program, calling for the AHIP to operate with the Apache as a hunter/killer team. The AHIP would locate targets, and the Apache would destroy them in a throwback to the traditional OH-58/AH-1 relationship in the Vietnam War.

The Secretary of the Army directed instead that the aircraft's armament systems be upgraded, based on experience with Task Force 118's performance operating armed OH-58Ds in the Persian Gulf in support of Operation Prime Chance, and that the type be used primarily for scouting and armed reconnaissance. (Ref. 3)

The armed aircraft would be known as the OH-58D Kiowa Warrior, denoting its new armed configuration. Beginning with the 202nd aircraft (s/n 89-0112) in May 1991, all remaining OH-58Ds were produced in the Kiowa Warrior configuration. During January 1992, Bell received its first retrofit contract to convert all remaining OH-58Ds to the Kiowa Warrior configuration. Other Kiowa Warrior Wars (Ref. 3)

Gulf War

During Operation Desert Storm, 115 deployed OH-58D helicopters participated in a wide variety of critical combat missions and were vital to the success of the ground forces mission. During Desert Shield and Desert Storm, the Kiowas collectively flew nearly 9,000 hours with a 92 percent fully mission capable rate. The Kiowa Warrior had the lowest ratio of maintenance hours to flight hours of any combat helicopter in the Gulf War.

Operation Just Cause and action in the 1990s.

During Operation Just Cause in 1989, a team consisting of an OH-58 and an AH-1 were part of the Aviation Task Force during the securing of Fort

Amador in Panama. The OH-58 was fired upon by Panama Defense Force soldiers and crashed 100 yards (90 m) away, in the Bay of Panama. The pilot was rescued, but the co-pilot died.

Afghanistan and Iraq

The U.S. Army employed the OH-58D during Operation Iraqi Freedom in Iraq and Operation Enduring Freedom in Afghanistan. Between a combination of combat and accidents, over thirty-five airframes have been lost, resulting in the deaths of thirty-five pilots. Their presence was also anecdotally credited with saving lives, having been used to rescue the wounded despite their small size.

In Iraq, OH-58Ds reportedly flew seventy-two hours per month, while in Afghanistan, the type flew eighty hours per month.[46] During April 2013, Bell stated that the OH-58 collectively accumulated 820,000 combat hours and had achieved a 90% mission-capable rate. [47] Pictures of OH-58D Kiowa Warriors in Afghanistan are shown in Figures 75A and 75B.

Figure 90A. OH-58D at Kandahar, 2011

Figure 90B. Group of Kiowa Warriors covered by snow at Bagram AB, 2013

Retirement

The U.S. Army's first attempt to replace the OH-58 was the RAH66 Comanche of the Light Helicopter Experimental program, which was canceled in 2004. Airframe age and losses led to the Armed Reconnaissance Helicopter program and the Bell ARH-70, which was canceled in 2008 due to cost overruns. The third replacement effort was the Armed Aerial Scout (AAS) program. Due to uncertainty in the AAS program and fiscal restraints, the OH-58F's planned retirement was extended from 2025 to 2036. (Ref. 3)

The Kiowa's scout role was supplemented by tactical unmanned aerial vehicles, the two platforms often acting in conjunction to provide reconnaissance to expose crews to less risk. The OH-58F had the ability to control UAVs directly to safely perform scout missions. (Ref. 3)

In 2011, the Kiowa was scheduled to be replaced by the light version of the Future Vertical Lift aircraft in the 2030s. In December 2013, the U.S. Army had 338 Kiowas in its active-duty force and thirty in the Army National Guard. The Army considered retiring the Kiowa as part of a wider restructuring to cut costs and reduce the variety of helicopters operated. The Analysis of Alternatives for the AAS program found that operating the Kiowa alongside RQ-7 Shadow UAVs was the most affordable and capable solution, while the AH-64E Apache Guardian was the most capable immediate solution.

One proposal was to transfer all Army National Guard and Army Reserve AH-64s to the active Army for use as scouts to divest the OH-58. The Apache costs 50 percent more than the Kiowa to operate and maintain; studies note that had it been used in place of the Kiowa in Iraq and Afghanistan, total operating costs would have risen by $4 billion, but also saved $1 billion per year in operating and sustainment costs.

UH-60 Black Hawks would transfer from the active Army to reserve and Guard units. The aim was to retire older helicopters and retain those with the best capabilities to save money. Retiring the Kiowa would fund Apache upgrades. The Army placed twenty-six out of 335 OH-58Ds in non-flyable storage in 2014. In anticipation of divestment, the Army looked to see if other military branches, government agencies, and foreign customers had any interest in buying the type.

The Kiowas were considered to be well-priced for foreign countries with limited resources; Bell had not yet agreed to support them if sold overseas. [53] Media expected OH-58s to go to foreign militaries rather than civil operators due to high operating costs.] By 2015, the Army had divested 33 OH-58Ds. By January 2016, the Army had divested all but two OH-58D squadrons.

The success story of the progression from a scout observation helicopter to the Kiowa Warrior is based on a "successful partnership" between industry and government over a decade of improvements and modifications. This partnership has been well documented by industry, Frank Harris, and Mike Kawa, and by government, Dan Schrage, COLs Walt Rundgren and Bud Forester, John McLaughlin, Jack Van Kirk, Darrell Harrison, Dave Weller, and Bill Lewis in the following references. (Refs. 2,3, 5, 6)

To quote Frank Harris about AHIP/OH-58D: "All the right people were in the right place at all the right times." (Ref. 2)

Who are/aren't the right people, and what/when place/time for government and industry?

Competent Technical Experts on the government and industry side who trust each other and support the overall objective at the right place and time.

*I*n Frank Harris' words, the overall objective for AHIP/OH-58D was:

"The No. 1 goal was to satisfy the pilot/co-pilot scout team."

PEO/PMO/Industry PMOs who don't understand, encourage, fund, or use the necessary technical expertise to support Development Assurance, not only Qualification Assurance Tradeoffs

The Air Vehicle Margins from the OH-58D Development Process led to successful further development to the OH-58D Kiowa Warrior as the Prime Chance armed OH-58D and the Multi-Purpose Light Helicopter (MPLH).

From Mike Kawa (Ref. 3) Contains a detailed documentation of the Prime Choice OH-58D Kiowa Warrior and the Multi-Purpose Light Helicopter (MPLH) development programs.

Program and Operational Highlights of the Armed Kiowa Warrior Development and Production

The successful launch of the OH-58D by an incredibly unique industry-government team was closely followed a timely production program that then launched an armed version in record time to fulfill an urgent military need in the Middle East.

Like the key AHIP/OH-58D Development Program document by Frank Harris's *AHIP: The OH-58D From Conception to Production,* (Ref. 2) a detailed documentation by Mike Kawa and Jack M. Van Kirk, "*Program and Operational Highlights of the Armed OH-58D Kiowa Warrior,*" (Ref. 3) provided the next step toward an evolution to a successful Future Attack and

Reconnaissance Aircraft (FARA) Program. An overview of this document will be provided, followed by lessons learned.

U. S. Army units operating in the field learned through experience that the basic production OH-58D helicopter is a premier scout helicopter. It had no equal as a night scout, operating quietly to detect distant targets and handing off to ground artillery, other Army killer aircraft, or U. S. Air Force strike forces. Until it was armed, OH-58D crews had to be content to hand off targets to others for final disposition, or at most, laser-designating the target for a killer aircraft. Now enter the fully armed OH-58D, as illustrated in Figure 75.

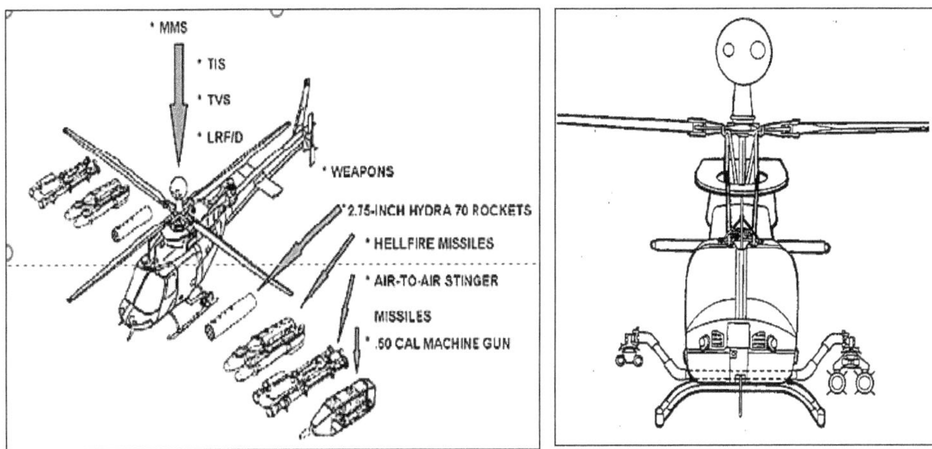

Figure 91A. The OH-58D Kiowa Warrior Figure 91B. Weapons Pylon

Conclusion: "Accomplishments in the OH-58D Kiowa Warrior program were a result of innovative engineering, aggressive management, and open-mindedness of the Bell/Army/Industry team working together. A model program, producing a unique aircraft in the world—inspiring now, but what's next for the OH-58D?" (Ref. 3)

THE OH-58D: THE MOST HONORED MILITARY AIRCRAFT

The U. S. Army's OH-58D, manufactured by Bell Helicopter Textron Inc., was the central figure in four major U. S. military aviation awards during

1989.These were: Bell Helicopter employees who build the OH-58D, the Army personnel who manage the OH58D program, and Army aviators who fly the Advanced Aero-Scout have been honored with the following awards: The Army Aviation Materiel Readiness Award; The Leich Award; the AAAA Department of the Civilian of the Year Award; and the Daedalian Weapon System Award (Colonel Franklin C. Wolfe Memorial Award). (Ref. 3)

The OH-58D Kiowa Warrior became equipped with two universal quick-change weapons pylons. Each pylon can be armed with two TOW or Hellfire missiles, seven Hydra 70 rockets, two air-to-air Stinger missiles or one .50-calibre fixed-forward machine gun, as in Figure 75.

A hostile gunboat presence at night in the Persian Gulf in 1987, created a need for a small, stealthy, armed helicopter for interdiction purposes. Teamwork within the U. S. Armed Forces and industry made it possible for the Army's OH-58D aircraft, built by Bell Helicopter Textron, Inc. (BHTI), to meet that need.

The chapter tells the story of how military and civilian members of the *Operation Prime Chance* team added armament to fifteen already superb aerial observation OH-58D helicopters, tested them, developed operational procedures, trained crews, and deployed them overseas in record time. The basic production OH-58D was developed under the Army Helicopter Improvement Program (AHIP), with initial deliveries beginning in December 1985.

The chapter identifies the differences between the basic production and the special armed helicopter, called the Prime Chance OH-58D, and what it took to prepare the aircraft for Army deployment. The impact of the Prime Chance OH-58D on the hostile gunboat activity was so great that shortly after they arrived in the Persian Gulf, there were no more mining or harassment incidents. Because of the obvious advantages of armament configuration and the successes in the Gulf, a program has begun to arm all OH-58Ds in Army inventory, starting with current production.

The paper discusses improvements being made in integrating the weapons systems into the modern, user-friendly OH-58D cockpit controls and displays. Planned qualification tests and schedules are described. The paper briefs another facet of the armed OH-58D picture with the description of the Multi-Purpose Light Helicopter (MPLH) development concept. In the MPLH configuration, the armed OH-58D was capable of many different missions, including rapid air deployment aboard military transport aircraft. And finally, armed operational aspects are covered, describing performance at the National Training Center (NTC), REFORGER exercises, and armed fielding.

Several appendices are attached to the paper, which includes engineering descriptions of the air vehicle and the subsystems that make up the mission equipment package (MEP).

Differences: Production OH-58D and OH-58D Kiowa Warrior: Air Vehicle Differences (Ref. 3)

The major differences in operation between the Prime Chance and the basic production OH-58D are listed below.

- -Weapons pylons added.
- -100% maximum continuous transmission.
- The rating is 510 hp at the main rotor mast (was 455hp).
- -Use of DOD-L-85734(AS) gear lube oil in transmission and gearbox (was MIL-L-23699 turbine engine oil).
- -Engine turbine gas temperature (TGT) allowable limit increased for takeoff at high ambient temperatures:
- -2-minute takeoff limit 797'C, 1468'F
- -30-second takeoff limit = 825'C, 1518'F (new)
- (Note: The two limits are not to be used in
- conjunction with each other.)
- -Aircraft lateral center of gravity (cg) limits

- increased to + 4. 0 inches (was 1. 0 left and 2. 0 right).
- -Maximum gross weight increased to 5,400 pounds (was 4500 lbs.)

Weapons Pylons. The weapons pylons for Prime Chance were similar to pylons developed on earlier BHTI 206-type helicopters and referred to as "tube in-a-tub" pylons (Figure 76). The weapons stores were mounted on racks at the ends of a long, aluminum tube which curves downward, enters the sides of the cabin below the avionics bay doors behind the flight crew and extends along the floor from side to side just forward of the fuel cell. Large fittings transferred weapons loads from the tube to the fuselage sides. Local sheet metal riveted doublers were installed to distribute loads from the fittings to the fuselage structure.

Special Allison T703-AD-700 Engine Consideration

Because of ambient temperatures in excess of 113' F. A killer aircraft. Now enter the fully armed OH-58D, as illustrated in Figure 22 (39'C), were encountered during summer operations in the Persian Gulf, the Prime Chance helicopters were allowed to operate at increased engine TGTs for takeoff and short-term hover at 5,400 pound gross weight. Engine Hot Section Factor. As a means of tracking the higher TGT allowable limits for the engines in the Prime Chance helicopters, a parameter known as the hot section factor (HSF) was automatically recorded by the Honeywell computers. The HSF can be accessed on the ground by calling up the Engine History page on the multifunction display. Use of the HSF is guided by the following (reference Allison Gas Turbine Operations):

1. When T5 is greater than 715, 6'C and less than or equal to 825. 6'C, HSFs are accumulated

$$\text{(Counts/Minute} = 10)^{\left(\frac{T5 - 715°C}{42}\right)}, T5 \text{ is } °C$$

2. To avoid excess HSF counts due to peak starting temperatures, HSF counts are not accumulated for thirty seconds after T5 reaches 300'C during an engine start. A curve of HSF counts per minute versus TGT is presented in Figure 92, showing how HSF accumulates at a much greater rate at higher engine temperatures than at lower temperatures.

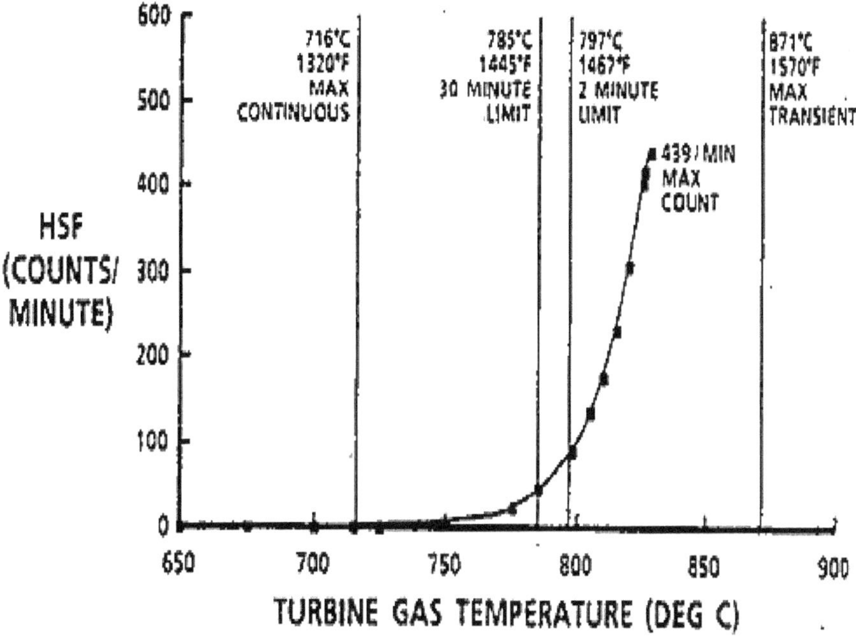

Figure 92. Engine HSF vs. Turbine Gas Temperature

Initially, Allison recommended removal of an engine for analytical teardown inspection at 131, 000 counts, or 300 hours, which ever occurred first. Later, after several engines had been inspected for specified hours of accumulated use, the inspection period was increased first to 600 hours and then to 1000 hours. The critical HSF count remains at 131, 000, which is the maximum allowable number of thermal stress cycles in the hot end section of the engine.

Mission Equipment Package Differences

The mission equipment package (MEP) selected for Prime Chance included subsystems that had not been previously installed on the OH-58D. These subsystems could have been integrated into the basic OH-58D 1553 multiplex data bus. However, the schedule precluded this concept. Consequently, each subsystem was added as a stand-alone installation. As stand-alone systems, dedicated control panels were required, completely filling the center console, and overflowing onto the instrument panel, as shown in Figure 93. (5)

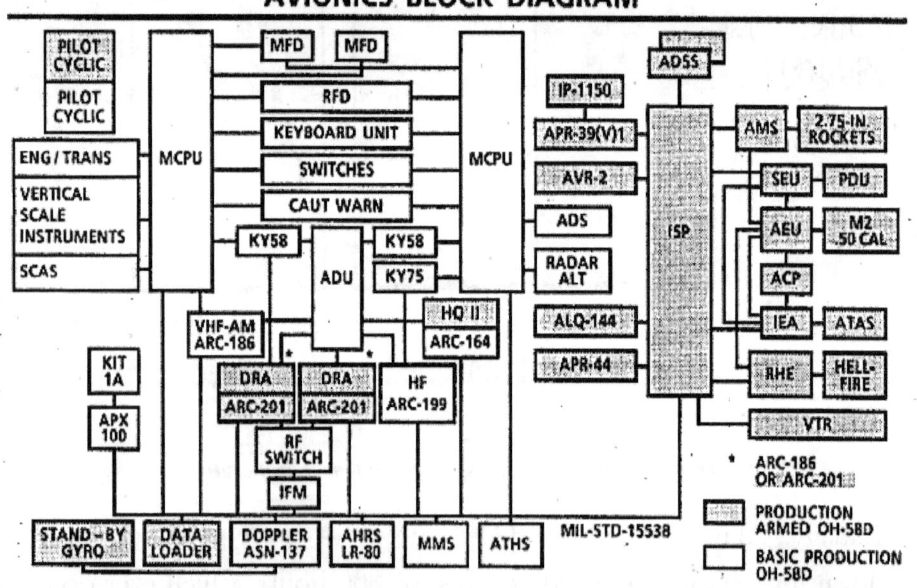

Figure 93. Prime Chance OH-58D Kiowa Warrior Avionics Block Diagram

As previously mentioned, an extensive armament subsystem was added to the OH-58D MEP giving the Prime Chance helicopter the capability to deliver air-to-air and air-to-ground munitions and fulfilling the purpose of the Prime Chance program.

HELLFIRE (HF) is an air-to-ground, laser guided, anti-armor missile system. The Prime Chance OH-58D can carry two HF missiles on each side of the aircraft. The HF system can be guided autonomously by the OH-58D through the laser designator in the Mast Mounted Sight (MMS) or can be guided by others. Ground-or air-based laser designators. Steering indicators from the MMS are displayed on a pilot steering indicator (PSI) installed as a part of the HF system.

Air-to-Air Stinger (ATAS) is an airborne, infrared, antiaircraft missile system. The Prime Chance OH-58D could carry two ATAS missiles on each side of the aircraft. The system uses passive IR detection to acquire and destroy targets. Target acquisition information is displayed on the Pilot's Display Unit (PDU), Figure 94.

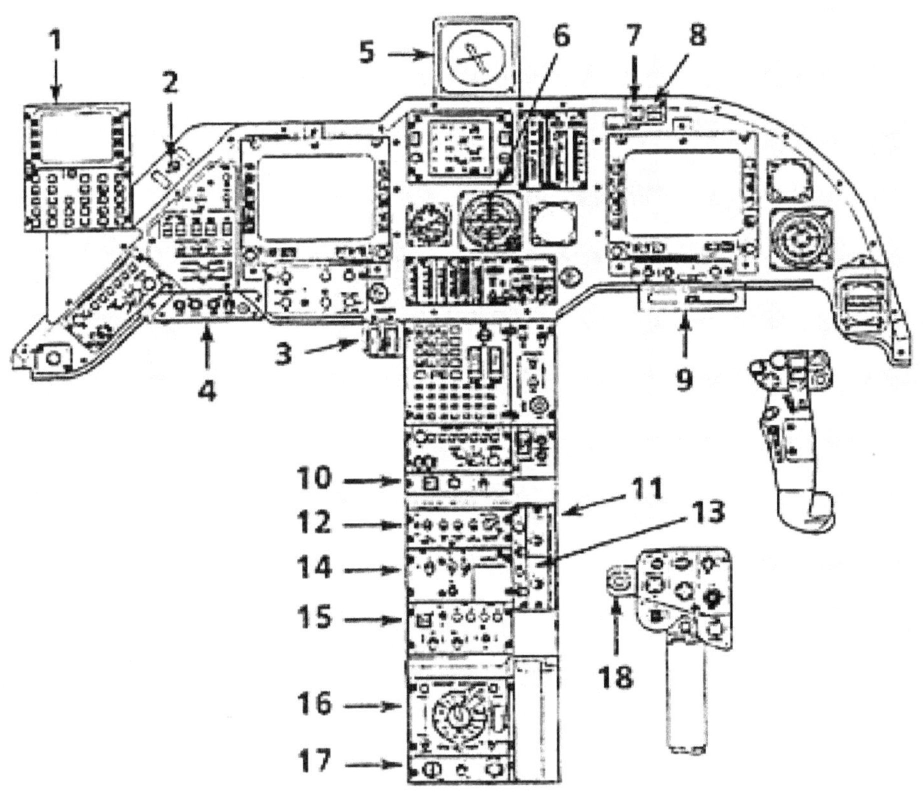

1. HELLFIRE control display unit (CDU)
2. Weight on gear switch interrupt
3. Jettison switches
4. Video recorder control panel
5. Bearing distance heading indicator (BDHI)
6. Attitude deviation indicator (ADI)
7. AN / APR-44 radar indicator
8. AN / ALQ-144 IR jammer annunciator
9. Pilot steering indicator (PSI)
10. AN / APR-44 radar control panel
11. AN / ALQ-144 IR jammer control panel
12. HELLFIRE control panel
13. AN/APR(V)-39 radar warning control panel
14. ATAS control panel
15. Armament control panel (ACP)
16. Armament management system (AMS) control panel
17. Gyromagnetic compass control panel
18. HELLFIRE fire switch

Figure 94. OH-58D Instrument Panel and Console Showing Weapons and ASE Controls/Displays

LESSONS LEARNED FROM AHIP/OH-58D KIOWA WARRIOR PROGRAM

Weapons Development

In 1985 AVSCOM was requested to examine the possibility of arming the AHIP. Mr. Charlie Crawford led the effort and recruited the design team. His comments were, "CPT Lewis, I want you to analyze the AHIP for weaponization. I want a roof-mounted sight incorporated into the TOW system. It also must carry two motorcycles in a utility configuration. You have two months." With this blessed little guidance, CPT Bill Lewis began the study.

The whole time CPT Lewis was working on the TOW design, he knew that the HELLFIRE was a better weapon. The AHIP had a laser designator and target tracker in the MMS. From his perspective, this simplified the design/operation considerably, but there were three problems. First, with a 4200-pound airframe, the weight of a fully loaded aircraft, including four missiles, exceeded the max gross weight.

Second, the lateral CG of 4.5 inches did not allow an asymmetric firing of the missiles. Lastly, Charlie Crawford wanted TOWs! As an augmentation to the TOW study, CPT Lewis examined what would be required to increase the GW by 300 lbs. and keep the loading within the lateral CG limits. Incorporating the HELLFIRE hardware and integration was far simpler than the TOW. CPT Lewis briefed the HELLFIRE option to Mr. Crawford on the plane enroute to the 9th ID brief to the commander. During the brief to the CG, he stated that he wanted a weapon that would blow the turret off a tank with one hit. Mr. Crawford then requested the AHIP HELLFIRE study brief.

The concept was viewed favorably, but not executed as a part of the 9th ID Aviation Study. The study was utilized as a basis for the development of the AH-58 for use in Operation Prime Chance in 1987. While the AH-58 configuration was never adopted by the Army, the incremental development of the weapons suite resulted in the OH-58D. The major upgrade to the OH-58D was the development of the universal weapons pylon allowing rockets, guns, STINGER, and the HELLFIRE to be incorporated. The OH-58D was then known as the KIOWA WARRIOR, or KW. was selected. Blast blankets were incorporated to accommodate the plume overpressure on the fuselage. Existing TOW hardware was integrated for firing. TOW tracking was accomplished from the observer seat. The final design was marketed as the 406 Combat Scout by Bell. ["Bell Model 406 CS Combat Scout". *Jane's All the*

World's Aircraft 1992–1993. Jane's Information Group, 1992. subscription article dated 15 July 1992],

The Backside of the V Diagram (Ref. 3)

In early 1986 the AHIP was in final DT testing at the Army Engineering Flight Activity (AEFA) at Edwards AFB. These were some of the final tests to achieve an operational flight release. One of the final tests was the testing of the aircraft during speeds below forth-five knots (low speed). With the aircraft operating at 235-degree relative azimuth (left, rearward flight) at approximately thirty-five knots, a severe compressor stall was experienced. Further flight tests proved the data to be repeatable. Due to the severity of the repeatable stall, the upper fuselage fairing (doghouse) was redesigned, as illustrated in Figure 95. This redesign took several months, and the follow-on testing even more. The result of this discovery proved the need for experimental flight testing of a production design. While today's detailed analysis methods were not available, the design was thought to be complete prior to the low-speed flight tests. Current tools may have been more capable in a first design success, but that is not certain. It is important to understand the need for component and final flight tests and realize that incremental changes will most likely be needed. It is not a reason to terminate/pause a program, but to plan for these needed changes as a matter of course.

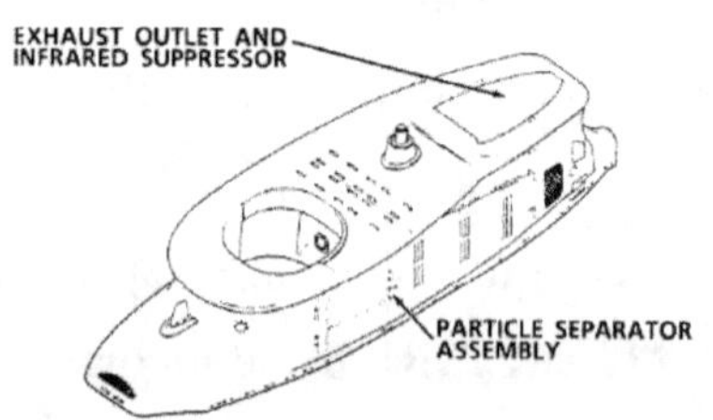

Figure 95. Upper Fuselage Fairing Redesign (4)

Simulation Verification

The development of the AHIP was paralleled by the integration of the STINGER onto the aircraft. STINGER was to be pylon mounted and fired from the pilot's station via a fixed drop-down HUD. The initial pylon development supported the STINGER on classical underside hard points.

The pylon development was started in parallel with the analysis. Typical structural and dynamic loads were used to design the mount, and the underside hard points were found to be suitable. A subsequent jettison analysis was conducted via a simulation and resulted in an unacceptable envelope due to the missile flying into the rotor. Immediately the redesign of the mount was initiated. It was discovered through the simulation that orienting the mounting points down and outboard at a 45-degree angle solved the jettison problem. This resulted in a heavier mount and delayed testing for approximately six months. A few months later, it was discovered that the simulation model was incorrect. The engineer had failed to recognize that there was a 250 pound holdback on the jettison kicker. Instead of modeling the jettison force as a step force starting at 250 lbs. to the max, it was modeled as a ramp input from zero to max force. This faulty assumption and subsequent model error resulted in high cost and schedules. Had the jettison force been implemented properly, the first iteration pylon design would have been acceptable.

AHIP Architecture (Ref. 3)

By today's standards, there was no formal architectural analysis performed for AHIP. At CDR 3, there were eighty-nine boxes and fifteen antennae to be incorporated. The marketing slide actually touted ninety boxes, not capabilities, as illustrated in Figure 81. Utilizing the existing form factors of the boxes resulted in a huge Space, Weight, and Power (SWaP) issue. They were located throughout the aircraft for purpose and CG. While using these boxes was cheaper than creating a new form factor, the weight trade was large.

Additionally, stuffing these boxes in small spaces created thermal issues for the aircraft. The boxes were connected via a 1553 data bus. Software developers were key to making the system operate. This was a federated integration at best. The savior of this approach was the software developers and the over design of the computers and related hardware. These computers had been designed for other DOD systems and mitigated many of the issues with throughput, timing, etc. In retrospect, installing radios meant for a tank into an aircraft was a poor idea. Requirements should be a list of capabilities, not part numbers.

The solution being proposed today is a Modular Open Systems Approach (MOSA). However, it really needs to be an expanded application of the Open Integrated Modular Avionics (IMA). This approach has been used in recent civil transport aircraft avionics certification DO-297 Integrated Modular Avionics (IMA), such as the Airbus 380 and the Boeing 787.

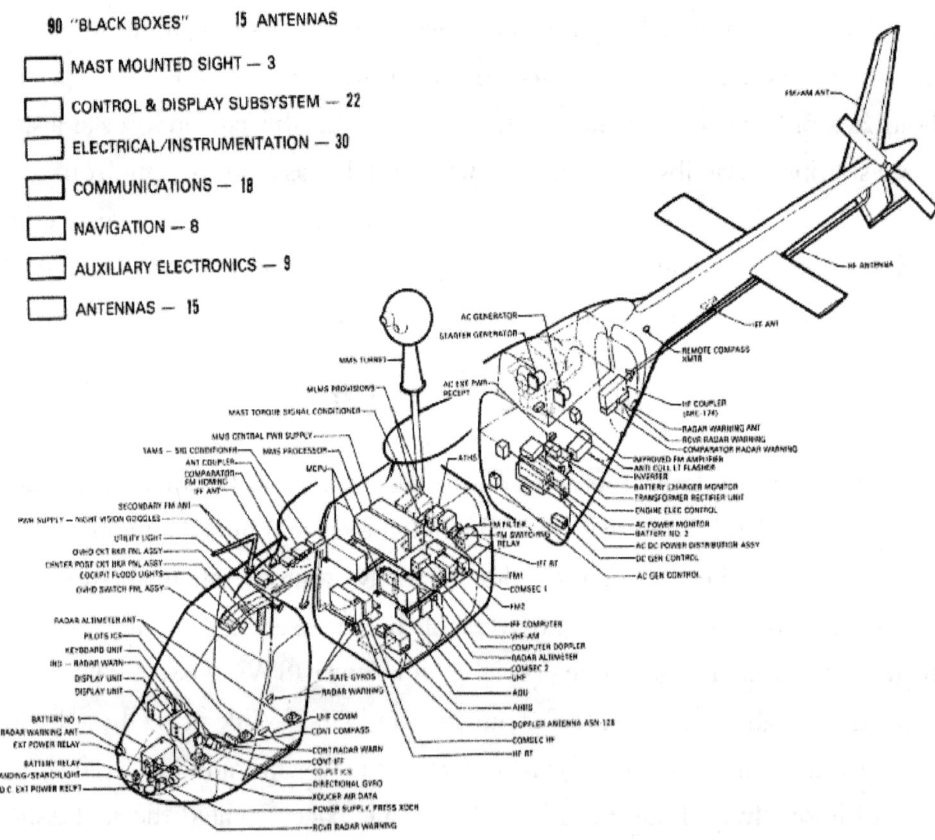

Figure 96. Major MEP 7 Major Systems w/90 Boxes

Endnotes

1. Porter, D. J., "Howard's Whirly Birds: Howard Hughes' Amazing Pioneering Helicopter Exploits", FONTHILL MEDIA, Printed 2013

2. Harris, F.D., "AHIP: The OH-58D From Conception to Production," Proceedings, AHS 42nd Forum, June 1986.

3. Kawa, M., and Van Kirk, J.M., "Program and Operational Highlights of the Armed OH-58D Kiowa Warrior," AHS 46th Forum, May 1990.

4. Freedberg, S.J., "FVl: The Army's 10-Year Plan for FARA Scout", March 26, 2020, AIRWARFARE

5. Forester, B, Weller, D., Harrison, Schrage, D.P., and Lewis, W.D., "AHIP/OH-58D Kiowa Section," Army Aviation-Foundations of the Modern Fleet, PEO Aviation, Public Affairs Office, Redstone Arsenal, AL, 2013

6. Schrage, D.P. "Use of Rotorcraft Computer Analysis for U.S. Government Design Assessment", Comp, & Maths. with Appls. Vol. 12A. No. I. pp. 1-9. 1986 009" Pergamon Press Ltd.

Chapter Six

ARMY MEDIUM AND HEAVY-LIFT CARGO HELICOPTERS

In 1956 the United States Department of the Army announced its intention to replace the piston engine-powered Sikorsky CH-37 Mojave with a new, gas turbine–powered helicopter. In 1957 Boeing Vertol (the former name of an American aircraft manufacturer now known as Vertical Lift division of Boeing Defense, Space & Security) began work on a prototype tandem rotor helicopter, designated as the Vertol Model 107 or V-107. In 1958, the US Army ordered a small number of V-107s from Vertol under the YHC-1A designation. The YHC-1A would be improved and adopted by the US Marine Corps as the CH-46 Sea Knight. (Ref. 1) Following Army testing, however, the YHC-1A was determined by some Army officials to be too heavy for assault missions and too light for transport purposes.

Fig. 97A. Vertol V-107. Fig. 97B.Sikorsky CH-37. Fig. 97C. Boeing CH-46

Seeking a heavier transport helicopter, the Army ordered an enlarged derivative of the V-107 with the Vertol designation Model 114. Initially designated the YCH-1B, the preproduction rotorcraft performed its maiden flight on 21 September 1961. In 1962, the HC-1B was re-designated *CH-47A* under the 1962 United State Tri-Service aircraft designation system Named after the Native American people of Oregon and Washington states, the CH-47 Chinook is a tandem rotor helicopter developed and manufactured by the American rotorcraft company Boeing Vertol. (Ref. 1)

About the same time, the single main rotor system, twin turboshaft engine CH-54 Tarhe "Flying Skycrane" was developed by Sikorsky Aircraft and had its first flight in 1962. The CH-54 was used in Viet Nam for transport and downed aircraft retrieval. It was gradually replaced in Regular Army aviation units by the CH-47 and subsequently retired from all National Guard units in the 1990s. The CH-54B had a payload of 20,000 lbs.

Originally fielded in the Vietnam War, the CH-47 Chinook has been the US Army's primary heavy supply and troop transport aircraft ever since. The CH-47A had a maximum payload of approximately 10,000 lbs. and has undergone a series of upgrades to increase lift and airworthiness in combat environments. The CH-47D Chinook is among the Western World's heaviest lifting helicopters. The CH-47D lift capability, however, is dwarfed by the huge Soviet-Russian, heavy-lift helicopters such as the Mil Mi-26, with 44,000 lb (20,000 kg) payload, and the experimental Mil Mi-12, with 55,000 to 88,000 lb (25,000 to 40,000 kg) payload.

For a long time, Boeing and the US military had the urge to match or top the Russian heavy lifters. From 1965, Boeing developed designs for machines with broad similarities to the Sea Knight and Chinook, but about twice the size of the Chinook in terms of linear dimensions. The proposed machines included the "Model 227" transport and the "Model 237" flying crane. Figure 98. From 1965, two Boeing Vertol Heavy Lift Helicopter concepts in model form, to scale with the Chinook (at far right). At far left is the Model 227 which carried loaded internally, in the middle is Model 237, designed as a flying crane, note though, that even though the Model 237 is designed to carry payloads externally the vehicle is so large that the reduced fuselage still has room for a substantial passenger load, windows and all.

Shown in Figures 98B and 98C are the Sikorsky CH-54 Crane and the Mil MI-26, the world's largest helicopter.

Figure 98A. Boeing Models 227 and 237 Compared to Boeing CH-47

Figure 98B. Sikorsky CH-54.

Figure 98C. The MIL MI-26

The U.S. Department of Defense (DoD) issued a request for a proposal (RFP) for a Heavy-Lift Helicopter (HLH) in November 1970. On May 7, 1971, the DoD announced the selection of Boeing Vertol to perform the first phase of HLH development. Following the award of an Army contract for an HLH prototype in 1973, Boeing did move forward on building an oversized flying crane prototype, the "XCH-62," as illustrated in Figure 99. (Ref. 2)

HLH

The XCH-62 prototype was in an advanced state of assembly in 1975, being readied for a planned initial flight in 1976. The XT701 engine had passed its 30-hour Prototype Preliminary Flight Rating Test (PPFRT) on March 12, 1975, and then passed a 60-hour Safety Demonstration Test (SDT)Aug 4. 1975

Figure 99, XCH-62 Prototype, 1975

Beginning in 1982 and ending in 1994, all CH-47A, B, and C models were upgraded to the CH-47D version featuring composite rotor blades, an improved electrical system, modularized hydraulics, triple cargo hooks, avionics and communication improvements, and more powerful engines and carry payloads up to 2o,000 lbs.— twice the CH-47A original lift capacity. (Ref. 1) The Army upgraded the CH-47D Chinook helicopters to the CH-47F standard, which is now primarily an avionics and digital systems upgrade not necessarily related to lifting performance.

Yet, today's heavy-lift helicopters must handle heavier equipment and climb to higher altitudes at hotter temperatures, as witnessed in Afghanistan. Boeing proposed the CH-47 Block II (see Fig. 100) combining new technologies, including Advanced Chinook Rotor Blades (ACRBs), redesigned fuel tanks, an enhanced fuselage, and an improved drivetrain-all aimed at increasing lift.

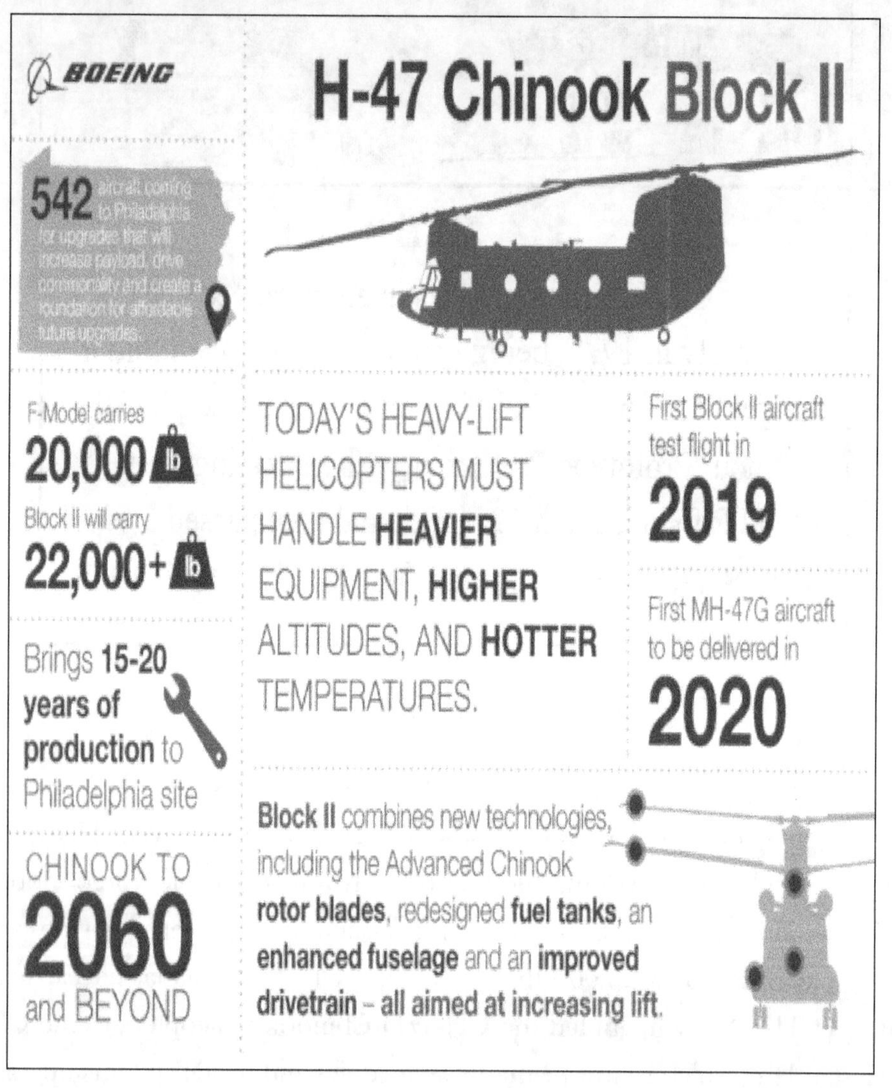

Figure 100. Boeing's Optimistic Chinook Future with Block II and Beyond Improvements

An early test flight of the CH-47 Chinook Block II configuration at the Boeing Philadelphia facility is illustrated in Figure 101.

Figure 101. Early test flight of CH-47 Chinook Block II configuration at the Boeing Philadelphia facility.

During first flight testing by Army test pilots and reviewed by the Pentagon's chief weapons tester, it was stated that "The new Advanced Chinook Rotor Blade, or ACRB, on the CH-47F Block II aircraft, produces excessive vibrations in ground hover and forward flight that may cause a safety of flight risk." The report stated, "Aircrews reported prolonged fatigue and other physiological conditions due to excessive vibrations following a developmental test flight using the redesigned ACRBs." (Ref. 3)

Subsequently, the Army did not include the ACRB in its first three CH-47F Block II buy. At the Army Aviation Association of America (AAAA) Convention in Nashville, TN, in April 2022,the Program Executive Officer (PEO) for Army Aviation, said that the US Army has decided to shelve the ACRBs it was developing as part of the Boeing-manufactured CH-47F cargo helicopter's latest upgrades. This was due to vibration issues that cropped up during testing.

However, the critical performance technology was the ACRBs and without a new hub design and blade frequency placement, the 6P vibrations (discussed in the following section) will still be the weakest length in the chain, and the Achille's heel of any Chinook improvement.

The tandem rotor helicopter design was pioneered by Frank N. Piasecki when the US Navy gave him a contract to design a large tandem rotor helicopter capable of carrying heavy loads. In 1945 Piasecki delivered the first HRP-1, nick named the "Flying Banana." His innovative designs laid the foundation for today's tandem rotor helicopters and in 1986 President Reagan honored him with the National Medal of Technology. Piasecki Helicopter Corporation was acquired by Boeing in 1960. (Ref. Boeing.com executive bio of Piasecki)

Figure 102A . Piasecki HRP-1 Rescuer "Flying Banana". Figure 102B. H-21 Army Vietnam Version

The shooting down of a CH-21 Shawnee near the Laotian-Vietnamese border with the death of four aviators in July 1962 were some of the U.S. Army's earliest casualties in the "Vietnam War." Despite these events, the Shawnee continued in service as the U.S. Army's helicopter workhorse in Vietnam until 1964 when it was replaced with the Bell UH-1 Huey. In 1965, the Boeing CH-47 Chinook was deployed to Vietnam and later that year, most CH-21 helicopters were withdrawn from active inventory in the U.S. Army and Air Force

Vibration has always been an issue on large tandem rotor helicopters, like those shown in Figures 103A and 103B, largely due to the wake from the forward rotor impacting the aft rotor. The Russians were never successful in developing tandem rotor helicopters due to this wake interaction.

Fig. 103A. Family Tandem Rotor Helicopters. Fig. 103B. CH-47 lifting HLH

Tandem Rotors and Helicopter Vibrations

Designing helicopters for low vibration levels is so difficult and complex that determined efforts should be made at the earliest possible stages in an aircraft's development to ensure that all the basic factors are in the proper range. This involves rotor blade, drive system, and fuselage natural frequencies, to provide adequate fuselage/rotor aerodynamic clearance. To ensure proper placement of fuselage modes and frequencies, finite element analyses, and correlative ground shake tests are mandatory. (Ref. 3)

The same may be said for blade designs where branched structures are encountered. Rotor blade natural frequencies should be calculated with all the known couplings, including aeroelastic effects represented. If a linear theory with small perturbations is assumed, they should always be taken about large, mean deflected positions. Nonlinearities, such as for example, post-buckling behavior in fuselage structures and geometric nonlinearities in rotor blade dynamics, should be incorporated, based on the kind of "ordering analyses" which are beginning to show which effects are commensurate with a particular level of accuracy.

The implication of composites is very strong for both fuselages and rotor blades, and in some respects rotor drive shafts. Designers, structural analysts, and dynamicists must be alert for those changing ratios of strength and stiffness which may throw rotorcraft structures, non-rotating and rotating, into new ranges of parameters for which there is little experience. (Ref. 3)

Main rotor flap-wise and edge-wise rotor natural frequencies have to be well removed to achieve acceptable vibratory loads. They often aren't, and don't meet the required separations identified by Mr. Ken Grina, former Boeing Vertol VP of Engineering in his rotorcraft design short course, as illustrated Figure 98. (Ref. 4)

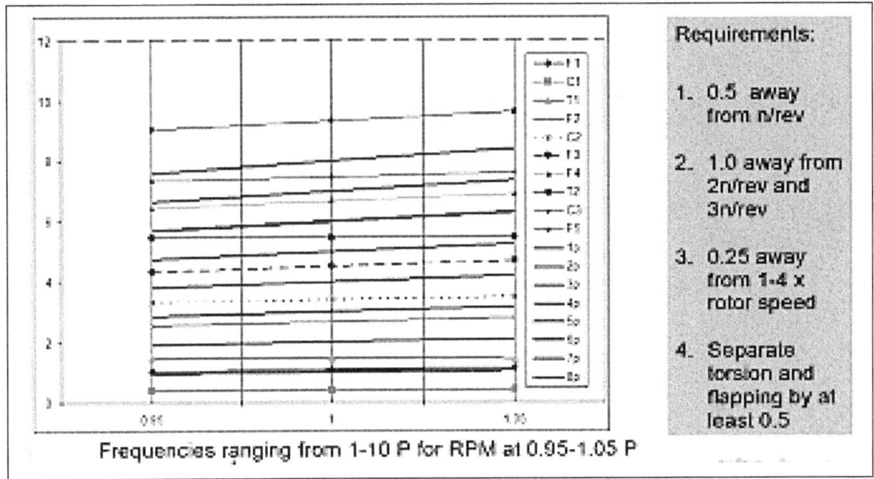

Figure 104. Rotor Frequencies Separation Requirements (Ref. 4)

The XCH-62 HLH program was officially canceled on August 1, 1973. At the time of cancelation, the XCH-62 prototype airframe was thought to be at 95 percent completion, and it needed about three months of final assembly and checkouts before rollout and installation for pre-flight testing. Vibration would have also been a problem on the XCH-62 HLH, as evidenced by the rotor flapping out-of-plane natural frequencies, in resonance with 5P and 6P, as illustrated in Figure 105 where the circles illustrate unacceptable resonances. (Ref. 5)

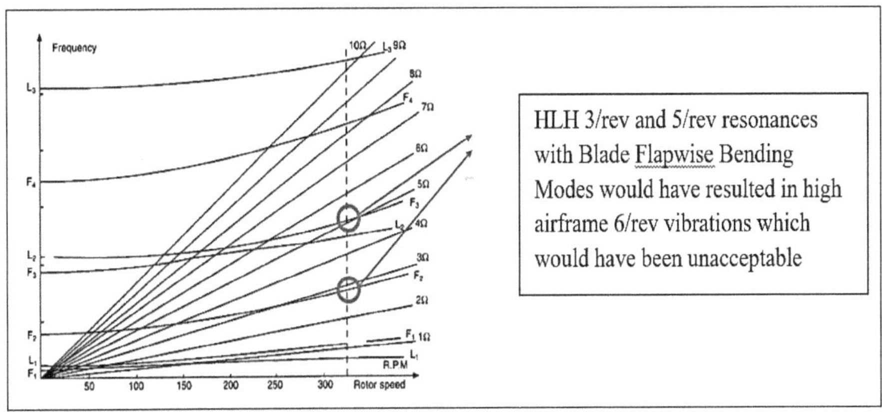

Figure 105. Frequencies Resonances on HLH XCH-62 Prototype (Ref. 5)

Also, failures in the spiral bevel gearing of the main transmission were experienced in tests because the method of analysis employed had not considered the effect of rim bending. (Ref. 2) Consequently, new gears with strengthened rims were designed and fabricated. For a more accurate prediction of the load capacity of the gears, an extensive Finite Element Method (FEM) system was developed. The US Army's XCH-62 HLH aft rotor transmission was finally successfully tested at full design torque and speed, but the US Congress cut funding for the program in August 1975. The designers of the Mil Mi-26 avoided similar problems by using a split-torque design for its single main rotor transmission. The Sikorsky CH-53K, a single main rotor design, also uses a split-torque gearbox design. (Ref. 2)

Chinook Upgrades to Improve Mission Success

The CH-47D Modernization Program has not been without some problems, mainly associated with the inclusion of composite rotor blades with new airfoils, the VR-7 and VR-8 for more hover lift. A description of the CH-47D/F Chinook Specifications is shown in Figure 106. (Ref. 1)

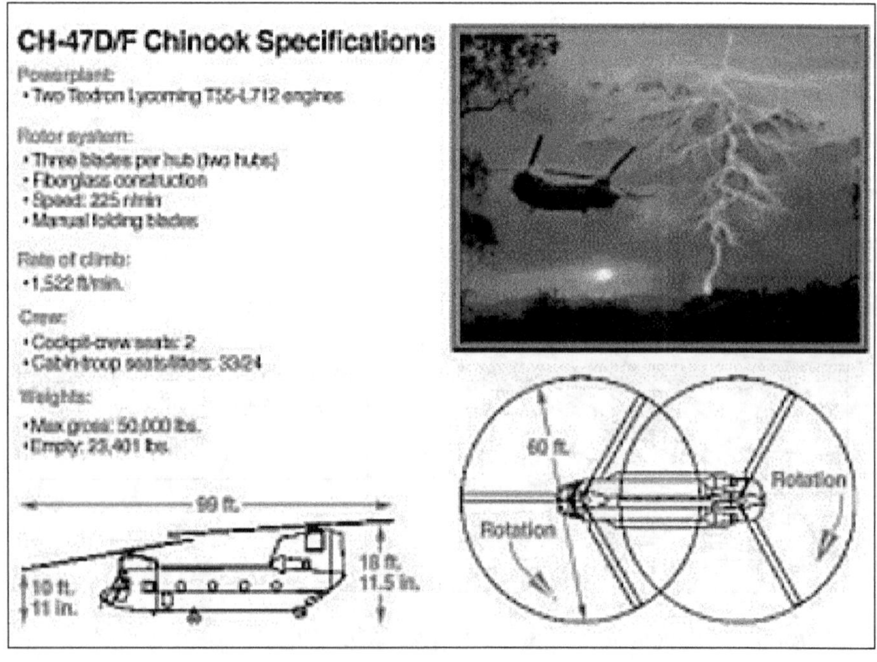

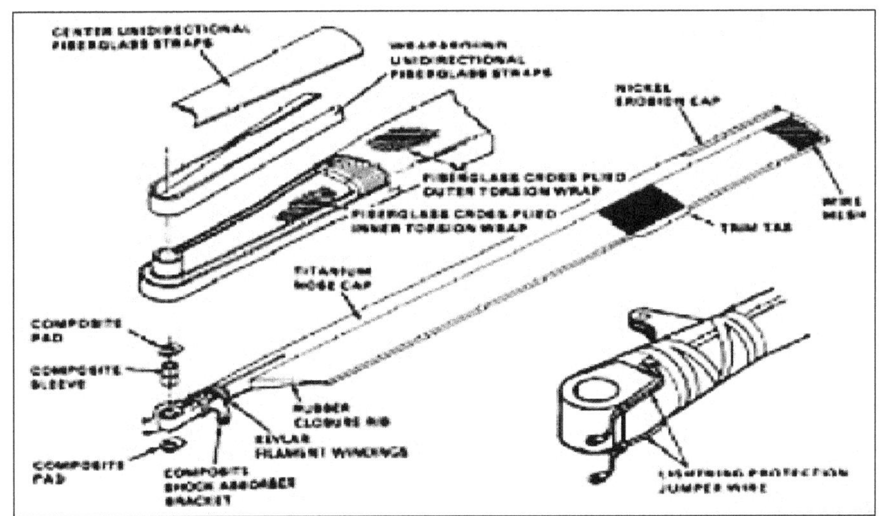

Figure 106. CH-47D/F Chinook Specifications and Composite Blade Layup

An upgrade program was planned to remanufacture 300 of the existing fleet of 425 CH-47Ds to the CH-47F standard. The 6P airframe vibration on the CH-47D/F has also caused a 6P vibration problem with the fiberglass blades (twice as high as CH-47C metallic blades) and now also on the new Advanced Chinook Rotor Blades (ACRBs).

The primary intent for the CH-47F Block II upgrade was to increase payload and to reduce operating costs on the battlefield so that the Chinook would continue to be an enabler of Army operations well into the future.

Boeing's primary emphasis with the new composite blades for the CH-47D was to increase hover lift capability, e.g., GW to 50,000 lb, and not to degrade forward flight speed. The rotor design environment to improve hover lift capability by reducing stall requires cambered airfoils and to reduce high-speed loads by slowing the rotor to reduce compressibility, as illustrated in Figure 107A. Figure 107B illustrates the aeroelastic analysis to design for both flight conditions. (Ref. 5)

Boeing also stiffened the fuselage in the back end, so the helicopter can accommodate bigger engines and can increase the lift capability of the aircraft by 4,000 pounds, according to the company.

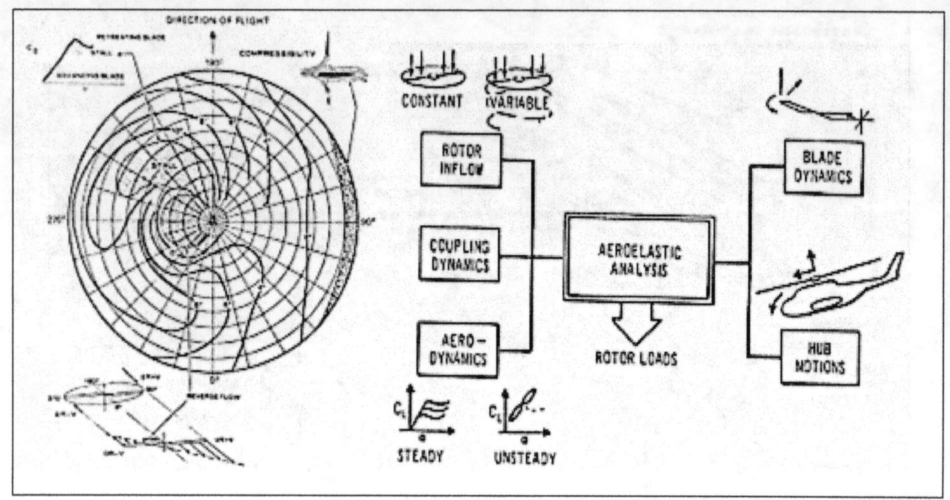

Figure 107A. The Rotor Design Environment. Figure 107B. The Required Aeroelastic Analysis{?}

The CH-47C used two different rotor speeds for hover, e.g., 245 RPM and 215 RPM, but planned for the CH-47D to use one rotor speed, 225 RPM. However, difficulty with the camber in the VR-8 at the tip led to Mach Tuck advancing blade loads which required redesign of pitch links and upper transmission cover. (Ref. 5)

Similarly, this has resulted in the ACRBs being removed from the CH-47F Block II technologies, which compromises any necessary increase in hover and forward flight performance from the CH-47F Block II or III re-manufacturing.

This again seems to be due to the rotor second flap-wise bending proximity with 5P causing hub pitching and rolling moments in the fixed system at 6P. (Ref. 5) This has resulted in high airframe and aft drive train 6P vibrations; especially in the aft end of the airframe and drivetrain. This resulted in two CH-47D aircraft being lost at Fort Sill, OK in the 1990s.

The Block II configuration has more in common with the Special Forces MH-47G model—especially the nose of the aircraft. The G model has a longer nose because Special Forces has more equipment than they store in that region. That extended nose, however, made the G model different from the F model. Going to a common structure means the F and G models can benefit from greater commonality and interchangeability, creating more flexibility and cost savings for the Army in maintenance. (Ref. 6)

Special operators have pushed the Chinook hard by operating them at a maximum gross weight of 54,000 pounds, which is greater than what Army CH-47F models operate. That has caused the maintenance and logistical needs of the special forces model to diverge from others in the Army's fleet.

Illustrated in Figure 108 is a summary of current Army helicopter rotor design parameters critical to performance and aeroelastic considerations for different rotor systems. As can be seen in the circles, Boeing used three airfoils, VR-7, VR-8, and VR-9, with decreasing thickness inboard to tip and trailing edge (TE) tab reflections for the YUH-61A UTTAS prototype. It can be seen that upward TE reflex deflection of -6 and -3 degrees were used for the YUH-61A. For the CH-47/D fiberglass rotor blades (FRB) with the VR-7 and VR-8 airfoils, circled, does not use the VR-9, but uses a TE reflex tab of -6 degrees for the VR-8 tip airfoil. (Ref. 5)

ROTOR SYSTEM	TIP SPEED (FT/SEC)	$M_{(1,90)}$ @ 0°R	CHORD (FT)	SOLIDITY	TWIST (DEG)	DISC LOADING (LBS/FT^2)	AIRFOIL DISTRIBUTION	M_{DD} @ 0°	T.E. TAB DEFLECTION (DEG)	TIP SHAPE	EST FIRST TORSIONAL FREQ. (PER REV)
YUH-61A (UTTAS)	734	.97	1.92	.100	-12	7.9	VR-7(.12)to.75R VR-8(.08)@.91R VR-9(.06)@R	.69 .84	VR7-6 VR8-3 C_{M_0} = .004	Rect.	
YUH-60A (UTTAS)	725	.94	1.73	.082	-18. (Highly Non-Lin)	7.2	SC1095R6 to .5R SC1095R8 @ .5R to .82R SC1095R6@.83R to R	.76	NONE C_{M_0} = -.015	Swept (20 Deg @.95R)	4.2
TAH-63 (AAH)	750	.95	3.55	.086	-12 (Non-Lin)	7.1	FX69-H1-083	.80	NONE C_{M_0} = .002	Leading Edge Sweep (70 Deg @.94R)	2.81
YAH-64 (AAH)	726	.96	1.75	.093	-9	7.9	HH-02 (Mod 63A410.5) 64A006 @ R	.81	-5 C_{M_0} = .025	Swept (20 Deg @ .95R)	3.66
K747 (AH-1S)	746	.92	2.5	.063	-10	6.25	VR7 to .67R VR8@.85R to R	.69 .84	VR7 -4.6 VR8 -2.08 C_{M_0} = 0.0	Tapered (3:1) .85R toR	3.23
CH-47 FRB	707	.93	2.67	.085	-11	7.0	VR7 to .85R VR8 @ R	.69 .84	-6 C_{M_0} = 0.0	Rect.	4.78

Figure 108.Summary of Recent Rotor Design Parameters Critical to Performance & Aeroelasticity

A summary of recent rotor design parameters critical to performance and aeroelasticity are included in Figure 109. (Ref. 5) It illustrates negative pitching moments for VR-7, VR-8, and VR-9 cambered airfoils and the impact using trailing edge (TE) tabs set at 0 degrees, -3 degrees, and -6 degrees, with negative being reflexed up. While the YUH-61 rotor blade used all three airfoils, the CH-47D used only two, the VR-7 and VR-8 airfoil sections as illustrated in Figures 109. Without the thinner VR-9, the CH47D ran into a problem for advancing blade pitching moments, called Mach Tuck, which resulted in high advancing blade pitch link loads, which jeopardized the use of single rotor RPM, e.g., 225 RPM, for both hover and cruise speed. It is believed that Boeing ended up using 225 RPM by improving transmission cover design.

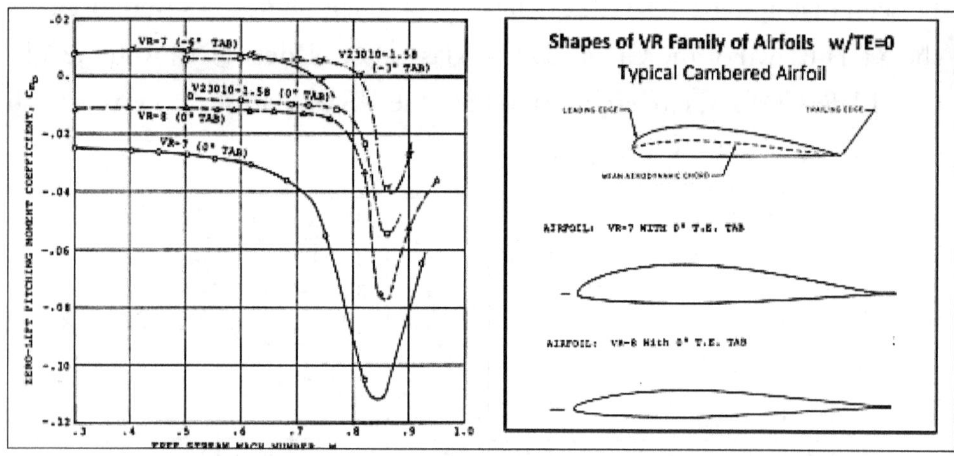

Fig. 109. Compressibility Effects on Pitching Moments and Typical Camber & VR7&8 w/TE Tabs

Chinook Rotor System Management & Technical Lessons Learned

In the 1979-1981 timeframe, I served as Chief, Structures and Aeromechanics Division, Development and Qualification Directorate, Aviation Research and Development Command (AVRADCOM and the technical lead for overseeing the technical development and airworthiness qualification of the CH-47D Modernization Program. Improvements to the

CH-47C standard included upgraded engines, composite rotor blades, a redesigned cockpit to reduce workload, improved and redundant electrical systems and avionics, and the adoption of an advanced flight control system.

In the 1990s, the Director of Army Aviation Research, Development, and Engineering (AVRDEC) asked me in a consultant capacity to evaluate and address CH-47 vibration problems. My contention was that there was a close resonance in the rotating system between the 5P rotor harmonic and the composite blades second flap-wise bending moment. This resulted in hub pitch and roll vibratory moments at 5P at the flap hinge offset which fed into the aft airframe and drive system as 6P vibrations and excited the aft airframe and drive system. Once again, to fix the problem, I recommended they modify the rotor hub with less rotor flapping hinge offset or redesign the rotor blades so that the second flap-wise bending moment was not as close to resonance with 5P. However, neither of these recommendations were implemented, and vibrations, especially 6P, have continued to plague all CH-47 derivatives.

In 1995-97 a three-year, CH-47D Airframe Tuning Program, Figure 110, was initiated to help solve the vibrations problems, both 3P and 6P, in the airframe and aft gearbox. I confirmed my opinion to the Boeing engineers that the source of the vibration problems was in the rotor system. I relayed to them once again that they needed to either redesign the rotor blades to move the 2nd flap-wise bending mode further away from 5P, or reduce the flapping hinge offset, which would result in lower pitch and roll 6P vibrations in the airframe.

On January 11, 1996, I received a large three ring binder from a Boeing engineer, with a letter providing a detailed weight break-down of a CH-47D, that I had requested. He stated that they still owed me a 6P vibration reduction story, but as he understood my plan, he wanted to get the baseline aircraft done first. I never received a 6P vibration reduction story.

Airframe Tuning Program

A Four Phase Program Has Been Defined:

- **Phase I:** - Baseline Flight Test Program
  - Baseline Shake Test Program
  - Installation of Forward Pylon Stiffening ⎫ Reduce 3P
  - Shake Test of Forward Pylon Stiffening ⎭

- **Phase II:** - Integration of the Advanced Vibration Suppressor (AVS) Units into the Fuselage
  - Installation of the Aft Pylon Stiffening ⎫ Reduce 6P
  - Shake Test Forward and Aft Pylon Stiffening ⎭

- **Phase III: U.S. Army Flight Testing of the Aircraft**
  - Forward & Aft Pylon Stiffening Alone
  - Forward & Aft Pylon Stiffening - With Various AVS Configurations
  - Forward & Aft Pylon Stiffening - With Various 3P Absorber (STVA) Configurations
  - Best Aircraft Configuration

- **Phase IV: Refurbish Aircraft**

Figure 110. Boeing CH-47D ICH Airframe Tuning Program, FY95-97 (Ref. 1)

However, by this time, I realized the only solution was to redesign the hub design, rotor blades, or both. A trade-off determination (TOD) was conducted by WESTAR and Boeing to determine what technologies could be remanufactured into the CH-47D ICH. I believed that the rotor and hub designs would have to be a key to future aircraft capabilities and performance, but redesigning the rotor and hubs was the most difficult and expensive. This is also especially true if the CH-47F Blocks II and III are going to have increased speed to remain closer to the future advanced rotorcraft, such as a tilt rotorcraft. It appears the Advanced Chinook Rotor Blades (ACRBs), as Block II solutions, have not been successful. They used similar rotor blade airfoils to what was on the RAH-66 Comanche without vibration success. (Ref. 6)

The WESTAR/Boeing TOD hub and rotor designs identified some potential solutions but most of the following, if any, were not implemented.

1. Improved Lubricated Main Rotor Hubs - The existing CH-47 wet t design could be improved by incorporating design improvements aimed at recognized problem areas. Areas of concern affecting the life of the hub are excessive spline wear and low retirement life of horizontal pin bearings. By increasing the bearing diameter by 25", the retirement life of the horizontal pin bearings would be at least doubled.

2. Low Maintenance 3-bladed Main Rotor Hubs - The CH-47 rotor system can be redesigned to eliminate all lubrication at the hinges. Existing roller bearings would be replaced by elastomeric bearings and/or non-lubricated (Kaydon for example) bearings. All hinges, flap, lag, and pitch, would remain at their existing locations, thus preserving existing rotor geometry and dynamic responses, including vibrations.

3. Low Vibration 4-bladed Main Rotors - The hub would be of low maintenance design, like the three bladed system. However, the lead-lag damper is replaced with an elastomeric spring to accommodate a short hinge offset and to manage blade to blade clearance. The hub and blades are designed together to achieve targeted low vibrations characteristics. The rotor change requires changes to the rotor masts, flight controls, and aft transmission due to the additional blade and increased loads. The main rotor diameter must be reduced to maintain blade separation; rotor speed is increased to recover hover performance. The rotor alters both the aerodynamics and dynamics of the aircraft. The Automatic Flight Control System (AFCS) would require reprogramming.

I have continued to follow the CH-47D, CH-47ICH and CH-47F vibrations problems over the years. My lessons learned for what kept the CH-47D from achieving its full potential included: (Ref. 5)

1. By using the VR-7 and VR-8, Boeing got the required hover lift without blade stall, cross-sections illustrated in Figure 109, to meet the maximum hover gross weight of 50,000 lb. However, without the thin

VR-9 airfoil near the blade tips, the rotor pitch link loads were quite high from Mach Tuck, which resulted in a need to redesign of pitch links and the transmission cover.

2. Another problem with Mach Tuck was the desire to convert the CH-47D to a single rotor speed in hover and forward flight at 225 RPM. This could not be achieved, at least initially. On the CH-47C, two different rotor speeds had to be used; 245 RPM for Hover and 215 RPM for cruise speeds. In my opinion, by beefing up hub components, the CH-47D eventually led to using a constant rotor speed, i.e., 225 RPM.

3. A third problem was that the 6 P vibrations with the CH-47D composite rotor blades were more than twice as high as they were on the CH-47C, especially on the aft rotor hub, airframe and drive system, Figures 111A and 111B.

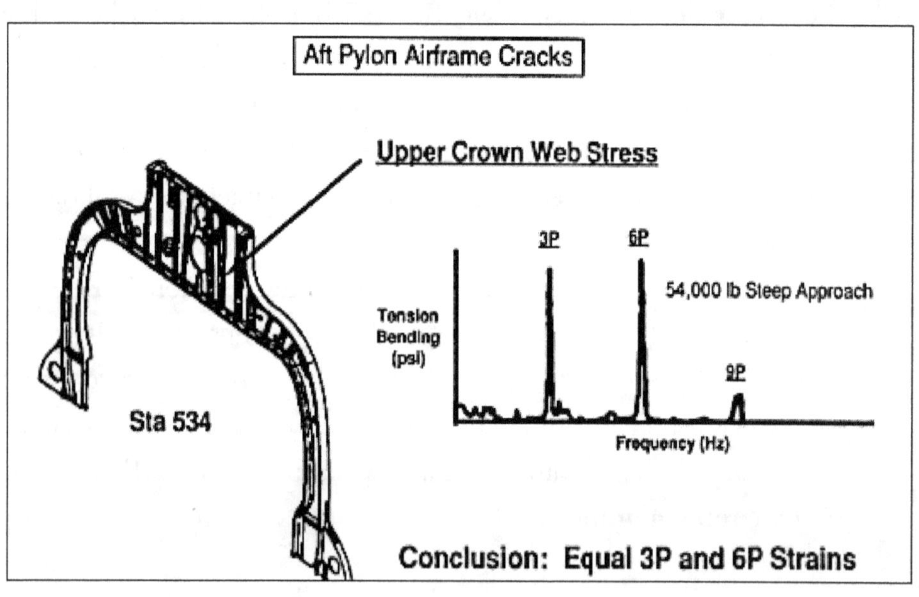

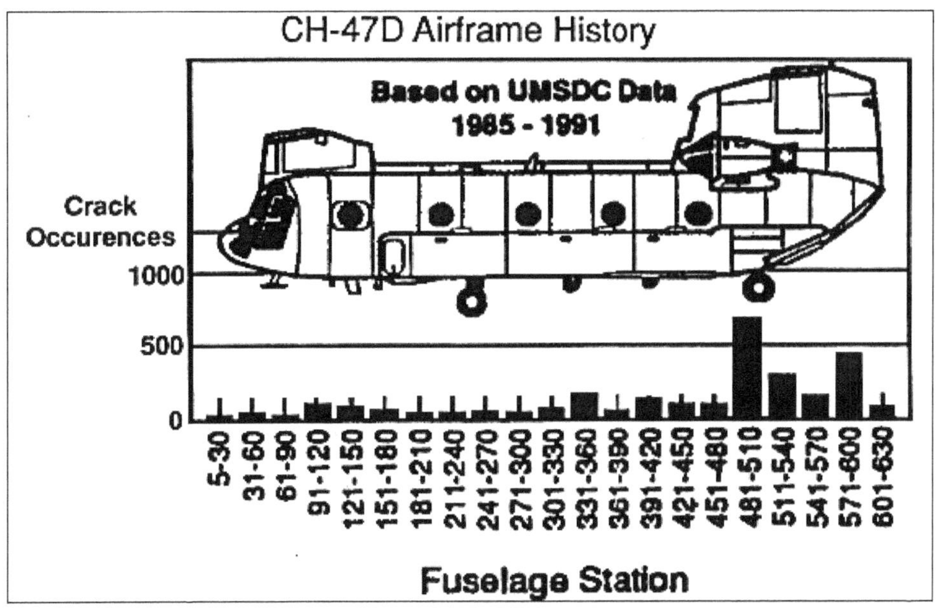

Figure 111A. 6P Vibration is an Unsolved Problem   *Figure 111B. Airframe Cracking*
Greatest Aft

4. A problem confronting the CH-47 PM was that there was only a 3P vibration requirement, and not a 6P requirement, in the CH-47D System Specification. Charlie Crawford, then the Director of Development and Qualification, and I, tried to convince the CH-47 PM at the time, to have Boeing solve the 6P problem, without much success. We warned the PM the problem would arise again, as it did when two CH-47D helicopters had short shaft and transmission failures due to 6P vibrations at Fort Sill, OK in the 1990s.

5. The vibration problems continued on the proposed Improved Cargo Helicopter (ICH) version of the CH-47D in the mid-1990s. Boeing began a three-year airframe tuning finite element modeling (FEM) program through a Cooperative Research and Development Agreement (CRADA) with the Army for FY '95-'97. The Government Accounting Office (GAO) questioned whether stiffening the airframe, as Boeing proposed, would solve the vibration issues or be a potential showstopper.

Army Future Vertical Lift (FVL) Program

As disruptive technologies emerge and adversaries adopt them, the Army is going to need a family of affordable Future Vertical Lift (FVL) aircraft to replace its current fleet. FVL is an Army-led multi-service initiative, focused on enhancing vertical lift dominance through the development of next generation capabilities. FVL increases reach, protection, lethality, agility, and mission flexibility to successfully dominate in highly contested and complex airspace against known and emerging threats. (Ref. 8)

The FVL seeks to have the same aircraft be able to perform multiple missions. The Army would then need fewer types of aircraft. That means a smaller number of parts will be needed to sustain the fleet, and a shared pool of maintainers and maintenance equipment. That would reduce logistics costs. The FVL could come in different sizes, depending on the mission it will perform, enhancing engine, drive train, and cockpit component commonality. FVL multi-role capabilities enabled by avionics will support a wide range of joint missions, including aerial reconnaissance, anti-submarine warfare, anti-surface warfare, special operations, amphibious assault, undersea warfare, surface warfare, air assault, medical evacuation, Intelligence Surveillance Reconnaissance, Search and Rescue, Command and Control, Combat Search and Rescue, attack, logistics, homeland security and cargo operations.

FVL's signature modernization efforts have approved requirements. The top priorities include the Future Attack Reconnaissance Aircraft (FARA), Future Long-Range Assault Aircraft (FLRAA), Future Unmanned Aircraft Systems (FUAS), and Modular Open System Approach (MOSA).

FVL is the acquisition follow-on to the Army-led Joint Multi-Role Technology Demonstrator Program (JMR-TD). Prior to JMR-TD, the Army produced over 35 years of incremental improvements to its rotorcraft fleet without a revolutionary increase in capability. The Army is leading FVL development by maturing a next generation of capabilities. Initially, the Army and other services worked together to refine requirements for next generation reconnaissance, utility, medical evacuation, and attack aircraft. The JMR-TD conducted ground and flight demonstrations of advanced rotorcraft designs for a revolutionary increase in capability. JMR-TD also developed a modular open systems approach (MOSA) to provide a common digital network capability and an open architecture that is portable across multiple platforms, thereby shortening integration timelines of critical new capabilities to the warfighter.

The FVL program should have shaped the United States military's helicopter fleet of the future; however, the cancellation of the FARA is a big setback. Led by the U.S. Army, FVL will replace aging helicopters across the DoD services through a family of five different categories, or capability sets. The over-arching Joint FVL efforts span a range of five classes of future aircraft, ranging from light helicopters to medium and heavy-lift variants and an ultra-class category designed to build a new fleet of super-heavy-lift aircraft. The ultra-class aircraft will be designed to lift, transport and maneuver large vehicles around the battlefield such as Strykers and mine-resistant, ambush-protected vehicles known as MRAPs. The ultra-class variant, described as a C-130 type of transport aircraft, is part of an Air Force led, Army-Air Force collaborative S&T effort called Joint Future Theater Lift, or JFTL.

Initially, planning focused on four capability sets, but later added a fifth, JMR-Ultra (Figure 112). The winning Bell FLRAA tilt rotor falls under CS2 and CS3; while the FARA would be under CS1 or possibly CS2. (Ref. 8)

Capability Set		speed knots	radius NM	pax	Payload lbs internal	Payload lbs external	legacy	objective	IOC
JMR-Light	CS1	200-250	170-230	6	2-2.5k	TBD	OH-56	FARA	2028-30
JMR-Medium	CS2	170-270	300-440	8-10	3.5-5k	6-8k	MH-60		
JMR-Medium-Heavy	CS3	270-350	300-450	10-12	3.4.5.5k	6-8k	UH-60	FLRAA	2030
JMR-Heavy	CS4	250-300	325-420	24-32	12-20k	12-20k	CH-47		2035-60
JMR-Ultra	CS5	270-350	750-1200	45-54	24-50k	30k+	C-130J	JFTL	2025!?

Figure 112. FVL Capability Sets (CS) 2018

1. <u>JMR-Light:</u> Scout version to replace the OH-58 Kiowa, with introduction planned for 2030. The FVL Capability Set (CS) 1 air vehicle is the smallest, most agile air vehicle in the FVL Family of Systems (FoS). The CS 1 air vehicle will conduct reconnaissance, light attack, and light assault/lift operations in support of Army and Joint forces.

2. <u>JMR-Medium:</u> Utility and attack versions to replace the UH-60 Blackhawk, with introduction planned for 2027-28. The FVL CS 3 air vehicle(s) is intended to be a versatile medium lift air vehicle in the FVL Family of Systems (FoS). The CS 3 air vehicle will conduct Assault, Urban Security, Attack, Maritime Interdiction, Medical Evacuation (MEDEVAC), Humanitarian Assistance/Disaster Relief (HA/DR), Tactical Resupply, Direct Action (DA), Non-combatant Evacuation Operation (NEO) and Combat Search and Rescue (CSAR) operations in support of Army and Joint forces.

3. <u>JMR-Heavy:</u> Cargo version to replace the CH-47 Chinook, with introduction planned for 2035, although Boeing expected 2060;

4. <u>JMR-Ultra:</u> New ultra-sized version for vertical lift aircraft with performance similar to fixed-wing tactical transport aircraft, such as the C-130J Super Hercules, with introduction planned for 2025.

A summary of light, medium and heavy concepts, missions, and potential military customers are illustrated in Figure 113 (Ref.12 www.vtol.org)

Light	Medium			Heavy
• Cockpit • FACE/JCA • Training	• Requirements • Reduced overhead • Mission flexibility	All Air Vehicles have common...	• Sustaining • Maintaining • Repair parts and components	
Capability Set 1 **Missions:** • Reconnaissance • Attack • Security • CCA/CAS • Surface Warfare • Direct Action • Maritime Interdiction Operations	**Capability Set 2** **Missions:** • Reconnaissance/Attack • Security • CCA/CAS • MEDEVAC • Surface Warfare • Direct Action • Anti Submarine Warfare • CSAR • Maritime Interdiction Operations • Mine/Counter Mine	**Capability Set 3** **Missions:** • Mine/Counter Mine • MEDEVAC • Air Assault • Logistics • HA/DR • Amphibious Assault • NEO	**Capability Set 4** **Missions:** • MEDEVAC • Air Assault • Logistics • HA/DR • Amphibious Assault • NEO	**Capability Set 5** **Missions:** • MEDEVAC • Air Assault • Logistics • HA/DR • Amphibious Assault • NEO
• Army • Marines • US Special Operations • Navy • Coast Guard • (DHS)	• Army • Marines • US Special Operations • Navy • Coast Guard • (DHS)	• Army • Marines • US Special Operations • Navy • Coast Guard • (DHS)	• Army • Marines • US Special Operations • Navy	• Army • Marines • US Special Operations • Navy

Figure 113. Summary of FVL Requirements (Ref. 12)

The fair and open competitive Other Transaction Authority (OTA) approach for FVL preserves competition equal to or greater than what the Army experiences in traditional acquisition programs of this level. The program solicits industry innovation through a series of competitive selections over the course of six years. Ahead of schedule, the Army awarded five initial FARA design contracts in April 2019. The Army intends to select two vendors for final design and competitive prototypes. November 2022 was the target date for the prototypes to first take flight followed by a year-long flight demonstration competition; however, delays in availability of the ITEP engines has pushed FARA flight demonstrations and down selection to 2024 and beyond. The FARA competitive prototype program goal was to reach Milestone C in 2028 and First Unit Equipped in 2030; however, these timelines have slipped. (Ref. 8)

Completion of the Future Long Range Assault Aircraft (FLRAA) analysis of alternatives (AoA) and substantial progress beyond the JMR-TD allowed the Army to accelerate the FLRAA approach using the knowledge and experience gained from the aforementioned JMR Program. Army plans to reach Milestone B in FY26, and Milestone C in FY30, which appears to be on schedule.

FVL FLRAA and FARA (now cancelled) promise to deliver a variety of advanced capabilities including speed, lift, lethality, range, survivability, and low sustainment costs to replace the already aging fleet of Black Hawk and Apache helicopters. Thus far, the FVL concepts are advanced helicopter such as the Bell FARA prototype or Compounds and Tilt Rotors as shown in Figure 114. (Once again, the cancellation of FARA jeopardized FVL)

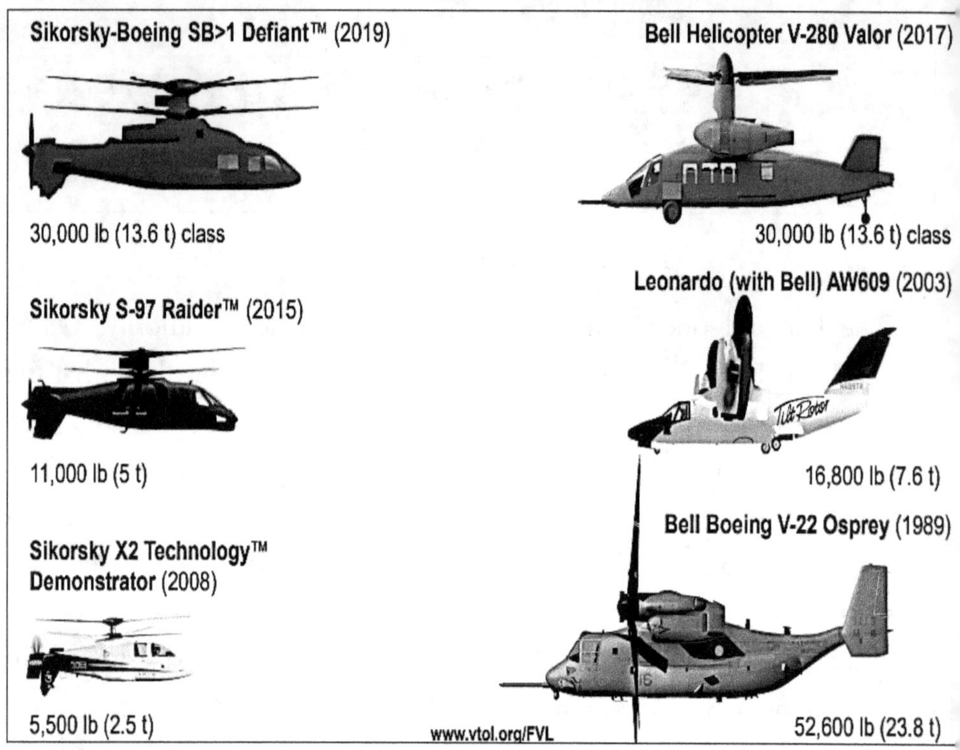

Figure 114. Compounds and Tilt Rotors

In summary, listed below are just some of the reasons why the Army needs to move to an advanced rotorcraft for a Future Heavy Lift (FHL) and possibly combined with a High Speed VTOL Rotorcraft (HSVTOL):

1. The Army's choice of the Bell Tiltrotor as its Forward Long-Range Assault Aircraft (FLRAA) replacement for the UH-60 (Ref. 13) opens the door for satisfying the three fundamental capabilities identified and required for Army Aviation:

- See—which addresses reconnaissance and security;
- Move—which includes air assault, air movement, and aeromedical evacuation; and
- Strike—which is close support and interdiction.

2. The CH-47F Block II and G lack FLRAA Bell V280 forward speed and lift capability. Rotor Loads & 6P Vibrations are still unacceptable on the CH-47F Block II with the Advanced Chinook Rotor Blades (ACRB), Figure 115. (Ref. 6)

Figure 115. CH-47F Block II w/ACRB Blades Not Acceptable (Ref. 6)

3. The Army Science Board (ASB) Fiscal Year 2015 Study, *Army Science and Technology for Army Aviation 2025-2040, Final Report, February 2016,* (Ref. 11) top two recommendations are shown in Table 1. They can help initiate a new, more appropriate FHL Future Heavy Lift (FHL) coupled System of Systems (SoS) Operational Analysis and Conceptual Design Assessment to initiate an Affordable FHL Program.

Topic	Findings	Recommendations
Context: Character of War in 2025 and Role of Army Aviation		
1. System of Systems Operational Effectiveness Analyses	Increasing threat sophistication and proliferation (e.g., missiles, UAS, cyber, and directed energy) pose critical concerns. (Classified annex provides additional details) Platform-centric survivability must evolve to a manned-unmanned system-centric approach with new CONOPS and TTPs. A system of systems architecture should include manned-unmanned teaming, supervised autonomous systems, and secure communications.	**TRADOC:** <u>Conduct operational effectiveness analyses</u> of potential system of systems concepts in a cost-constrained environment that address capability gaps for Army aviation in 2025 and beyond in complex threat environments. <u>Concepts</u> should include holistic air-ground approaches, high/low mixes of collaborative manned/unmanned systems, FVL performance characteristics, higher levels of autonomy, PNT in denied GPS environments, attritable UAS assets and enhanced lethality of DE. <u>Develop CONOPS and architectures</u> for the most cost effective concepts.
2. Affordability of Heavy Vertical Lift	Heavy vertical lift (20-30 stons) is required by the Army CONOPS for expeditionary and operational maneuver, validated by JROC (documented in the FVL and JHL ICDs) and supported by Unified Quest 2014, and "2012 Gaining and Maintaining Access: An Army-Marine Corps Concept." However, development of a new system is cost prohibitive within likely future Army modernization (RDA) funding before 2040. Interim solutions able to provide more limited capability are available (e.g., CH-53K at 18 stons, and Joint Precision Air Drop System – JPADS) and could provide viable options.	**TRADOC:** Assess interim and future HVL options for meeting Army CONOPS and documented JCIDS capability needs for expeditionary and operational maneuver and recommend road ahead • If development of HVL is cost prohibitive, consider alternatives. • If there are no plans for HVL, do not assume it is available in analyses, wargames, and exercises.

Table 1. ASB 2015 Top Two Recommendations for Future Army Aviation (Ref. 11)

This can be achieved through a Framework for Developing an Army Aviation Coupled System of Systems Operational Effective Analysis with an Affordable HLR Concept Design and Analysis (CDA), as illustrated in Figure 116. This approach could help get HLR FVL interest in the JCIDS Process (Ref. 10)

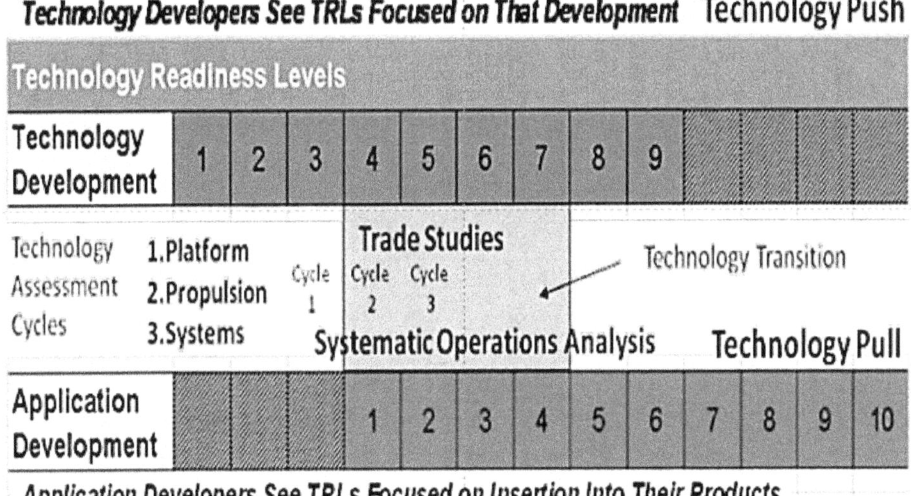

Figure 116.Framework for Execution of Army Aviation System of Systems Operational Effectiveness Analysis with Emerging Rotorcraft Design and Life Cycle Cost Assessment

The Need for a Joint Service Future Heavy Lift (FHL) /HSVTOL Concept Design and Analysis (CDA) Study

Potential V/STOL concepts such as rotor, propeller, ducted fan and turbofan/turbojet and their method of transitioning from hover to forward flight, e.g. tilting aircraft, tilting thrustor, vectored thrustor and separate thrusters are illustrated in Figure 117A (Ref. 10) Illustrated in Figure 117B is disk loading versus installed power with some V/STOL concepts from Carter Copter Slowed Rotor/Compound (SRC)to Ryan XV-5A ducted fan are identified. Low disk loading will be essential for FHL/HSVTOL. The best potential solution for an FHL/HSVTOL is achieved by combining the best features of a helicopter and a fixed wing aircraft. The best feature for FHL is low disk loading which provides the lowest hover power required and lowest downwash velocity. The best feature for high speed is the highest lift over drag (L/D) and efficient hybrid propulsion.

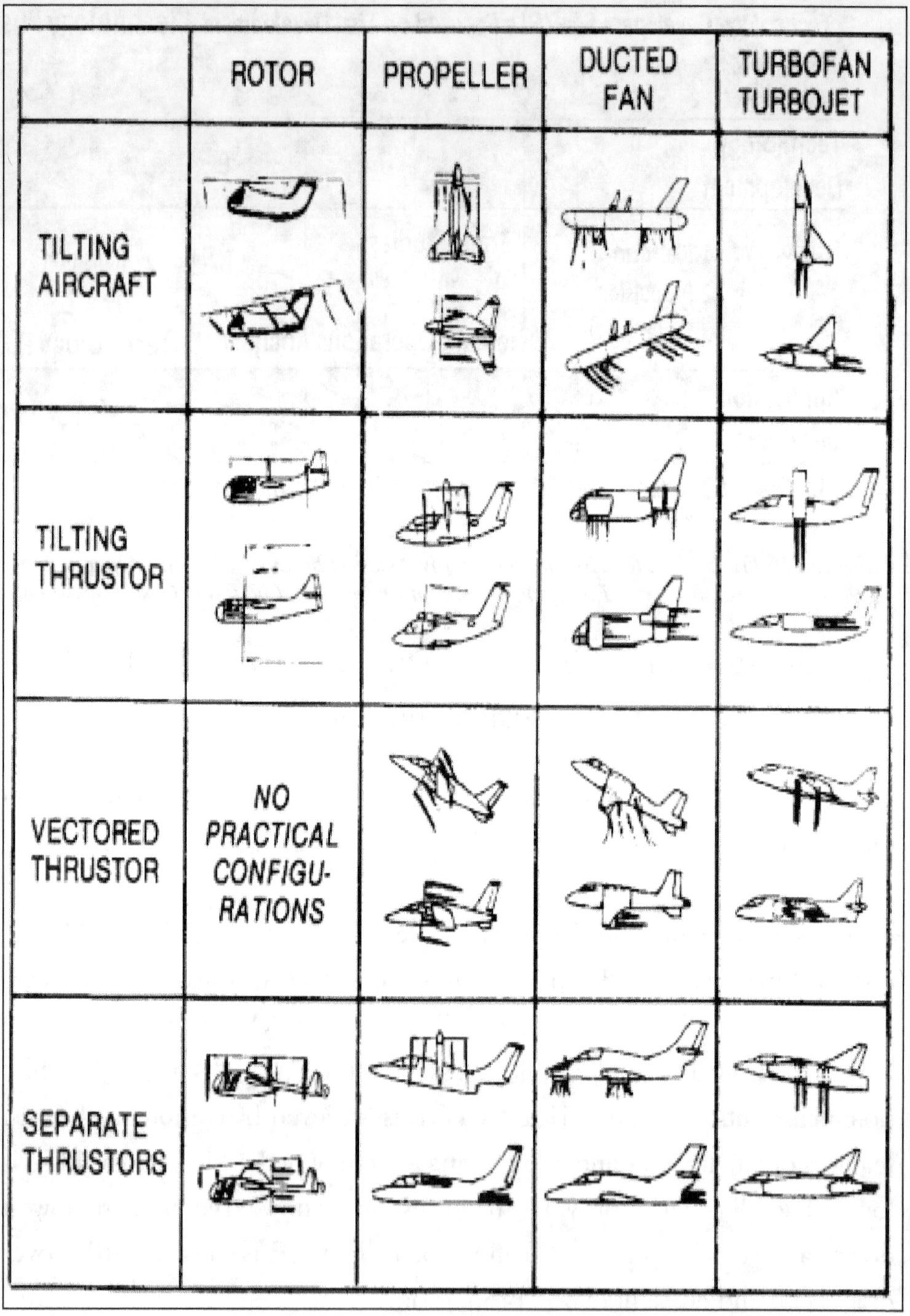

	ROTOR	PROPELLER	DUCTED FAN	TURBOFAN TURBOJET
TILTING AIRCRAFT				
TILTING THRUSTOR				
VECTORED THRUSTOR	NO PRACTICAL CONFIGU-RATIONS			
SEPARATE THRUSTORS				

Figure 117A. V/STOL Concept vs Tilt (Ref. 16)

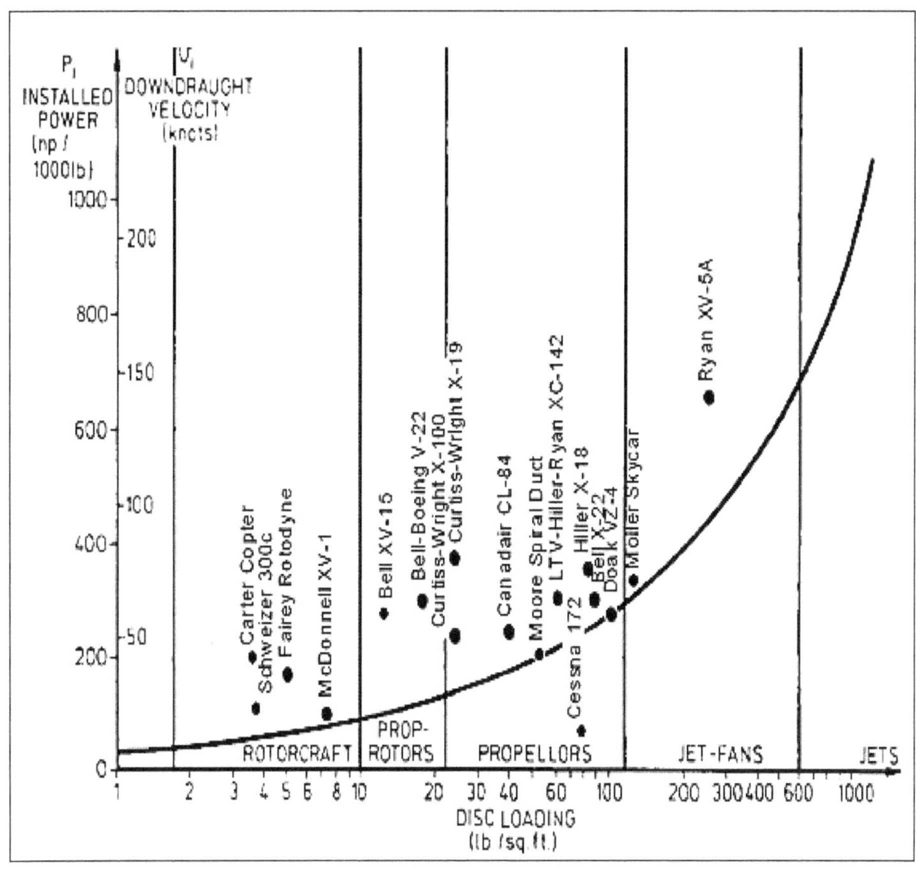

Figure 117B. Power Available vs Disk Loading (Ref. 10)

The Vertical/Short Takeoff and Landing (V/STOL) Wheel of Fortune, Figure 118, authored by Michael Hirschberg (Ref. 12) was an attempt to categorize almost all of the V/STOL aircraft designed, built and tested, mostly as prototypes in the 1960s-1980s. The inner circle identifies the type of rotor or propulsion systems for various categories. It also illustrates that very few prototypes ever made it to production. Those that have, have not been very successful, e.g., the Vectored Thrust Harrier, the Ducted Fan XV-5A and the Tilt Rotor V-22 have all demonstrated limitations. The 'What's Missing ??' annotation refers to a solution that is missing and could provide a solution for the necessary FHL/HSVTOL concept for the future. However, for the FHL/HSVTOL concepts to meet the missions required for Mission

Capability Sets 4 and 5 in Figure 113. They must include efficient hover and low-speed flight, high speed L/D efficiency and hybrid propulsion. In addition, for the Future there will be the requirement for autonomous aircraft. The initiation of a concept for an all-electric eVTOL aircraft can be traced to Mark Moore's participation as a team member on the Georgia Tech winning design in the 2007 AHS Student Design Competition and then as his follow-up at the NASA Langley Research Center. This was then picked up by Uber with the establishment of Uber Elevate which lead to the eVTOL Wheel of Fortune in Figure 119. The Air Force Research Laboratory (AFRL) then launched the Agility Prime Program in April 2020. This is being followed with the Autonomy Prime Program. (Ref. 17)

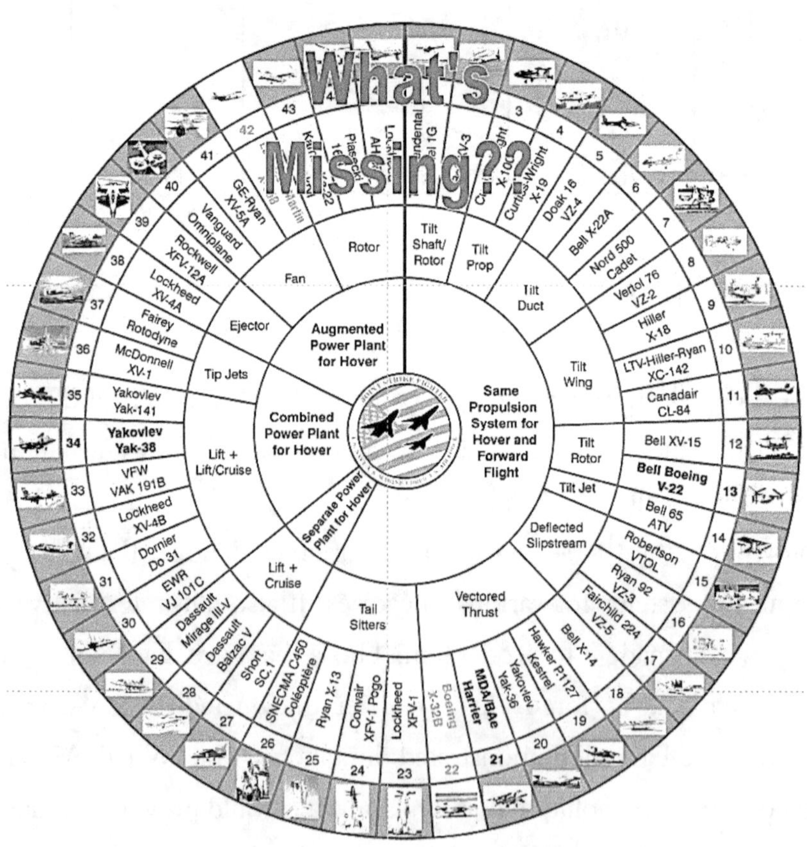

Figure 118. Searching for a FHL/HSVTOL Solution (Ref. 12)

While many electric eVTOL aircraft concepts are being designed as illustrated in the eVTOL Wheel generated by Mike Hirschberg, Figure 119, it is very doubtful that any can provide all the necessary electric power for FHL/HSVTOL in the next 20 years. Therefore, a hybrid VTOL aircraft will be required.

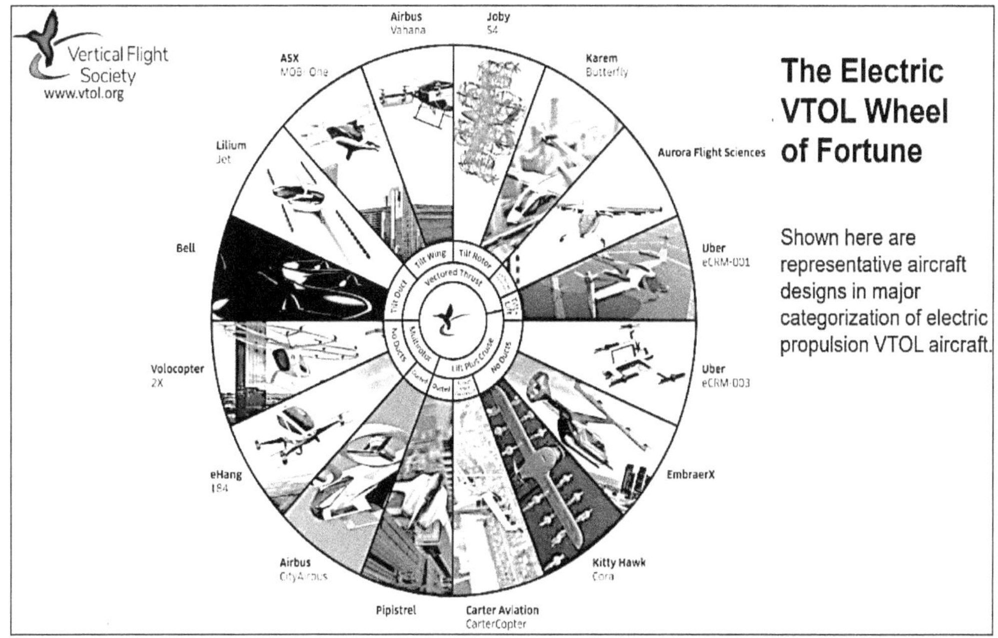

Figure 119. The Electric VTOL Wheel of Fortune (Ref. 12)

Uber Elevate was formed primarily to introduce eVTOL aircraft for Air Taxis. As was mentioned, Mark Moore, a former Georgia Tech student, was a key member of the Georgia Tech graduate winning team for the 2007 American Helicopter Society (AHS) Student Design Competition (SDC) for a Special Forces indigenous submarine mission. The cover of their design is illustrated in Figure 120A and illustrates the three designs the Team had to develop to win the competition. Illustrated in Figure 120B is the Dragonfly (UEV) eVTOL Information and its transition from Hover. (14)

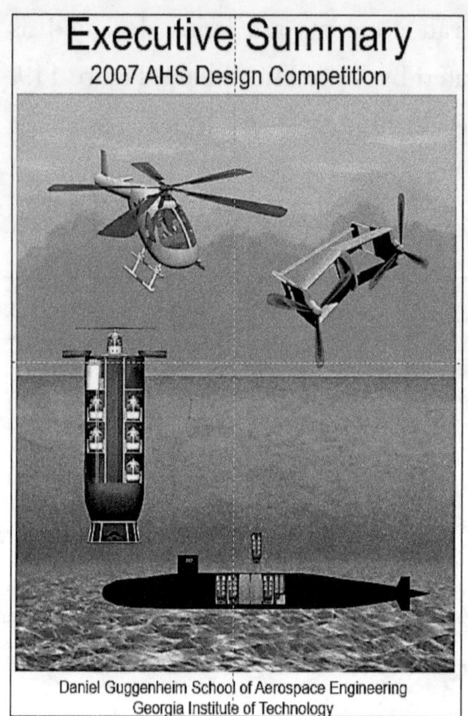

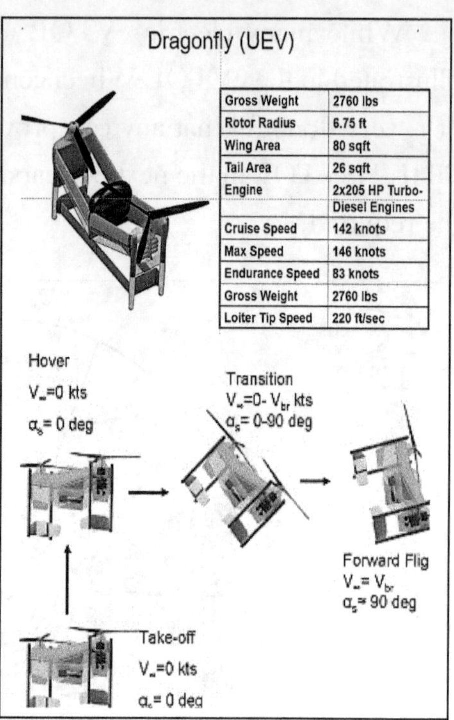

Gross Weight	2760 lbs
Rotor Radius	6.75 ft
Wing Area	80 sqft
Tail Area	26 sqft
Engine	2x205 HP Turbo-Diesel Engines
Cruise Speed	142 knots
Max Speed	146 knots
Endurance Speed	83 knots
Gross Weight	2760 lbs
Loiter Tip Speed	220 ft/sec

Figure 12oA. Winning Design Team Cover Figure 120B. Dragonfly eVTOL

Moore came up with the design for the electric powered, 12-foot (3.7 m) long, 14.5-foot (4.4 m) wingspan personal air vehicle as part of the coursework for his doctoral degree. Then Langley's creativity and innovation and revolutionary technical challenges funds paid for much of the research. How the Puffin rocketed from esoteric erudition to web sensation is a classic case study in the power of the viral nature of the web.

Upon returning to NASA Langley Research Center Mark Moore led a team that came up with the design of an eVTOL called the Puffin, Figure 121/

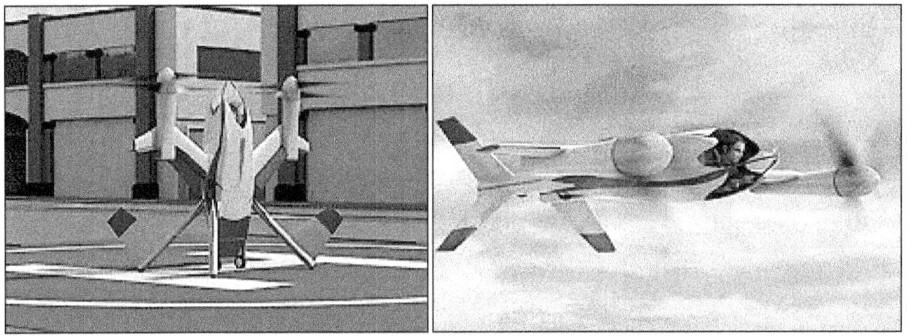

Figure 121. Moore's Puffin eVTOL Concept excited the World (Red. 15)

Mark Moore then left NASA and became the Uber Elevate point man for eVTOL development as Air Taxis. Uber Elevate identified six industry partners as shown in Figure 122 which has expanded worldwide as identified by the many in the eVTOL Wheel of Fortune in Figure 119.

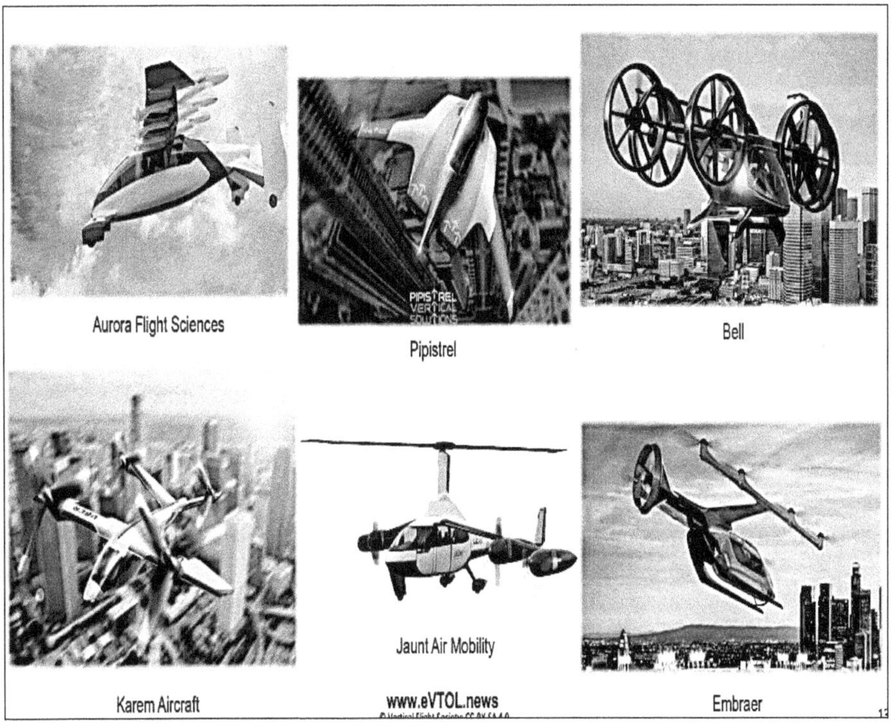

Figure 122. Uber Elevate Six Industry Partners (Ref. 16)

The Next Generation of Aircraft Dependent on Hybrid Propulsion

A recent White Paper on Hybrid & Electric Propulsion Systems for Sustainable Aviation was developed by ANSYS (Ref. 17). The commercial aviation industry is at the threshold of a decades-long transformation to dramatically improve its environmental sustainability. The success of this quest will largely depend on the development of new, more efficient propulsion systems. Unfortunately, current propulsion technologies are insufficiently advanced to enable a singular, comprehensive leap to solve the problem. Instead, commercial aviation companies must segment their approach, simultaneously developing multiple propulsion systems on separate timelines. This strategy will enable manufacturers to improve sustainability in the short term — through the development of hybrid-electric systems that use a combination of batteries and traditional combustion — while working toward hydrogen-powered and all-electric systems.

The military aviation industry and government must be aware of this transformation, especially as they prepare for future vertical lift (FVL) aircraft and especially for a FHL/HSVTOL solution. The Army and Department of Defense (DOD) has begun to accept from Commercial Aviation the Modular Open Systems Approach (MOSA) that Commercial Aviation has pioneered with Open Integrated Modular Avionics (IMA) to reduce Space Weight and Power (SWaP) and Time/Cost, as illustrated in Figure 123A (Ref. 18). Failure on the F35 Avionics Integration resulted in the Air Force taking IMA away from the Prime and now has given it to GE Systems. AirBus with its AB380 Commercial Transport and Boeing with its B787 initiated Open IMA ten years ago with a first generation, IMA1G, and are now moving to IMA2G which expands Sustainability beyond Open Avionics IMA, as illustrated in Figure 123B. (Ref. 19)

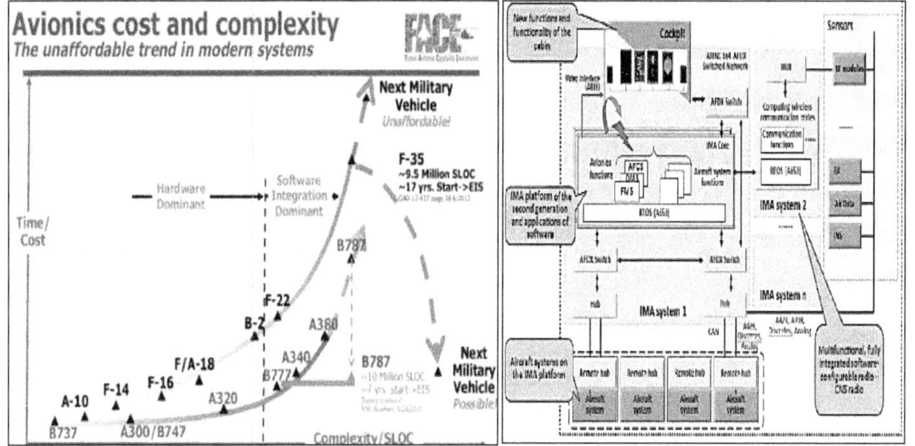

Figure 123A. Progression of Time/Cost for Civil & Military Aircraft. Figure 123B. Notional IM2G (Ref. 19)

Commercial Aviation Hybrid Propulsion Timeline (Ref. 17)

A Multi-Track, Multi-Timeline Approach Commercial aviation companies initially examined the feasibility of moving directly from traditional combustion engines to all-electric propulsion systems. They quickly realized that doing so in the near-term was impossible given the current limitations of battery technology. Instead, manufacturers decided to take a segmented approach that will enable them to pursue meaningful breakthroughs in sustainability for combustions engines while maintaining a longer-term focus on developing even more efficient hybrid-electric, all-electric, or hydrogen-based systems. Developing three distinct propulsion systems on distinct timelines offers another important advantage for the industry. As aviation companies develop these advanced propulsion systems for commercial aircraft, they can also apply their findings to other applications and markets, such as the civil and military aviation sector. As aviation companies contend with the technical challenges of sustainable propulsion systems, they must also establish a business case for each, and navigate the industry's extensive certification process. Regulators are establishing recommendations that will guide manufacturers as they develop the new systems. The FHL/HSVTOL can follow and take advantage of this effort.

DRIVERS OF NEW PROPULSION SYSTEM DEVELOPMENT

provide a Multi-Track, Multi-Timeline Approach as illustrated in Figure 124. (Ref. 20) They will explore:

• The factors driving aviation manufacturers to make commercial flights more sustainable.

• The industry's segmented approach to propulsion system development.

• The characteristics and development timelines of the three new propulsion systems.

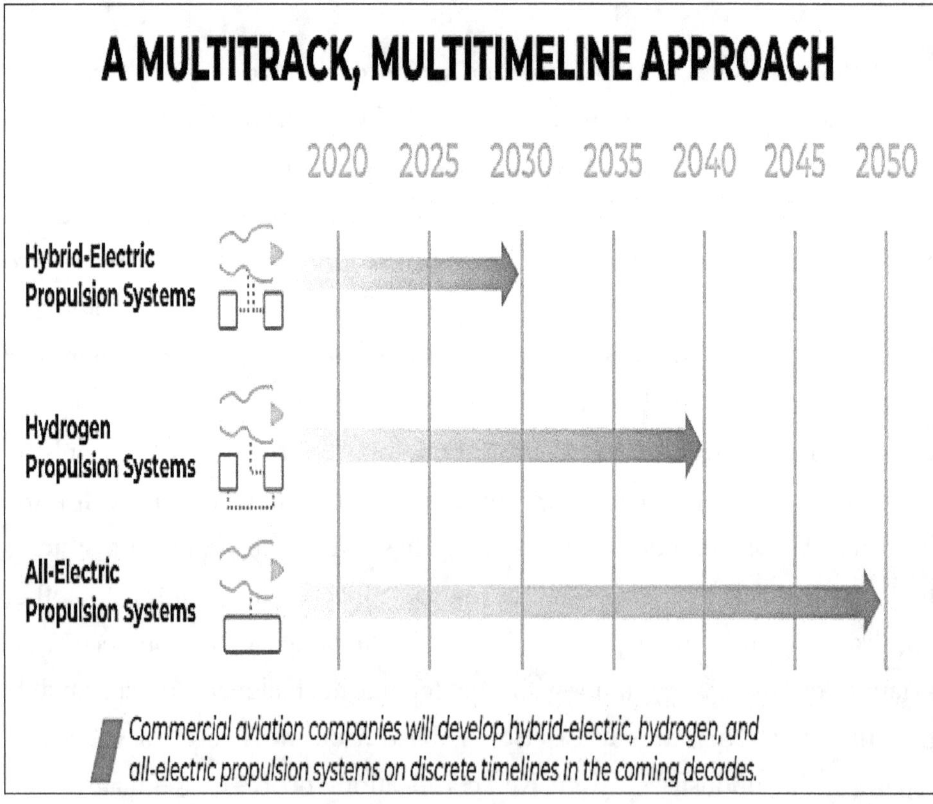

Figure 124. The Commercial Aviation Multi-Track, Multi-Timeline Approach

Impact of Hybrid Propulsion on FHL/HSVTOL Concept Development

This is an additional critical consideration for augmented installed power as a hybrid system including a combination of electrical and mechanical power which should be available from Commercial Aviation in the 2030s. An example rotorcraft tradeoff study of hybrid electric power is in the Epilogue to this Book. A potential solution is the Carter Slowed/Rotorcraft Compound (SR/C) aircraft which is now being developed as the Jaunt Journey eVTOL aircraft. It is being developed for FAA rotorcraft certification. It combines a separate lifting rotor that can be slowed, plus wing and propulsion devices with the attributes identified in Figure 125.

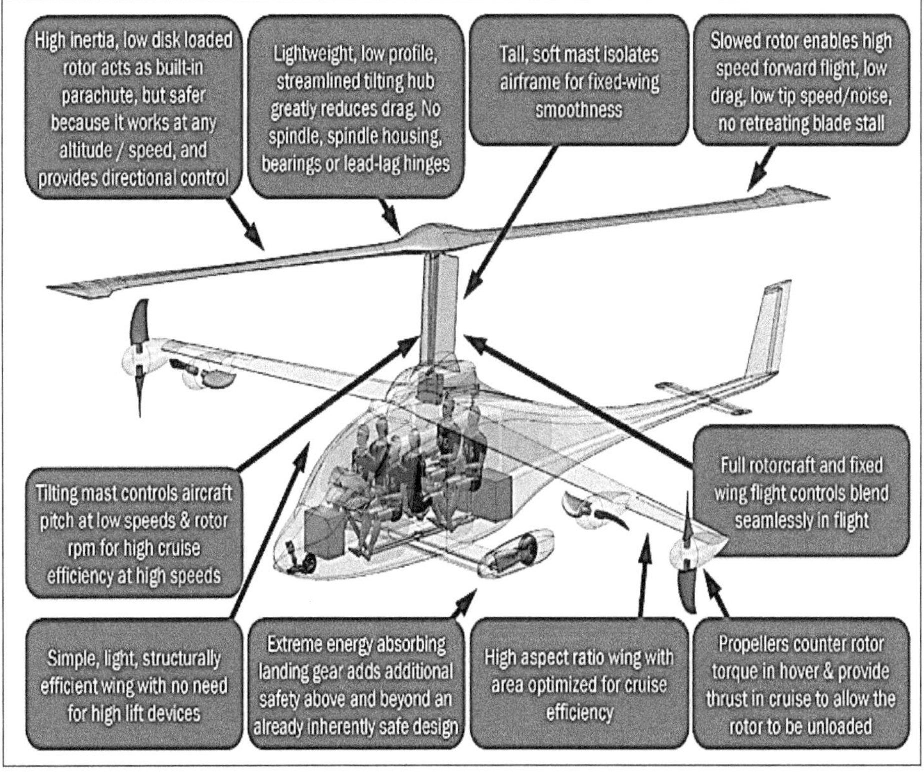

Figure 125. Unique Features of the Carter Aviation SRC (Ref. 20)

It is scalable to larger versions as hybrid aircraft. See Figures 126 A and B. Jay Carter, founder of Carter Aviation Technologies, has a FHL/HSVTOL hybrid solution, Figure 125C. He has described how a hybrid electric rotorcraft could be a viable solution as a FHL/HSVTOL aircraft.

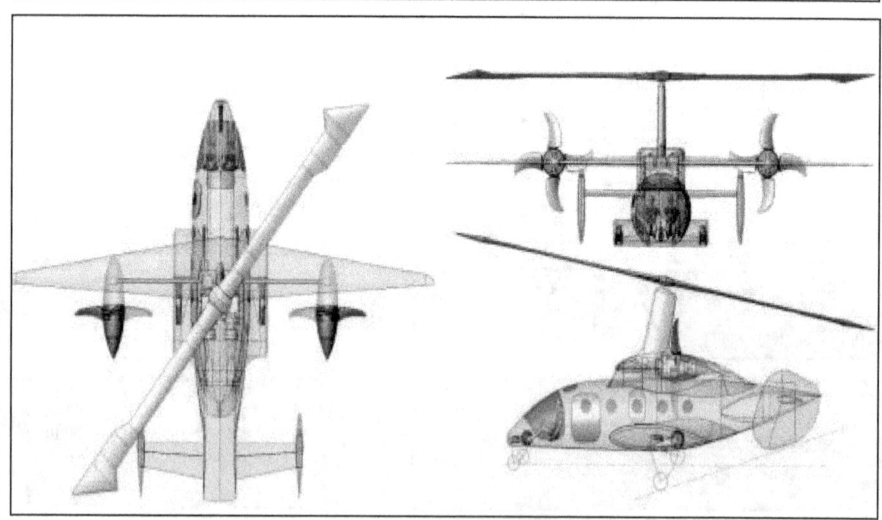

Figure 126A.Carter SR/C Configuration and 3-View Presentation

Figure 126B. Unmanned Cargo Transport 10,700 lbs Useful Load, 600 nm Range w/Cargo, 250 kts Cruise

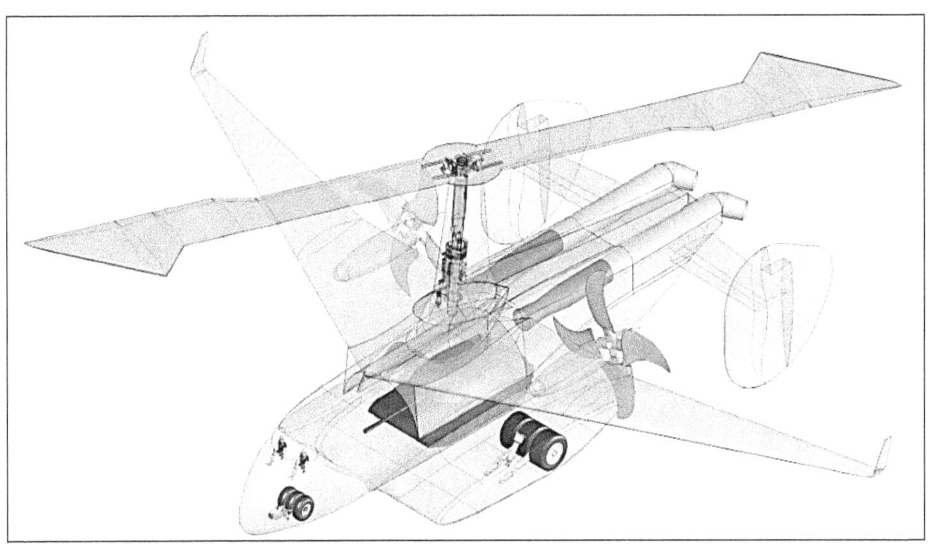

Figure 126C. Heavy Lift Transport, Full Hover-Anti-Torque from Vectored Jet Exhaust, 400k lbs. GW, Forward Speed 417 kts

The Carter SR/C prototype was the first rotorcraft to ever achieve an advance ratio, μ, greater than one. Carter Aviation Technologies CEO Jay Carter received the American Helicopter Society (AHS) Paul E. Hauter Award in 2013 for his achievements in SR/C aircraft designs capable of not only achieving a μ>1, but for his achievements in slowed-rotor compound aircraft designs capable of providing unprecedented improvements in rotorcraft operational flexibility, efficiency, speed, and safety. Carter's contributions culminated during flight-testing in 2013 when the four-place second generation Slowed-Rotor/Compound (SR/C) VTOL demonstrator consistently achieved level flight lift-to-draft values of nearly 12, exceeding values of conventional helicopters by a factor of 2.5. Carter Aviation Technologies was sold to Jaunt Air Mobility in 2021. The Jaunt Journey SR/C eVTOL aircraft was selected as one of six Uber Elevate concepts for Urban Air Mobility (UAM) Air Taxis, as was chosen in Figure 122.

Advantages of the Carter Aviation SRC for FHL/HSVTOL

Slowed Rotor Compound, or SRC, aircraft are a hybrid between an airplane and a rotorcraft. Carter pioneered this technology and has now transferred the intellectual property to Jaunt Air Mobility. (Ref, 20). It includes:

- SRC enables the speed and efficiency of an airplane with the vertical takeoff and landing of a helicopter.
- Large rotors are the most power-efficient configurations for hovering and vertical takeoffs and landings.
- The constantly turning, high inertia rotor can function as a built-in parachute at any altitude or airspeed, and allow the pilot to choose their landing spot for a soft, zero roll landing even in the event of a complete power failure.
- The low rotor rpm throughout the flight envelope reduces noise to levels barely perceptible above normal city traffic.
- The low drag rotor mounted on a tall mast supported by soft supports makes for a ride as smooth as a fixed-wing airplane.

Army/Air Force/DARPA/Navy have some Interest in combined Future Heavy Lift and High-Speed VTOL Aircraft (FHL/HSVTOL) aircraft development. These concepts were initially included in Army FVL Capability Sets #4 and #5, although CS4 is being re-defined. Army requirements for the evolving *Joint All Domain Command and Control* (JADC2), Multi-Domain Operations (MDO) and the *Great Power Competition* can provide Concepts of Operations (CONOPS).

The Air Force AFWERX Office is also trying to accelerate agile and affordable capability transitions through HSVTOL Aircraft Concepts. A new DARPA Speed and Runway Independent Technologies (SPRINT) X-Plane Program plan is also initiated to address HSVTOL. The Air Force with AFWERX and DARPA have interest in HSVTOL technologies, Therefore, a joint service approach is possible.

Army is moving to a Future Long Range Assault Aircraft (FLRAA) which exceeds a Heavy Lift Helicopter (HLH) speed and range capabilities.

The Army Science Board (ASB) Fiscal Year 2015 Study on *"Army Science and Technology for Army Aviation 2025-2040"* top two priorities were:

1. <u>A System of Systems (SoS):</u> The ASB Study Team found that the FVL family of systems represents the current planning for future of Army aviation manned (and optionally manned) vertical lift systems. FVL is currently an initiative within the Army Aviation Science & Technology (S&T) portfolio and is not expected to become a Program of Record (POR) until after the current Program Objective Memorandum (POM). It was noted that the cost of the heavy-lift option may preclude it from being a viable option. The team found that several significant technology initiatives within the Army Aviation S&T portfolio apply to both legacy and future manned systems. This technology of technologies includes ITEP/FATE engine insertion, Value Engineering capabilities, CBM/PHM capabilities, ASE improvements, and greater MUM-T. Current Army aviation UAS assets consist of the Raven, Puma, Shadow, and Gray Eagle systems.

215

These systems are all fixed-wing, have limited autonomy and limited MUM-T capability, and are essentially dedicated to one or two mission areas.

2. <u>The Affordability of Heavy Vertical Lift</u>: The ASB recommended that Training and Doctrine Command (TRADOC) assess interim and future heavy vertical lift options for meeting Army Concept of Operations (CONOPS) and documented in JCIDS capability needs for expeditionary and operational maneuver and recommend the road ahead. If the development of heavy vertical lift capabilities proves to be cost-prohibitive, alternatives must be considered (e.g. even the CH-53K at 18 tons, Figure 127. If there are no plans for HLR, do not assume it is available in analyses, wargames, and exercises. However, a FHL/HSVTOL Concept is required for the necessary speed, range and payload.

Figure 127. The Sikorsky CH-53K King Stallion

Previous Joint Heavy Lift (JHL) Efforts

The Aviation Applied Technology Directorate (AATD) solicited in April 2005 technical and cost proposals for conduct of Joint Heavy Lift (JHL)

Concept Design and Analysis (CDA). This Broad Agency Announcement (BAA) constituted the total solicitation. There was no formal Request For Proposals (RFPs), other solicitation requests, or other information regarding these requirements. (Ref. 21)

The JHL Concept Refinement (CR) phase was intended to support a Milestone A decision. The CDA portion of the overall JHL CR was not to exceed 18 months beginning in September 2005.

The CDA focus was on the identification and technical substantiation of viable design concepts that not only populate the desired trade space but have a reasonable chance of achieving a Technology Readiness Level (TRL) of 6 by 2012. The CDA was to identify the program and technical risks, cost, schedule, and critical path technologies.

The purpose of the JHL CDA was to first and foremost inform the requirements trade process with valid performance and design options that are achievable within the projected time env elope. It will assess the impact of trade parameters affecting total system suitability and affordability and define any technology hurdles requiring dedicated science and technology investments. Lastly, it was to establish credible cost, schedule, and risk estimates for the most probable options within the selected design envelopes.

Three design cruise speed bands of 160-200, 200-250 and 250-300+ knots were addressed by the CDA. A model performance specification has been developed to provide a representative example specification that was to be refined and expanded by the JHL CDA contractors to reflect the details of their concept design as it evolves. The five concept vehicles chosen for this effort, listed in order of their design cruise speeds, were:

- Sikorsky X2C, X2 Technology Crane - coaxial rotor (165 knots);
- Boeing ATRH, Advanced Tandem Rotor Helicopter (165 knots);
- Sikorsky X2HSL, X2 Technology High Speed Lifter - advancing blade compound (245 knots);
- Bell Boeing QTR, Quad Tilt Rotor (275 knots); and
- Frontier Karem Aircraft OSTR, Optimum Speed Tilt Rotor (310 knots).

These awards were for eighteen months and represented over \$30M of Government and Industry contribution. They include the delivery of the designs with substantiating data, a specification document, a technology development strategy, and cost/schedule estimates for a Component and Technology Demonstration phase to achieve TRL 6 in an appropriately large-scale flight vehicle. Award of any future JHL development activity, should it occur, was separate and independent of this BAA. However, the 2005. Army Joint Heavy-Lift (JHL) Concept Design and Analysis (CDA) (Ref. 21) was mostly a Technology Push rather than a Technology Pull. It was also unaffordable, not joint, and assumed unrealistic tiltrotor technology for the Army in that timeframe, Figure 128. A two-rotor tilt rotor in the 130-150K lbs. Gross Weight, as proposed by the Army and Kareem Aircraft was not feasible. More realistic solutions by Bell and Boeing were quad rotors as illustrated in the center of Figure 129.

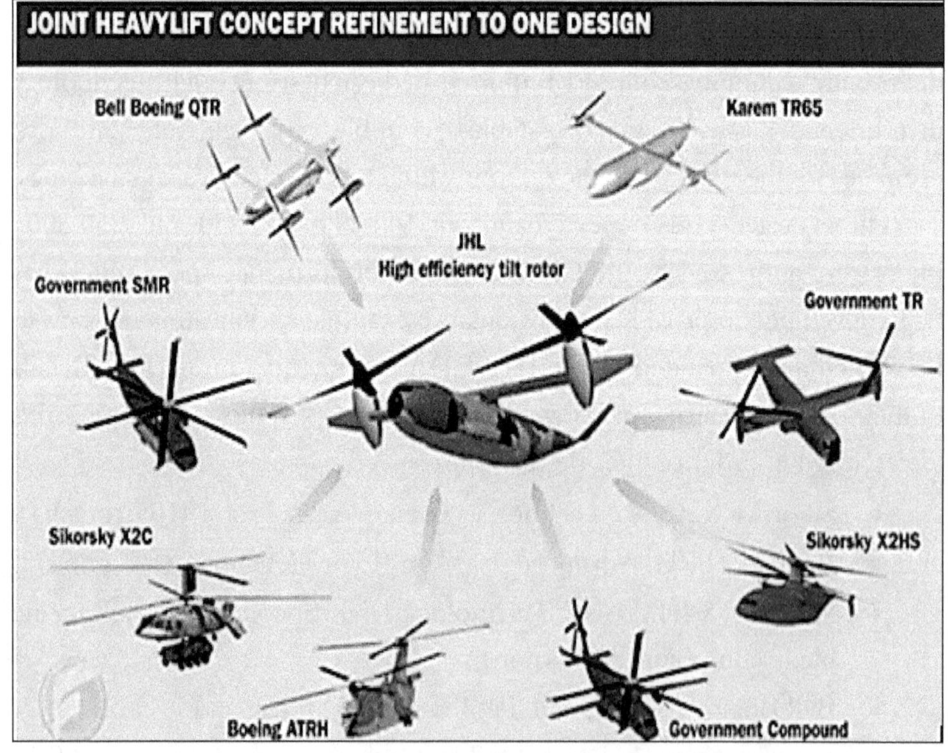

Figure 128. Army/Kareem JHL Two Rotor Tilt Rotor

218

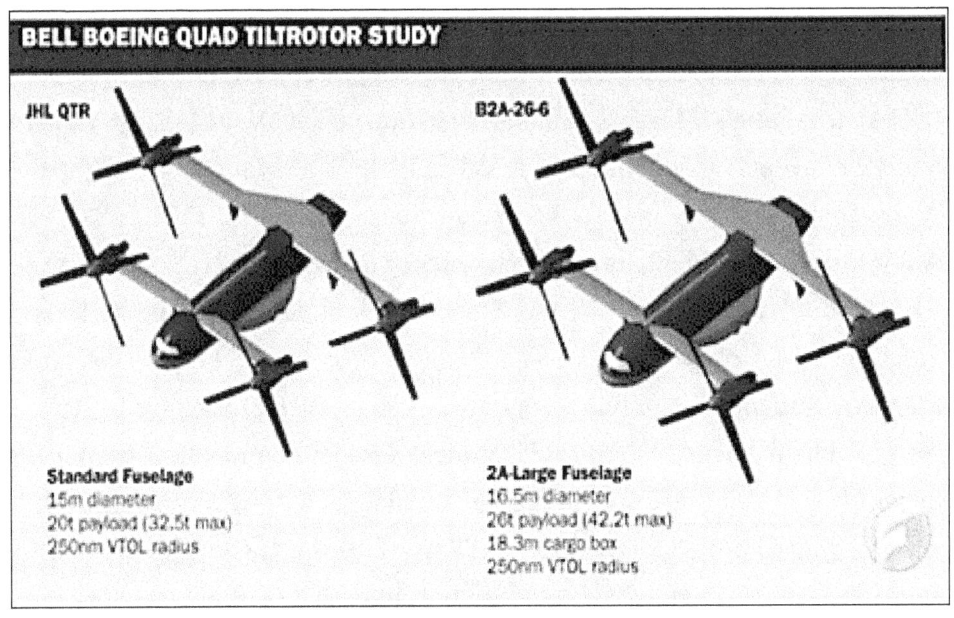

Figure 129. Bell and Boeing Quad Rotors

Bell Helicopter's Chief Engineer told me they would never build a two rotor tilt rotor on an aircraft greater than 100K lbs. In two of the Georgia Tech award-winning student design competitions two tilt rotor concepts were successfully designed as illustrated covers in Figure 130A and 130B. (Ref. 22)

Figures 130A and B. Georgia Tech Successful Quad Rotor Designs

In conclusion, the Army needs to work with the other military services, DARPA, FAA and NASA in conducting a Concept Design and Analysis (CDA) for Future Heavy Lift and High Speed VTOL (FHL/HSVTOL) aircraft concepts, as identified in a What, Why and How Section. For the Army this should include a variety of these advanced FHL/HSVTOL Aircraft which are being identified by other services, DARPA, and NASA. They should also include hybrid electric aircraft as well. The following describes some of the F-HL/HSVTOL aircraft concepts being considered for development in different programs. A kick-off CDA Study using the Vertical Flight Society (VFS) 2024 Student Design Competition (SDC) for initial FHL/HSVTOL Concept Design & Analysis (CDA) is recommended. The VFS 2023 SDC Competition was for a HSVTOL CDA sponsored by Lockheed Sikorsky. In 2024, it should be sponsored by the Army Research Lab. Perhaps three separate design solutions could be sought with varying lift capabilities and air speeds as identified in the previous Joint Heavy Lift (JHL) Program. (Ref. 21)

Evolving Hybrid and eVTOL Aircraft Concepts

Eleven companies were chosen to complete a Phase I contract award from the U.S. Air Force's AFWERX program in 2022. The market research program challenged applicants to design a high-speed VTOL aircraft concept and received about 230 submissions. Bell Textron, Horizon, Jaunt Air Mobility, Whisper, and Jetcoptera were some of the other companies to receive a Phase I contract award.

Two modern versions from those at the ends of the spectrum in *Power Available vs Disk Loading*, in Figure 117B are reviewed Modern versions of Carter Aviation and Ryan XV-5A will be discussed as follows.

The Jaunt Air Mobility Journey is the closest concept to a helicopter and is being certified to FAA Rotorcraft Type Certificate Standards, FAR Parts 27 and 29. It is shown in Figure 131. The author of this Book 2 was largely responsible for putting the Jaunt Journey Team together.

Figure 131. Jaunt Air Mobility Journey Slowed Rotor Compound (SRC)

For nearly a century, civil vertical flight has been nearly exclusively the domain of helicopters or gyroplanes suspended beneath a single, large, spinning overhead rotor. However, experts have long known that a "lift-compound" rotorcraft (i.e. with a wing) could achieve greater cruise speeds and efficiency than a traditional helicopter if the rotor was slowed to reduce drag during wing borne flight, but there were many technical and other challenges to overcome. Jaunt Air Mobility was created in 2019 to bring the world's first electric Slowed Rotor Compound (SRC) aircraft to market, leveraging recent advances in electric motors and controllers, lithium-ion batteries, and fly-by-wire flight controls to overcome some of these challenges. Jaunt acquired the intellectual property of Carter Aviation Technologies, which had spent 25 years refining all the systems and components required to operate as a highspeed and energy-efficient compound helicopter, part airplane and part helicopter — which Jaunt says is the most efficient electric vertical takeoff and landing (eVTOL) configuration. It can also come in a hybrid-electric version for a FHL/HSVTOL concept.

At the other extreme is the Horizon Aviation eVTOL Aircraft electric multi-ducted fans concept shown in Figure 132. (Ref. 23) While its high speed and short takeoff and landing capability may be good for the Air Force, it can't meet the Army's need for sustained hover and low speed flight for personnel and cargo delivery.

Figure 132. Horizon Multi-ducted Fan eVTOL Aircraft

It completed the AFWERX Phase I contract on June 30 and allowed the Horizon team to accelerate development of the Cavorite X5 prototype, a hybrid-. Canada-based Horizon Aircraft. It has finished building a half-size

222

prototype of its hybrid electric vertical takeoff and landing (eVTOL) aircraft, dubbed Cavorite X5. The nine-month-long Phase II contract will be starting in Fall 2023 and will offer companies significant non-dilutive financing to support further research and development of high-speed VTOL aircraft. Following that, Phase III of the AFWERX challenge will finance a 30-month program in which a functional full-scale prototype will be developed.

The DARPA Speed and Runway Independent Technologies (SPRINT) X-Plane project is a joint DARPA/U.S. Special Operations Command effort that aims to design, build, and fly an X-plane to demonstrate the key technologies and integrated concepts that enable a transformational combination of aircraft speed and runway independence. The SPRINT X-plane is intended to be a proof-of-concept technology demonstrator and its flight test program seeks to validate enabling technologies and integrated concepts that can be scaled to different size military aircraft. The goal of the program is to provide these aircraft with the ability to cruise at speeds from 400 to 450 knots at relevant altitudes and hover in austere environments from unprepared surfaces. A summary of the DARPA SPRINT X-Plane Demo Program is summarized below in Figure 133.

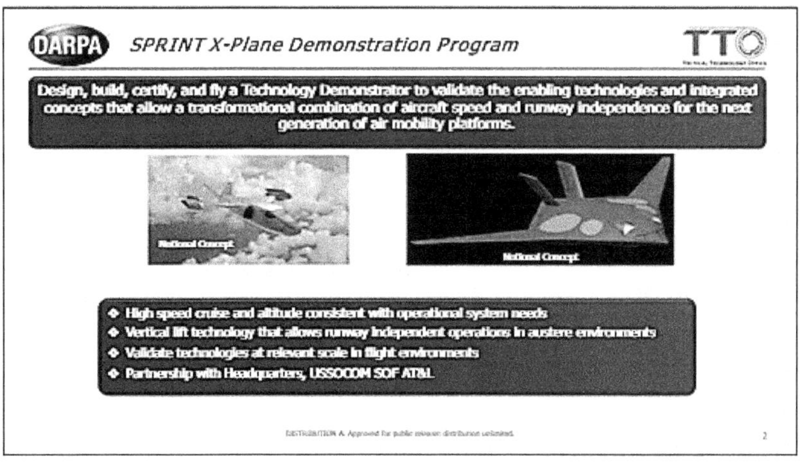

Figure 133. The DARPA SPINT X-Plane Demonstration Program

A summary of a DARPA Tactical Technology Office (TTO) Program that the author successfully lead was the DARPA Heliplane TipJet Program. The Heliplane Concept is shown in Figures 134A&B. (Ref. 24)

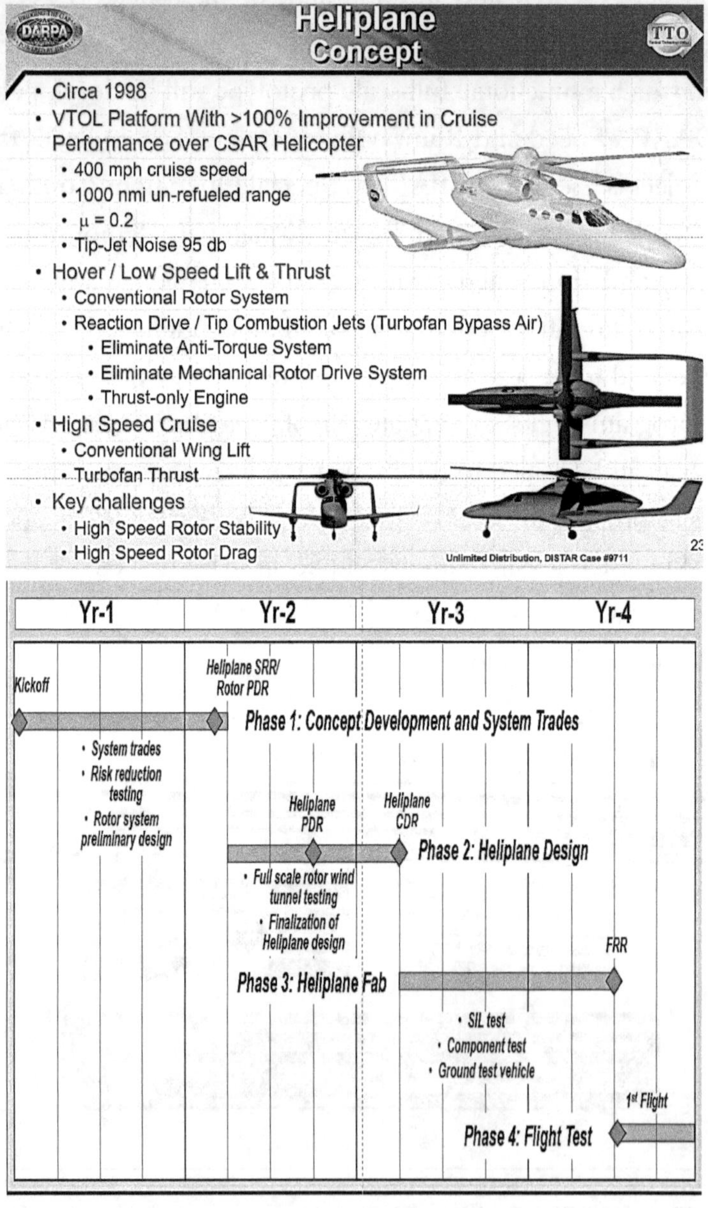

Figure 134A. Heliplane Concept. Figure 134B. Heliplane Schedule

The initial Phase I effort was led by Groen Bros. Aviation (GBA) as illustrated in Figure 135A with Georgia Tech, Williams International, and Adam Aircraft Technologies as partners. Illustrated in Figure 135B is a picture of the Heliplane Team outside Georgia Tech with an early tip jet rotor behind them which had been built and tested in the 1940s.

Figure 135A. Heliplane Phase I Organization. Figure 135B. Heliplane Team

Groen Bros. Aviation being a small company didn't have the relative cost sharing resources to execute the remainder of the Heliplane Program beyond Phase I. Since the Program was progressing it was determined by DARPA to add a Phase IB to complete the Preliminary Design Phase. Georgia Tech was given the Principal Investigator lead for Phase IB. The author was given the lead responsibility, as shown in Figure 136A and B.

Figure 136A. Heliplane Overview & Status. Figure 136B. Key Program Goals

As summarized in Figure 137A Key Performance Parameters (KPPs) were met. Figure 137B summarizes that Preliminary Design and potential of Heliplane was achieved. A picture of the Heliplane is shown in Figure 138.

> ## Phase 1 KPP's:
> - HOGE at 4kft/95degF at Range GW with 10% powe margin
> - 1,000nm unrefueled payload with 4k/95 HOGE at mission start and vertical landing at the destination
> - 400mph at max. continuous engine power for >30min
> - DL ≤ 10lb/ft² (similar to Blackhawk)
> - Achieve an effective perceived noise level of <95EPNdB
>   - This requirement was established by the DARPA director to meet a desire to "operate the aircraft from commercial airports"
>   - This noise was limited to rotor noise, including noise from the tip jets
>
> ## DARPA directed design constraints:
> - Reactionless (tip-jet) rotor drive
> - Rotor in auto-rotation at high speeds

- Preliminary design closes on top level system goals:
  - 400 mph cruise speed
  - 1000 nm range with 1000 lb payload
  - VTOL performance (including 4K95)

- Heliplane Has the Potential to Revolutionize VTOL Aviation
  - VTOL and hover capability of a helicopter combined with high speed cruise capability of a fixed-wing airplane
  - Heliplane has the unique potential to replace the helicopter
    - VTOL and hover capability
    - High payload fraction
    - Efficient hi-speed cruise at speeds 2-3x of today's helicopters

- Team is in place to complete the necessary Technical Data Package through a Phase 2 Effort

Figure 137A. KPPs Summary. Figure 137B. Preliminary Design & Potential

Figure 138. Heliplane Potential Capabilities

The Skyworks Global Company has collaborated with Scaled Composites to develop a VertiJet demonstrator based on Heliplane Technology

Conclusion

It is evident that the US Military has Vertical Take-Off and Landing (VTOL) Capability Gaps as illustrated in Figure 139. (Ref. www.vtol.org)

Figure 139. Future Vertical Flight, Mike Hirschberg, www.vtol.org

Army Aviation leaders are seeking a potential FVL Heavy-Lift Solution beyond the CH-47F Block II Solution. This effort needs to be implemented.

While Army aviation has consolidated its Aviation Development Directorate (ADD) into the Technology Development Directorate (TDD), DEVCOM Aviation & Missile Center, it would seem that a FHL/HS VTOL Solution could be started there. Finally, the ASB would seem to be an effective advocate for initiating the much-needed timely system-of-systems operational analysis with TRADOC and Concept Design and Analysis (CDA) for an affordable heavy lift solution.

Endnotes

1. CH-47D,F, Military.com
2. HLH XCH-62 Helicopter, Military.com
3. R. Loewy, AHS Nikolsky Lecture, 1984
4. Grina, K. Rotorcraft Design Short Course
5. Schrage, D.P. and O'Malley, J., "Performance and Aeroelastic Tradeoffs on Recent Rotor Blade Designs, AHS Rotor Design Specialists Meeting, October 1980.
6. CH-47F Block II and ACRBs Info, Braking Defense, 2022
7. Army Heavy Lift: An Enduring Capability for the Great Power Competition" Breaking Defense, 2021
8. Army FVL Program, Breaking Defense, 2021
9. GT-Liverpool NASA Quad Rotor Design, AHS SDC, 2011.
10. Schrage, D.P. Rotorcraft Design Courses, Georgia Tech, 1985-2019
11. Army Science Board (ASB) Report, January 2016, *Army Science and Technology for Army Aviation 2025-2040*
12. FVL Mission Capability Sets, 2018 AHS Website, www.vtol.org
13. FLRAA Selection, Breaking Defense, 2023
14. Georgia Tech 2007 AHS SDC Winning Design
15. NASA Puffin Program
16. www.vtol.org Mark Moore
17. ANSYS White Paper: Hybrid & Electric Propulsion Systems for Sustainable Aviation, 2023
18. FACE Website
19. Chuyanov, G.A., et al, "Advanced Avionics Equipment on the Basis of 2nd Generation Integrated Modular Avionics", ICAS 2014, St. Petersburg)
20. Carter Aviation Technologies Website
21. JHL Concept Design and Analysis (CDA): The Aviation Applied Technology Directorate (AATD) solicited in April 2005 technical and cost proposals for conduct of Joint Heavy Lift (JHL) Concept Design and Analysis (CDA)
22. Georgia Tech Tilt and Quad Rotor Winning NASA and AHS SDCs
23. Horizon Multi-ducted Fan eVTOL Aircraft Website
24. DARPA SPRINT HSVTOL Program, 2023
25. DARPA Georgia Tech Heliplane Program, 2009.

CHAPTER SEVEN

PROMOTION TO SENIOR EXECUTIVE SERVICE (SES) as Director of Advanced Systems (DAS) Aviation

Research and Development Command (AVRADCOM) & LHX Program

This provided an opportunity to lead Concept Formulation for LHX and help integrate technologies from the Army Aviation Labs. Also, I served as interim Chief Scientist for the Combined Arms Center (CAC), for six months, 1983)

The US Army Aviation Research and Development Command (AVRADCOM) Mission were illustrated in Figure 140, and the Research and Development Organizations for carrying it out are shown in Figure 141A, e.g. **MG** Story Stevens, Four Key Directors and Organizations across the USA Figure141B.

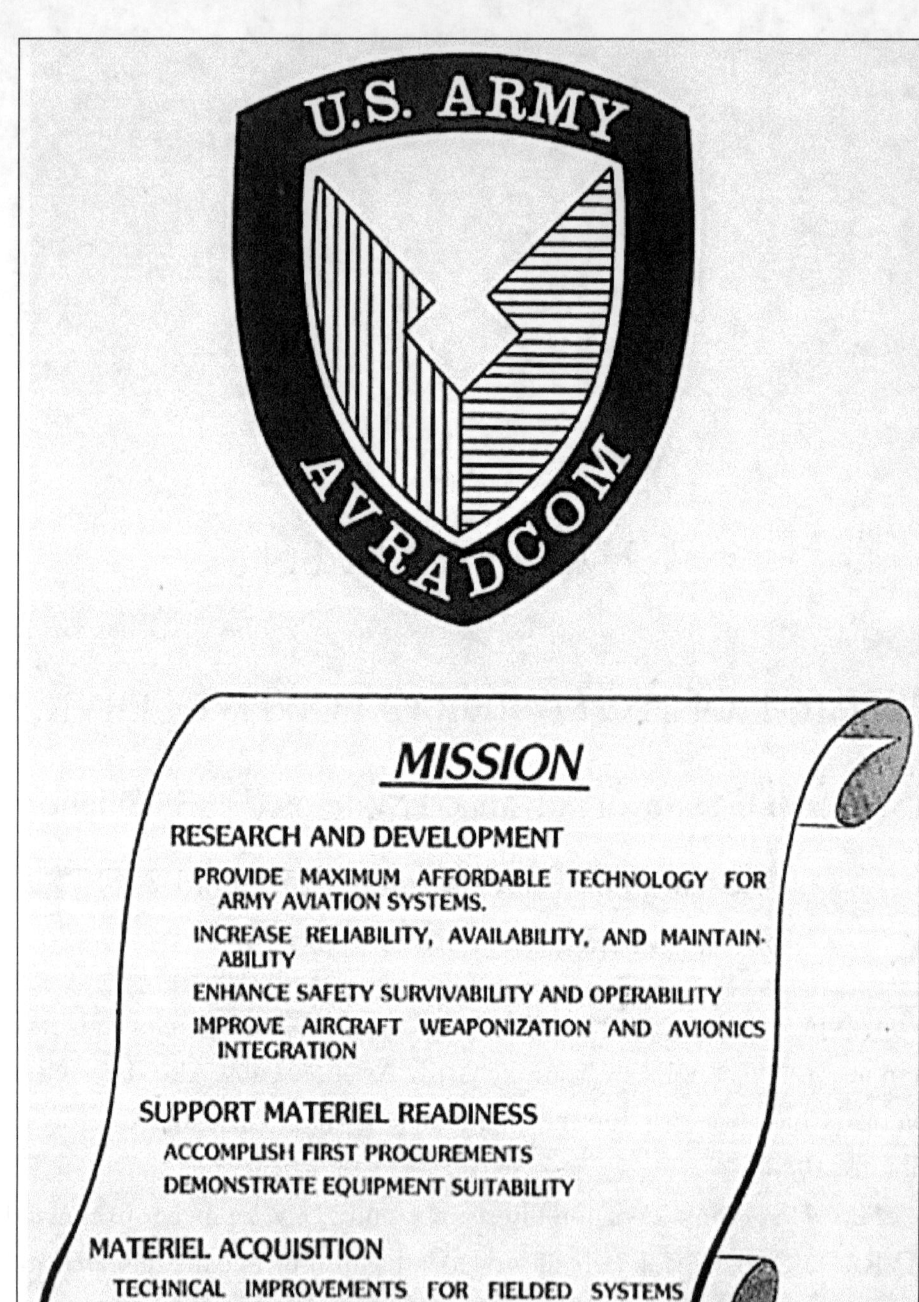

Figure 140. AVRADCOM Insignia and Mission w/Research & Development Focus

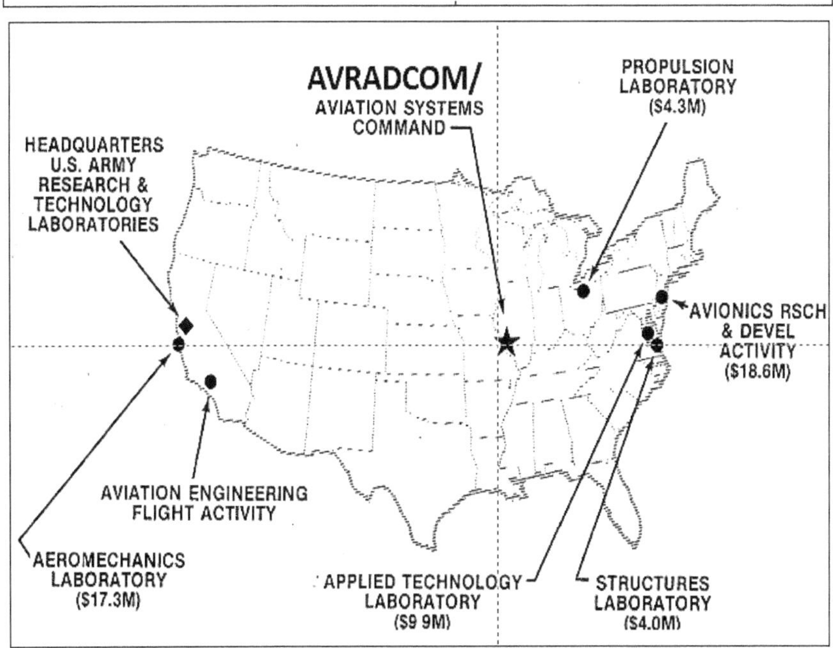

Figure 141A. MG Stevens with Four Key AVRADCOM Directors. Figure 141B. AVRADCOM HQs & Labs

In December 1981, Dean Borgman left the government and his position as the Director of Advanced Systems (DAS) to take a senior position as Director for Research and Development for the McDonnell Douglas Helicopter Company (MDHC). I was then selected as a Senior Executive Servant (SES), as the Director of DAS, and the AVRADCOM Associate Technical Director for Science and Technology (S&T) from 1981-84. This included overseeing and coordinating AVRADCOM's Research and Technology Labs (RTLs), co-located with the NASA Labs at Ames Research Center, Langley Research Center, and Lewis Research Center. Also, this included the Army Aviation Applied Technology Directorate (AATD) at Fort Eustis, VA, and the Avionics Research and Development Activity (AVRADA) at Fort Monmouth, NJ.

The Light Helicopter Experimental (LHX) program was initiated as a 1980s United States Army helicopter procurement project to replace the AH-1 Cobra, UH-1 Iroquois, and OH-58 Kiowa /OH-6 Cayuse helicopters.

An Army Aviation Mission Area Analysis (AAMAA), completed in 1982, identified deficiencies in current materiel and doctrine that needed to be addressed to meet the requirements of the new Air Land Battle Doctrine, Figure 142. Air Land Battle was the Army's doctrine during the First Persian Gulf War. It not only achieved widespread acceptance throughout the Army but was widely heralded both within and outside the Army as revolutionary to the American way of war and as the best fighting doctrine that this nation had produced.

The AAMAA study concluded that the current fleet of Army light helicopters would be unsupportable and non-survivable on the future battlefield. The Army then planned to develop and buy 2,000 Scout Attack Systems (SCATS) and 2,500 light utility helicopters. The specific material deficiencies were to be resolved through the LHX program. In 1981, a new Under Secretary of the Army, Secretary James R. Ambrose, was appointed by President Reagan, and he served until 1988. He wanted to expedite the LHX Program for an Initial Operations Capability (IOC) in 1994, which was years faster than the Army planned IOC. This acceleration was disruptive and unrealistic based on the necessary technologies needed for development, e.g., such as the Advanced Technology Engine (ATE) which became T-800 engine.

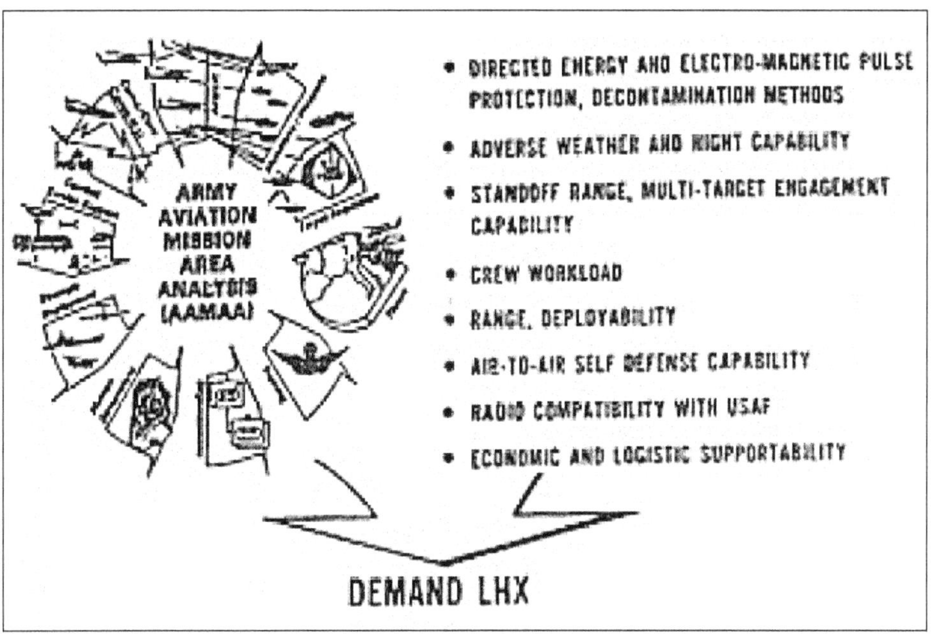

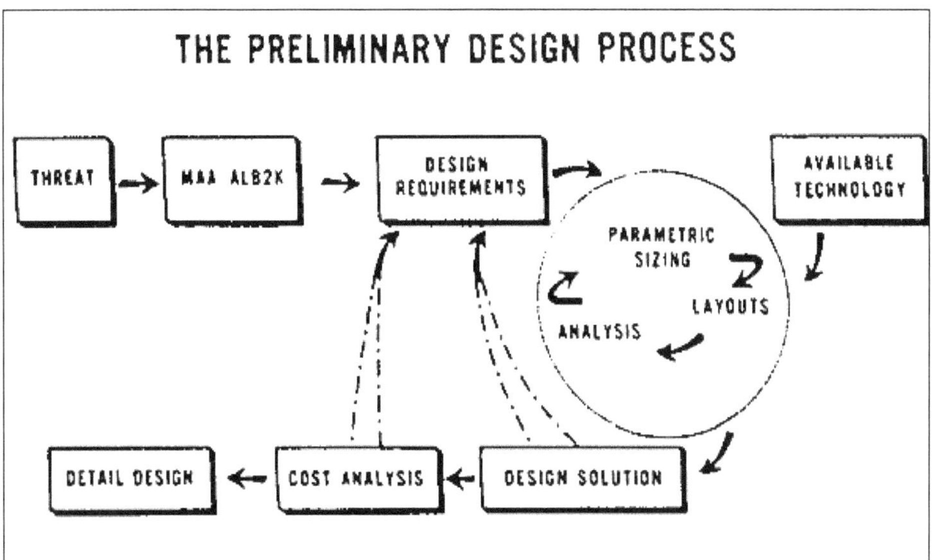

Figure 142. The AAMAA set the Demand for LHX and Preliminary Design Process

The Successful Concept Formulation Approach (CFA) used for UTTAS and AAH Development in the 1960s, 1970s, and 1980s was initiated. The current movement to JCIDS for FLRAA and FARA has been less effective and doesn't adequately include necessary timely user-developer interaction.

In the case of the LHX, the CFA was stretched out too long, due to uncertainty of it being an advanced concept, as opposed to a conventional helicopter; whether a single-pilot vs. two pilots. It was also due to lack of funding which kept it in the Dem/Val Program too long.

- Trade-Off Determination (TOD) (Material Developer Leads w/Science & Technology (S&T) baseline concept & Industry Concepts)
- Prepared by Material Developer to identify the technologies available and several technology alternatives. Usually, pre-design contracts were awarded to potential industry competitors. Government engineers/designers provide their own conceptual design assessments, as well as reviewing the industry proposed conceptual designs.
- Trade-Off Analysis (TOA) Combat Developer Leads w/Material Developer/Industry support.
- Prepared by the Combat Developer (usually the Training and Doctrine Command (TRADOC) organization) to assess technology alternatives from the TOD, and to determine whether a material solution was needed.
- Best Technical Approach (BTA) (Material Developer Leads with Combat Developer Support) Taking the TOA and industry competitors conceptual designs, the Material Developer determines the recommended BTA which was then provided to the Combat Developer for conducting a Cost and Operational Effectiveness Analysis (COEA).
- COEA w (Combat Developers/Cost Analysis Independent Group (CAIG) Conducted by the Combat Developer with the assistance of TRADOC Operational Research/Systems Analysis

(ORSA) community and modeling & simulation expertise. The COEA determined the best alternative for recommendation for the DoD Acquisition Milestone A decision.

SUMMARY OF THE LIGHT HELICOPTER EXPERIMENTAL (LHX) DEVELOPMENT

The Light Helicopter Experimental (LHX) combat helicopter program was initiated in 1983 to replace the Army's rapidly aging fleet of UH-1 Iroquois, OH58A Kiowa, OH-6A Cayuse, and AH-1 Cobra light utility and attack helicopters. At the time, as many as 6,000 helicopters were envisioned. The Scout/Attack (SCAT) was the design driver; the utility was to be derived from the SCAT. This was a reversal from the UH-1 to the AH-1 in the 1960s. To fulfill the missions undertaken by the existing fleet, different LHX models were to be equipped with a large variety of new technologies and mission equipment packages (MEPs). A light, survivable, lethal, armed reconnaissance helicopter, the LHX combat helicopter was to have greater capabilities, new sensors, millimeter wave radar, and other advanced features in a high visibility cockpit.

Maximum use was to be made of computer aided design of the LHX and systems as an integrated whole. A Scout/Attack version would be configured as a helicopter which can be tailored to perform both light attack and armed reconnaissance missions. The Scout/Attack version would replace the aging AH-l, 0H-58A/C and OH-6A helicopters. It would complement the OH-58D.

The SCAT was the more sophisticated of the two, and it was to be a single- seat helicopter whose armament would be included. Armament would include Hellfire antitank missiles, air-to-air missiles, and a gun system.

The utility version of LHX would have extensive commonality with the Scout/Attack version to include the same dynamic components (engine, transmission, rotor, etc.), and many common subsystems and mission equipment. The utility version would transport troops, unit commanders, and cargo. It would complement the UH-60 squad-carrying troop assault helicopter and replace the UH-1 Huey helicopter. The utility version would have two seats; it would carry air-to-air missiles for self-defense but would not have the target acquisition equipment of the SCAT. The Army planned to buy 2,000 SCATS and 2,500 utility helicopters. The Army had identified the need for the LHX in its Army Aviation Mission Area Analysis (AAMAA), completed in January 1982. An Army Aviation Systems Program Review in March 1982 endorsed the Mission Area Analysis and made a recommendation to replace portions of the current fleet of Army helicopters with the LHX. A LHX special working group was formed in January 1983 to develop the framework of the LHX Program. Six Army organizations were represented in the special working group.

The Department of the Army approved the Operational and Organizational Plan and the Justification for a Major Systems New Start in May 1983 and included the necessary LHX program funding in its fiscal year 1985 budget and out year funding profile. During the concept exploration phase, the acquisition strategy was developed, system alternatives were proposed and examined, and the material requirements document was refined to support the subsequent acquisition phases.

To support the Army's concept exploration activities, competitive preliminary design contracts were awarded in September 1983 to the four major helicopter firms—Bell Helicopter Textron, Sikorsky Aircraft, Boeing Vertol, and Hughes Helicopter to define specific aircraft system configurations (point designs). The four proposed LHX concepts are illustrated in Figure 143.

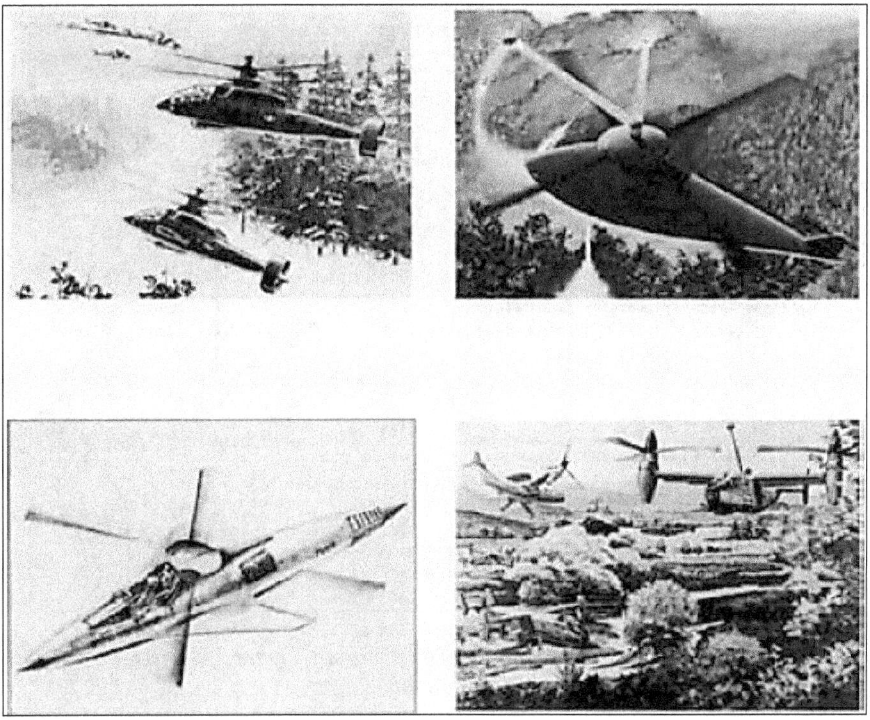

Figure 143. Proposed Four Industry LHX Concepts

From Upper Left Clockwise: Sikorsky, Boeing, Bell, and MDHC (10). The LHX Concept Formulation effort led to considerable Science & Technology (S&T) funding support, as illustrated in Figure 144A, was available. It also included funding the Advanced Rotorcraft Technology Integration (ARTI) Program, Figure 144B.

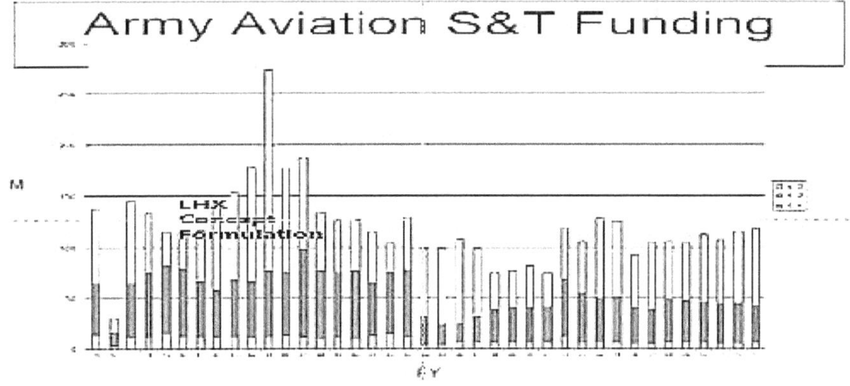

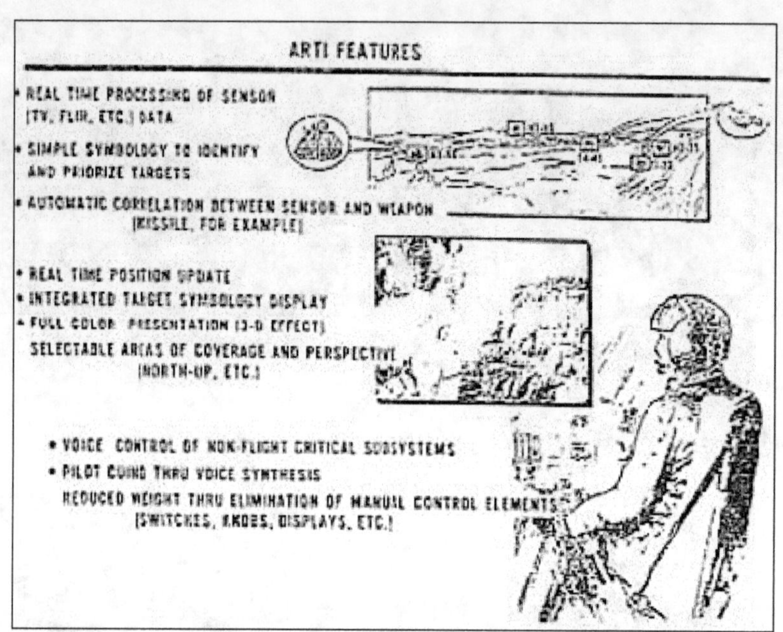

Figure 144A. Rise in S&T Funding for LHX Program. Figure 144B. ARTI Program Elements

Contracts were awarded in December 1983 for the Advanced Rotorcraft Technology Integration (ARTI) Science and Technology (S&T) Program to define the advanced/integrated cockpit design and architecture. It was used to demonstrate the feasibility of a single-pilot LHX Scout/Attack through full-mission simulations. These contracts were awarded to the four major helicopter firms and to IBM.

In parallel to the LHX Concept Formulation Effort a multi-service Joint Vertical Experimental (JVX) Technical Assessment (JTA) was being conducted at the same time at the NASA Ames Research Center, led by Charlie Crawford, Director of AVRADCOM's Development and Qualification Directorate.

However, the Army leaders decided that the required $5B was too much for the Army to continue leading the JVX development for its Special Electronics Mission Aircraft (SEMA) aircraft, which were usually modifications to commercial aircraft. The US Navy and Marine Corps took over the leadership, while the Army's JVX development leadership sought to push a tiltrotor solution for the LHX. In 1984 the Bell Advanced Tiltrotor (BAT) was proposed as a lightweight advanced, single seat, tiltrotor aircraft proposed by Bell in response to the US Army's LHX Program, Figures 100A and 100B. While the BAT had good capabilities, its lateral size was not good for tight landing zones (LZs). Also, its weight and cost would exceed conventional helicopters.

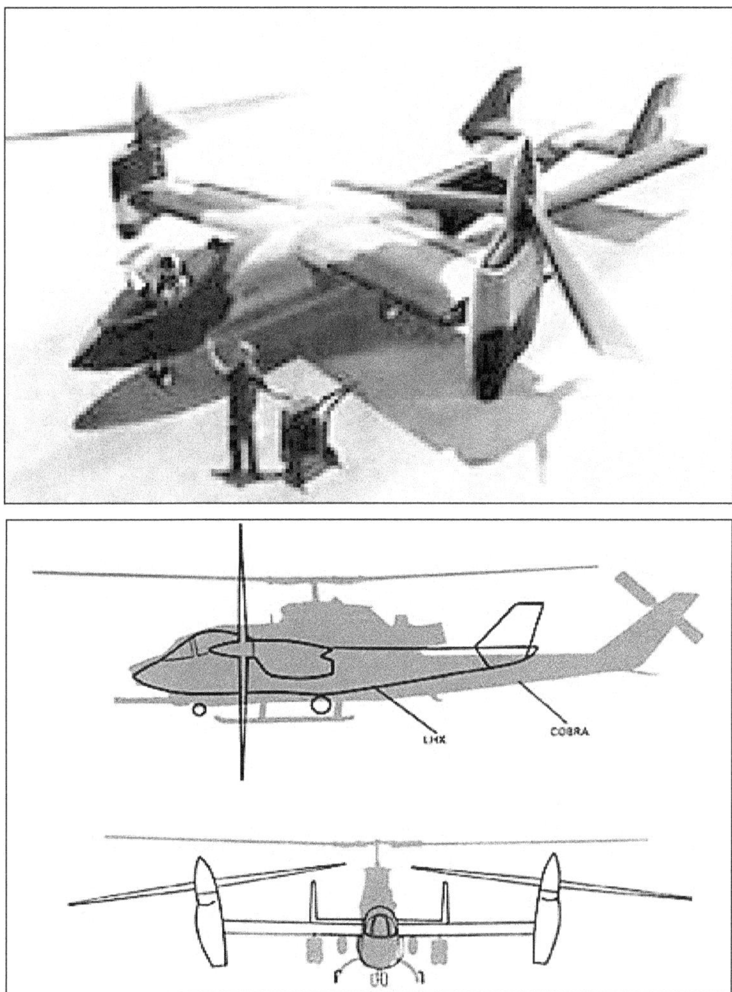

Figure 145A. Bell BAT Design Size. Figure 145B. Bell BAT Lateral Size vs. AH-1 Cobra

Development of the BAT was abandoned due to new LHX requirements, especially in the matter of weight (LHX had to be under 3150kg) and its lateral size. The BAT was a small aircraft, with a butterfly tail unit. In June 1985, Sikorsky and Boeing teamed to begin studies in the Army's Light Helicopter Experimental (LHX) competition. The Sikorsky (S75) Advanced Composite Airframe Program (ACAP) was an all-composite Sikorsky early LHX proof of concept aircraft, as illustrated in flight in 1984, Figure 146.

Figure 146. Sikorsky S-75 ACAP in Flight Test

Composite materials were used to replace metal to provide greater strength, lighter weight, lower manufacturing costs, and reduce maintenance costs. The Advanced Composite Airframe Program (ACAP) was undertaken to demonstrate the advantages of the application of advanced composite materials and structural design concepts to the airframe structure on helicopters designed to stringent military requirements.

The primary goals of the program were the reduction of airframe production costs and airframe weight by 17 and 22 percent respectively. The ACAP effort consisted of a preliminary design phase, detail design, and design support testing, full-scale fabrication, laboratory testing, and a ground flight test demonstration.

With the completion of the flight test demonstration programs, follow-on efforts were initiated to more fully evaluate a variety of military characteristics of the composite airframe structures developed under the original ACAP advanced development contracts.

NOTE

The author, Dr. Daniel P. Schrage, left his position as the Director of the AVRADCOM Directorate of Advanced Systems (DAS) at the end of 1983 to accept a faculty position at Georgia Tech in the School of Aerospace Engineering. The faculty position was as the Rotorcraft Design Professor and Associate Director of the Army sponsored Center of Excellence in Rotary Wing Technology (CERWAT). However, I continued to be involved with the LHX/LH/RAH-66 Programs illustrated below, as a consultant to the Army and industry. His assessment will be continued in this Chapter.

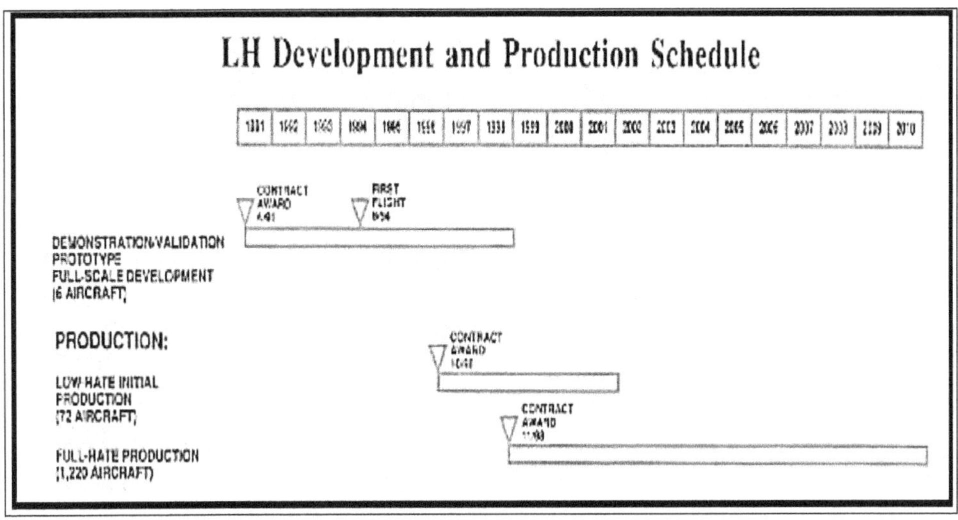

The LHX speed requirement was not defined early on. However, the speed issue was resolved when the Army decided to develop a conventional LHX helicopter. The Army Chief of Staff decided on March 4, 1985, to develop a conventional helicopter and eliminate other design concepts previously considered.

The principal factor prompting this decision was the risk involved in developing the other design concepts (compound, advancing blade, and tiltrotor). Other factors were cost, weight, and design needed to meet required performance capabilities.

The original concept of the LHX program was to produce a one-man helicopter that could do more than a two-man aircraft. Having a single-crew cockpit necessitates a higher level of automation than does having a two-member crew, as there must be greater interaction between sensors and more highly automated flight controls to reduce the workload of the single crew member.

However, it appeared that the LHX's design risks and sophistication are more heavily influenced by the mission requirements than by going to a single-member crew. The Sikorsky (S-76) Helicopter Advance Demonstrator of Operators Workload (SHADOW) had a single-pilot advanced cockpit grafted to its nose for the ARTI S&T Program. Figure 147.

Figure 147. Sikorsky S-76 SHADOW for LHX /RAH-66 Single-Pilot Evaluation

The purpose was to study the MANPRINT or human engineering interface between the pilot and the cockpit controls and displays. The cockpit was the prototype of a single-pilot cockpit designed for use on the prototype RAH-66 Comanche armed reconnaissance helicopter. The cockpit was designed so sensors would feed data to the pilot through helmet mounted

244

displays. The MANPRINT study determined that the single-pilot operation of the Comanche was unsafe and would result in pilot overload.

As result of this 1985 study, the Comanche was designed to be operated by a crew of two. Early program documents indicate that the Army identified the attack role (anti-armor) as the Comanche's primary mission. During development, the Army again emphasized the importance of the Comanche's attack capabilities.

For example, the Army's decision to switch to a two-seat aircraft was prompted largely because one pilot could not successfully carry out all the tasks related to the attack mission. Studies showed that available technology would not reduce the workload to an acceptable level so that a single pilot could fly the aircraft in combat, identify targets, and fire weapons at the same time.

The LHX baseline acquisition strategy approved by senior Army officials included competition for the LHX air vehicle in both the full-scale development and production phases. Two LHX development contracts were to be awarded, leading to a fly-off to select the winning design. The prototype to be used for the fly-off was to be an air vehicle produced during a pilot production program to be incorporated into the full-scale development phase. Full production competition would be introduced no later than production lot number three. If the winner of the fly-off was a team consisting of two firms capable of producing the LHX, these two firms would compete for the production contracts. In the event that the winner of the fly-off did not consist of a team with two firms capable of producing the LHX, the loser of the fly-off would become the alternate competitive production source. The winner of the fly-off would be responsible for qualifying the loser to produce the winning design for full production competition by lot number three.

By 1986, affordability considerations had caused the Department of the Army to significantly reduce its planned LHX research and development funding amounts in the Army's fiscal years 1987 to 1991 funding guidance submitted to DoD. This reduced budget profile did not provide adequate funds to implement the baseline acquisition strategy, which was to carry two

LHX air vehicle developers through the entire full-scale development phase, have a fly-off, and select the best design for production.

On July 19, 1985, Aviation Systems Command (AVSCOM) awarded two full-scale development contracts for engine development. Both contracts were awarded under teaming arrangements, one to the AVCO Lycoming/Pratt & Whitney team for $240 million, and the second to the Garrett/Allison team (Garrett Turbine Engine Company/Allison Gas Turbine Division of General Motors) for $264 million.

Each team was to compete for its design against the other during early development up to a preliminary flight rating. At that time, the Army would select one team— based on bench test demonstrations— to complete the engine's development. At the completion of the development phase, both engine manufacturers of the surviving team are to be capable of producing production engines. The Army then intended to require both engine manufacturers of the surviving team to produce approximately equal quantities of engines. The Army established unit flyaway cost goals for the LHX very early in the program. These goals were $6 million for the Scout/Attack (SCAT) version and $4 million for the utility version (in fiscal year 1984 dollars), or a fleet average of $6.3 million. The Army held fast to the unit cost goals, despite early indications that the goals could not be met without sacrificing capability.

The Army had conducted several requirements scrubs, including deletion of substantial amounts of mission equipment from the utility version, to keep estimates within the cost goals. The Army estimates were based on production buys of 480 aircraft per year, which preliminary assessments indicated were not affordable. Lower production rates would increase unit costs. At 480 aircraft per year, unit flyaway costs for the SCAT version in 1984 dollars are estimated at $6.02 million per aircraft; at 360 per year, $6.09 million per aircraft; at 240 per year, $6.42 million per aircraft.

As of February 1987, unit flyaway costs had increased for the utility version. SCAT unit costs increased primarily because of the need to increase aircraft weight to satisfy mission requirements. The overall reduced number of aircraft, and particularly the procurement of fewer SCATS, had also increased SCAT unit costs. The utility version's unit cost increase was due mainly to the Army's decision to outfit it with the same mission equipment as the SCAT, with some exceptions.

Research and development cost estimates remained fairly constant through 1986 since the first estimates in 1983. The 1984 constant dollar estimate has remained at $2.6 billion. The escalated research and development estimate rose from $3.1 billion in 1983 to $3.2 billion as of 1986. The increase was due primarily to delays in starting full-scale development because funds obtained for the Advanced Rotorcraft Technology Integration (ARTI) program were less than expected. By 1987, Research and Development costs had increased to $3.8 billion, due mainly to changes in the acquisition strategy suggested by DOD's Defense Science Board, which extended the LHX team competition through full-scale development to include a competitive flight test. The earlier strategy had called for selecting one contractor before beginning prototype fabrication. Also contributing to cost increases were the Army's decisions to build a two-seat SCAT prototype in addition to a one-seat prototype.

By 1987, the Army learned that the performance necessary from the optical sensors to automate the targeting function for the single pilot fully, might not be available for application to the initial LHX helicopters. The Army considered developing a radar sensor to complement the optical sensors available for the LHX to achieve full automation but determined the additional equipment to be too costly and heavy for inclusion on the initial LHX helicopters.

These factors, combined with an assessment by the Defense Science Board, raised some doubts about achieving the single-seat objective and have led to the addition of a two-seat version to the development program. In addition to automated targeting technologies, other areas where performance expectations had been lowered include the quality of the visual displays, digital map, automatic hover hold, and aircraft survivability equipment. Performance reductions reflect tradeoffs due to cost, weight, technical risk, or a combination of these. The original weight goal for the one-seat SCAT was 8,500 pounds. By late 1987, the goal for the same version was 9,500 pounds. A two-seat version would weigh more.

Summary Evolution from the LHX to the RAH-66 Comanche

- In 1982, the US Army initiated the Light Helicopter Experimental (LHX) program with the aim of producing a replacement for several existing rotorcraft, including the UH-1, AH-1, OH-6, and OH-58 helicopters.

- It was another six years until, in 1988, a formal request for proposal (RFP) was issued to various manufacturers, in time which the requirement had evolved into a battlefield reconnaissance helicopter.

- In October 1988, the Army announced that two teams, Boeing–Sikorsky and Bell–McDonnell Douglas, would receive contracts to further develop concepts.

- These were the Demonstration/Valuation Contracts. In the 1990s, the program's name changed from LHX to *Light Helicopter* (LH). In April 1991, the Army awarded the Boeing–Sikorsky team a $2.8 billion contract to complete six prototypes. Later that month, the helicopter received its official designation of *RAH-66 Comanche*

RAH-66 Comanche

Figure 148. RAH-66 Comanche's First Flight on 4 January 1996

LESSONS LEARNED FROM LHX/RAH66 COMANCHE DEVELOPMENT

RAH 66 Comanche Prototype Testing and Air Vehicle Lessons Learned

249

- The Army's decision to limit LHX alternatives to single main rotor helicopters was based on their belief it would be a lower risk than advanced rotorcraft concepts, such as tiltrotor, single main rotor compound, or coaxial compound, as illustrated in Figure 98.

- During November 1993, assembly of the first prototype commenced at Sikorsky's facility in Stratford, CT, and at Boeing's manufacturing plant in Philadelphia, PA.

- In December 1994, the DoD reduced the number of planned prototypes to two, as the services shifted budgets to pay for increased troop salaries.

- On 25 May 1995, the first Comanche prototype was formally rolled out at Sikorsky's production facility, after which it was transferred to West Palm Beach, Florida to commence flight testing activities.

- On 4 January 1996, the prototype Comanche, flown by test pilots Bob Gradle and Rus Stiles, performed its 39-minute maiden flight.

- On 30 March 1999, the second prototype conducted its first flight before joining the flight test program shortly thereafter.

- During testing, the Comanche was recorded as having attained a cruise speed of 162 knots (186 mph; 300 km/h), as well as having achieved a "dash speed" of 172 knots (198 mph; 319 km/h). Perhaps one of its more atypical capabilities was the demonstrated ability to perform a 180° turn in under five seconds.

Following the Cancellation of the Comanche Program in 2004, the PM Required All Program Office Integrated Product Teams (IPT's) to Develop Lessons Learned. The Comanche Government Team provided a series of Lessons Learned in the public domain, 2005-2006. They also consisted of RAH-66 Comanche Workshops and Technology Transfer as part of The University of Alabama in Huntsville (UAH) Professional Development. The following descriptions and results come from this material along with my comments. (Ref. 5).

A Comanche Advanced Subsystem Overview is provided in Figure 148. The Comanche Subsystem Technology Incorporation is in Figure 149.

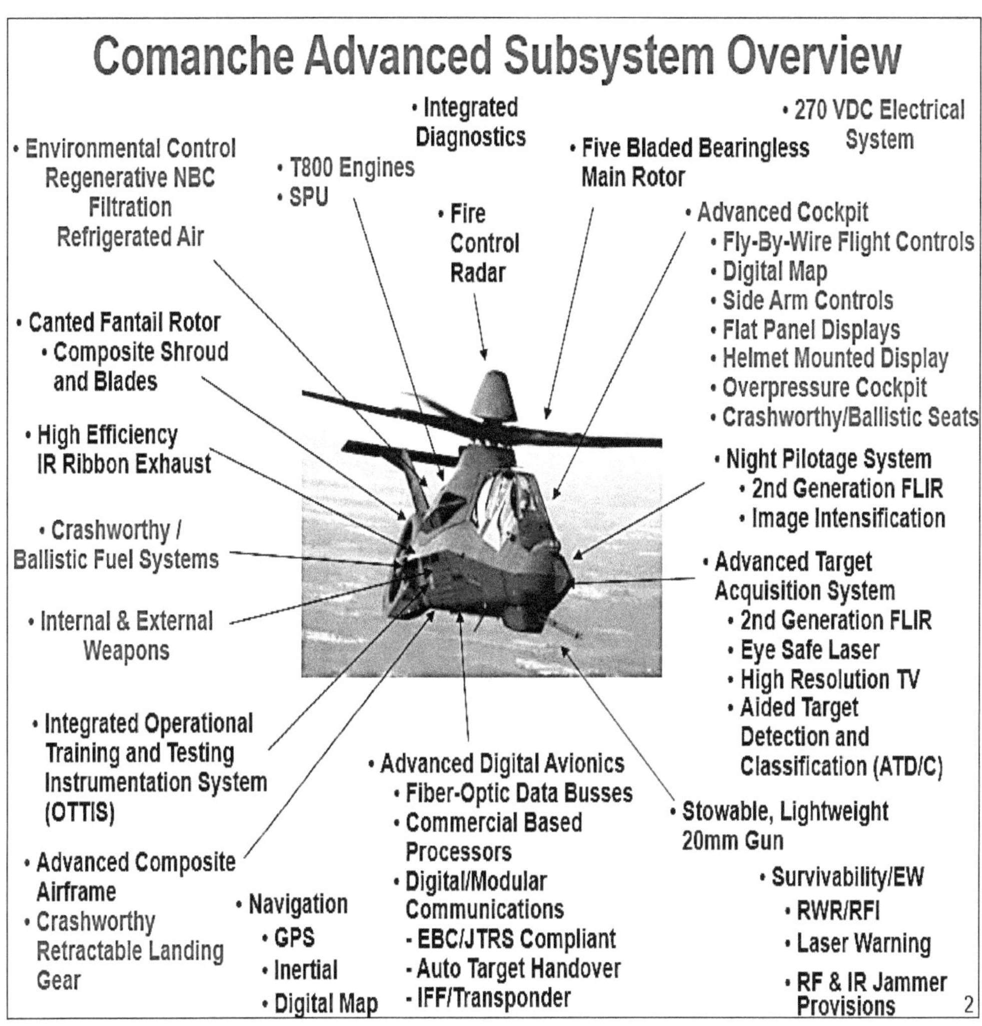

Figure 149 Comanche Advanced Subsystems

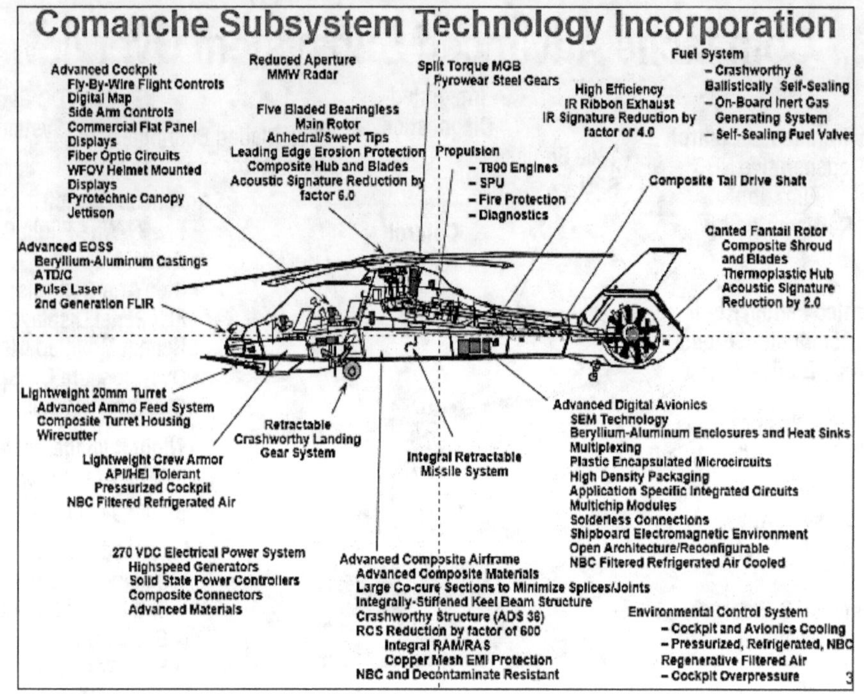

Figure 150. Comanche's Substantial Air Vehicle and Subsystem Technologies

The Air Vehicle Dynamic Systems for the RAH-66 Comanche are in Figure 151. (Ref. 3)

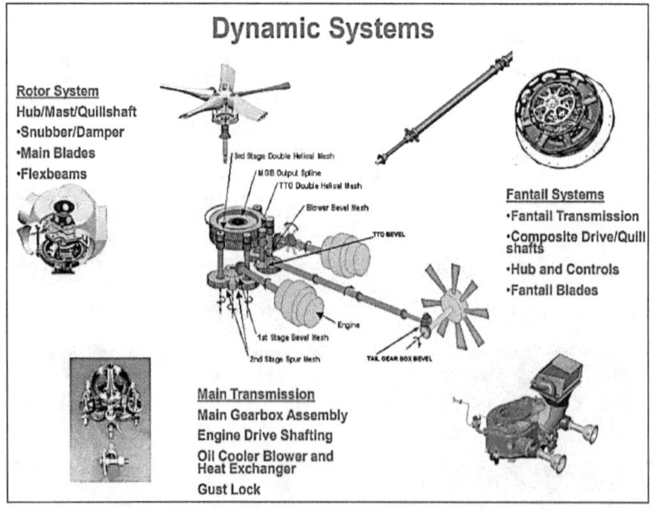

Figure 151. The Air Vehicle Dynamics Systems for the Comanche

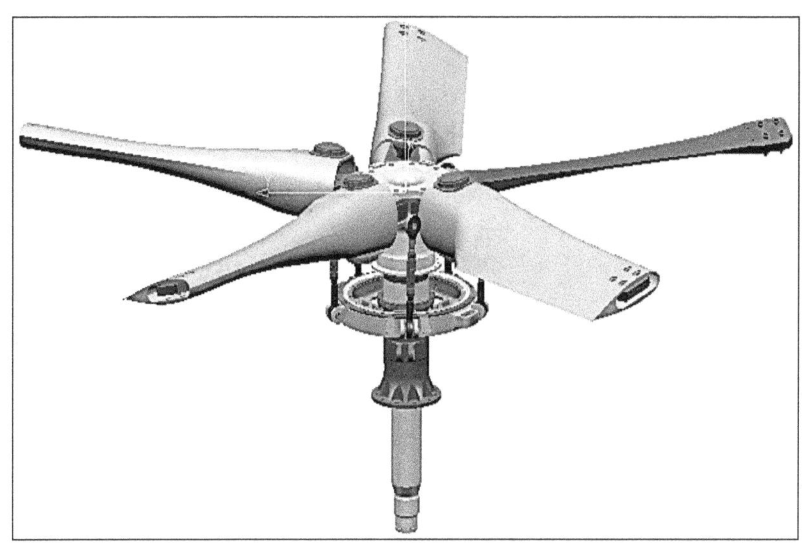

Figure 152. Comanche Five Bladed Bearing Less Main Rotor was its Achille's Heel

These were the comments I made as a consultant to Sikorsky in the mid-late 1980s on their selection of a Bearing less Main Rotor (BMR):

- Sikorsky's engineering experience with soft in-plane BMRs was very limited. Their participation in the Army's Integrated Technology Rotor (ITR) Program was with Evan Fradenburgh's gimbal rotor instead of a soft in-plane BMR. The only other Sikorsky experience I knew of was on their UTTAS Tail Rotor Test, where Bill Paul, the test engineer, ran into an instability in a UTRC wind tunnel test. Boeing had a good BMR design engineer, Pete Dixson, who had worked under Frank Harris. However, I don't believe he was involved at all in the Comanche. Frank Harris was then with Bell Helicopter Textron (BHT).

- I recommended the German manufacturer, Messerschmitt-IBoelkow-Blohm (MBB), based on my experience with their expertise in hinge less and bearing less rotors, that they should be on the First Team. I was good friends with Gunther Reichert, MBB Chief Engineer.

- Sikorsky put them on the First Team, but then dropped them after the First Team won the Army Demonstration Validation (DEMVAL) Full-Scale Development Program. Boeing had BMR expertise not utilized.

I resigned as Consultant for the First Team and became a Consultant for the Super Team. This was based on my association with COL(Ret) Bud Patnode and Dean Borgman, who had become President of McDonnell Douglas Helicopter Company (MDHC).

Additional insights I made at the Government First Team Weight Review, Jupiter, FL 2003.

Dr. Bill Lewis had been named in 2003 as the Army's Comanche Chief Engineer, and COL Cantor, Comanche Project Manager, a former student of mine at Georgia Tech. Bill and I had worked together at AVRADCOM, and I was his PhD Adviser at Georgia Tech. The weight of Comanche Block III was quickly approaching 200 lb over Spec Weight. In an effort to bring more focus on weight, the RAH-66 Comanche Government Program Weights Engineer, Frank Kirsch, a former colleague of mine at AVSCOM set up weekly Weight Review Board meetings. My comments at the first kickoff Review were limited after not having been involved with the First Team Comanche Design for a few years.

I had been a consultant for the Super Team in the mid to late 1980s. The Army RAH-66 Technical Manager, Mike Richey, kicked off the Weight Review in Jupiter by saying that "no company has successfully designed a hinge less or bearing less rotor with equivalent hinge offset of 10% or more." He was a good Mission Equipment Package (MEP) engineer who had worked with me on the AHIP/OH-58D Program. I asked him if he had ever heard of the Messerschmitt-Boelkow-Blohm (MBB) BO-105 or BK 117 helicopters, which have equivalent hinge offset well over 10 percent. He did not, but said he was concentrating on MEP as he felt Sikorsky and Boeing had the air vehicle covered, although the Army participation was limited. My other comments based on my previous experiences as a government technical expert were as follows:

- The Government Technical Team Participation in helping solve this problem was mostly missing. In the past, I had been the key government technical expert on rotor systems, such as the AH-1 IMRB, AAH, UTTAS, CH-47D, and the OH-58D, as discussed in other chapters in this book.

- Air Resonance and High Hub Loads were Comanche's biggest Air Vehicle Showstoppers, and it didn't appear that government personnel and First Team members understood the severity of the problem or how to solve it.

- Main Rotor System redesign resulted in the biggest weight increase in EMD going to Production, as shown in Figure 153. (Ref. 3)

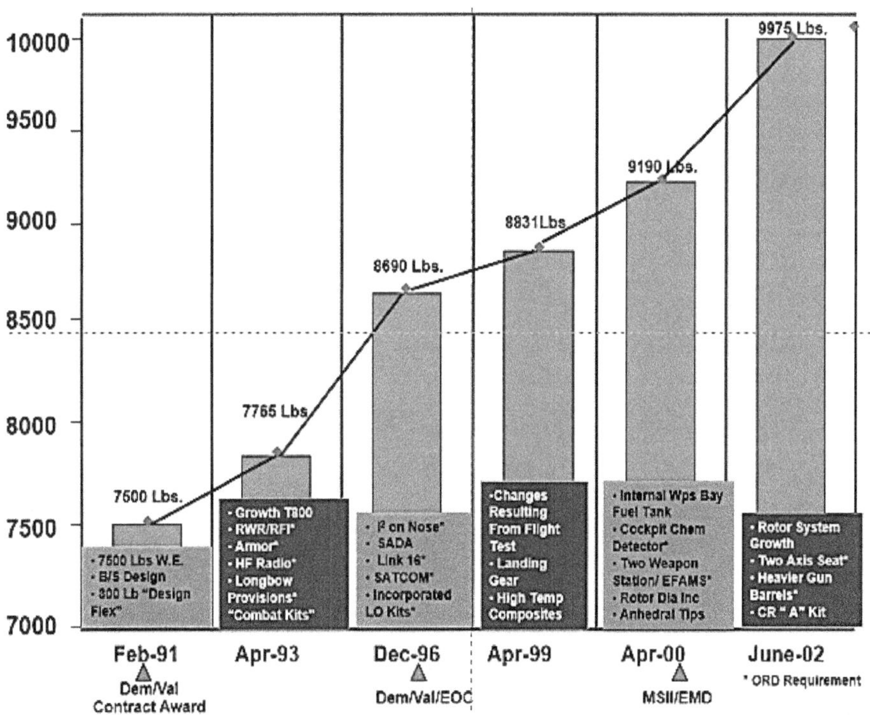

Figure 153. Rotor System Growth on RAH-66 resulting in Empty Weight Increase

The planned Production Comanche Main Rotor would make the design case even worse. The effective hinge offset was increased from 9-10 percent to 12 percent, as illustrated in Figure 154(Ref. 3)

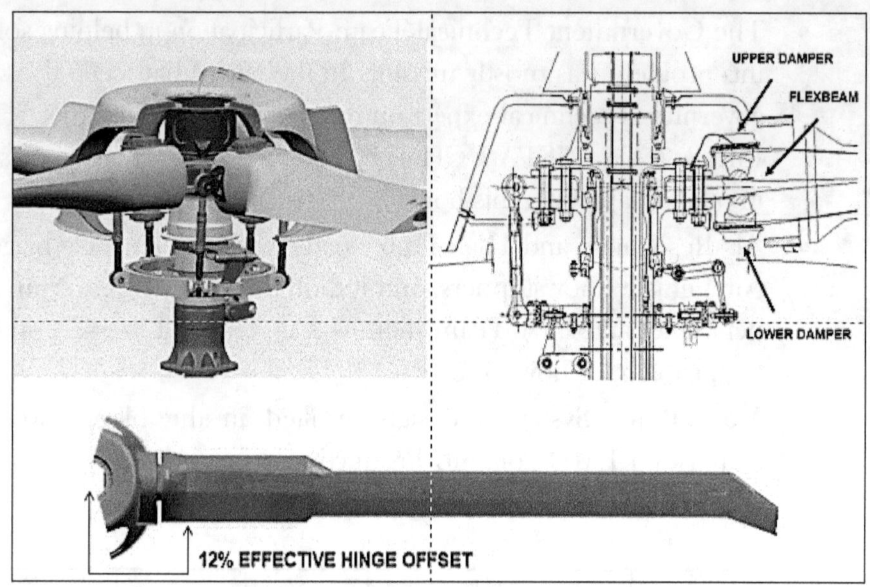

Figure 154. Planned Production Main Rotor System with 12% Effective Hinge Offset

Additional RAH-66 Main Rotor Blade Engineering Manufacturing and Development (EMD) Changes are identified in Figure 155.

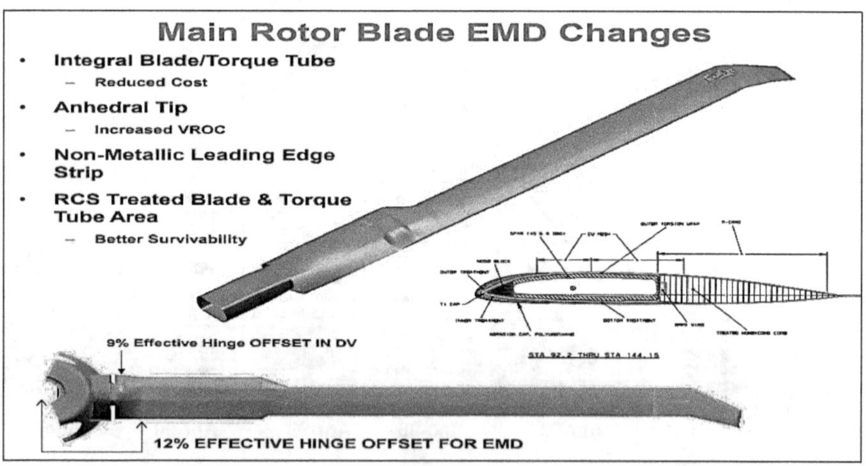

Figure 155. RAH-66 EMD and Production Main Rotor Changes.

Comanche Main Rotor Weight and Loads Growth with a larger Snubber Damper to hopefully increase its life from 50 hrs. to 1200 hrs. is shown in Figure 156. (Ref. 3)

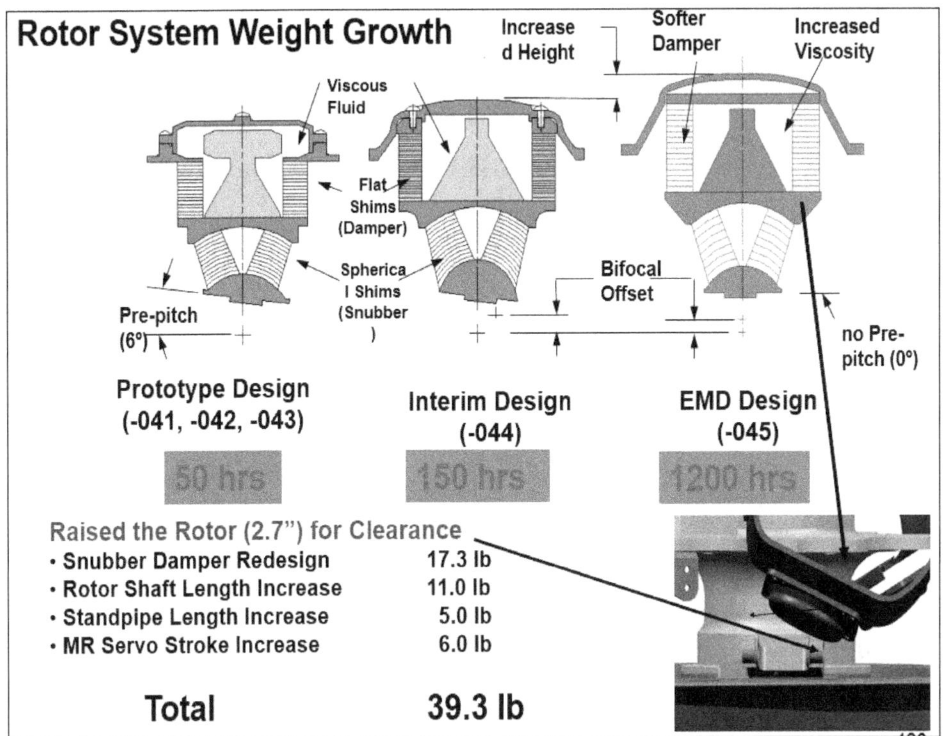

Figure 156. Comanche Main Rotor Weight and Life Growth for Snubber Damper

It was clear that the First Team didn't have the expertise or knowledge to solve the aeroelastic Bearing less Main Rotor (BMR) problems facing the RAH-66 Comanche for Production. Dr. Jerry Higman, who got his MS degree at the Naval Post Graduate School (NPGS), and his PhD in Rotorcraft Aeroelasticity and Higher Harmonic Control (HHC) with me at Georgia Tech in Aerospace Engineering, had a solution. Following his graduation, Jerry, and his wife, who was Japanese, moved to Japan to be close to her family. He worked for Fuji Heavy Industries (FHI) designing a new soft in-plane hinge less rotor system for the FHI Bell 412 helicopter. While there, he worked with Lord Kinematics on a subsystem snubber damper for the FHI 412 main rotor. This snubber damper was designed to suppress a potential Air Resonance lead-lag instability problem.

However, Jerry and his family returned to the USA. This was partly because he discovered that promotion to higher management positions at

FHI and other Japanese companies was unlikely if you weren't of Japanese descent. He then moved to Jupiter, FL, and was working for the Army on flight simulation support for Comanche flight testing in West Palm Beach, FL. He was also a member of the RAH-66 Comanche Main Rotor Integrated Product Team (IPT). The following is a description of Dr. Higman's RAH 66 Main Rotor Damper Design Snubber Solution. He provided this information to the Army and the First Team for assessment, and it was included in the First Team planned RAH-66 Production Design.

His following summary is in the public domain after Comanche Cancelation. It was part of a Lessons Learned Package for the RAH-66 solution in July 2006. Jerry's summary follows:

Main Rotor Damper Design

Comanche IPT: Main Rotor IPT

Author: Dr. Jerry Higman

Executive Summary: Advanced technology dampers are required to stabilize the Comanche's coupled main rotor/airframe modes and prevent helicopter air and ground resonance. The Comanche program team initially employed dampers that used a snubber concept to allow for the relative motion between the flex beam and torque tube.

In actual practice, the snubber disallowed or negated the required relative motion between the damping plate and the upper section of the damper. The net effect was little relative damper motion and a greatly diminished damping force. Upon recommendation by the US Army, a radically different fluid-elastic damper design during the EMD phase was implemented in which the snubber used a centering post to connect the top and bottom damper. With the centering post in place, both dampers are forced to move in unison with an optimal relative motion, thereby obtaining the maximum potential of the dampers. The flex beam was modified to accommodate the EMD damper with the placement of an elliptical hole at the 6 percent radial location through which the centering post passes.

Background: In 1992-1993, a detailed analytical study was performed by Boeing and Sikorsky to determine the aeromechanical stability characteristics of the Comanche. From this study, it was found that the proposed "all" elastomeric dampers (no fluid) would provide insufficient damping for the coupled rotor/fuselage modes for the conditions of low gross weight, high rotor weight, and extreme temperatures.

From this study, the concept of the fluid-elastic damper (with fluid) with a snubber was developed, which increased the dynamic damping margin by a factor between 4 and 5 over the "all" elastomer damper. Because this concept was a new application to the bearing less type rotor, there was no known measure of accuracy on how effective the snubber damper would be. The DemVal dampers, depicted in Figures 157A. and 157B., from upper left clockwise, attach to the upper and lower surfaces of the flex beam using a flexible elastomeric "spherical bearing" snubber. The body of the damper and spherical bearing is an elastomeric structure made up of stacked rubber polymer plies, as shown in Figures 157A and B. Bonded between each rubber polymer layer is a metal shim made of titanium or aluminum alloy, depending on the bending and shear stiffness requirements.

The DemVal dampers experienced two problems which could not be solved in manufacturing or flight tests, and ultimately resulted in the redesign of both the dampers and the flex beam. The problems encountered were:

1. inadequate damping of the coupled rotor/fuselage system

2. the separation or debonding of the laminated elastomeric plies from the titanium shims in the spherical bearing. The debonding, as it evolved, exacerbated the already inadequate damping. Regarding problem 1, the elastomeric snubber produced undesirable motions and allowed the damper to move with six degrees of freedom of motion. The effect was to diminish or cancel the required relative motion between the damper housing and damper plate. The diminished relative motion reduced the effectiveness of the damper such that the available system damping was insufficient to prevent the aggravation of the regressive lag mode (RLM). Large shear and tension stresses were created on the spherical bearing, which resulted in a debonding of the elastomeric plies

Discussion

Many facets of rotorcraft engineering are highly non-predictive, nonlinear, and often stochastic in nature. The prediction of aeromechanical stability is quite difficult and cannot be reasonably understood by analysis alone or even a low level of ground testing. Having said this, a redesign of the dampers and flex beam was required to ensure adequate system damping was available for all flight conditions and that the in-plane rotor modes were stabilized.

The primary issue of inadequate damping due to inadequate relative damper motion was eliminated with the removal of the snubber, as shown in Figures 157D. and 157E. The removal of the snubber also ended the occurrence of debonding of the elastomeric plies on the spherical bearing.

The solution to force the dampers to respond in a desired relative motion and to create a dependency between the upper and lower damper was to use a centering post. The centering post, shown in Figure 157C, forces the dampers to move in approximately equal displacements, or move as a system, thereby deriving the greatest damping efficiency from both dampers. To attach the upper and lower dampers via the centering post, an elliptical shaped hole was placed in the flex beam through which the centering post passes, Figure 157F. The centering post moves primarily in angular rotation in the lead-lag directions to accommodate the lead-lag motions of the dampers.

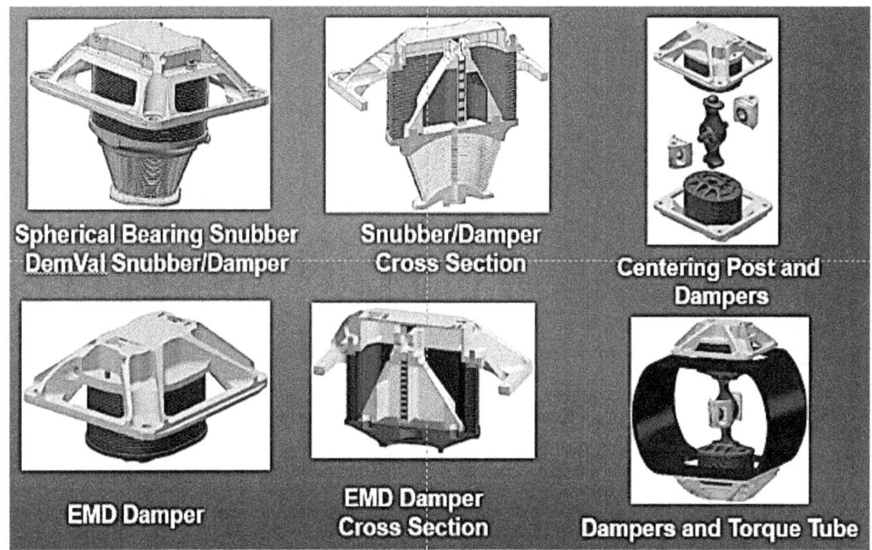

Figures 157 A-F Clockwise Evolution Snubbers/Dampers from DemVal/EMD to Production Dampers (Ref. 3)

Recommendations

Dr. Higman's lessons learned from his experience on the FHI 412 Main Rotor led to an improved damping system for the Comanche RAH-66 BMR. These lessons are as follows:

- The use of adequate analytical tools to adequately understand physical phenomena.

- To use proven methodologies/techniques developed by others.

- To understand, in detail, the loads and motions in which the component or parts will be subjected to

- Employ laboratory or ground testing to the greatest extent possible

- If required, use analytical or preliminary testing to better understand the conditions for the ultimate or fatigue tests.

- After initial test ramp-up, to test the components at the end points of the required operating cost.

Summary of RAH-66 Comanche Air Vehicle Lessons Learned for Production

The RAH-66 BMR was increased in size for EMD/Production primarily to solve the regressing lead-lag instability problem. This led to RAH-66 weight increase and higher loads from the Bearing less Main Rotor (BMR) System. This is something that the First Team needed to solve and redesign during LHX DemVal and EMD. Instead, they were doing a complete redesign, which drove up the risk, weight, and cost of the RAH 66 Production Design, which was never completed.

On 23 February 2004, the US Army announced the termination of the Comanche program, stating they had determined that the RAH-66 would require many upgrades to be viable on the battlefield and that the service would instead direct the bulk of its rotary systems funds to renovating its existing attack, utility, and reconnaissance helicopters. The Air Resonance instability hadn't been completely solved for production with the newer and larger snubber dampers, as with the RAH-66 Program.

Also, addressed in the Winter-Spring 2004 was the demonstration and use of Active Control to address the RAH-66 Air Resonance instability. The active controller block diagram used is illustrated in Figure 132, although not completely verified for production it identified a workable solution.

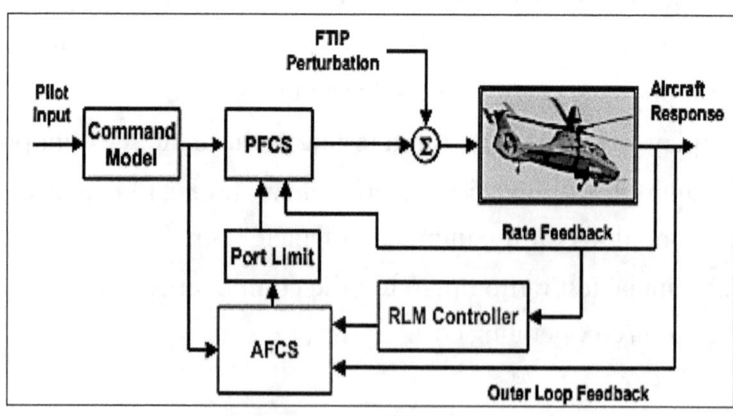

Figure 158. Active Controller for RAH-66 Air Resonance

Successful development and demonstration of this active controller mitigated somewhat the risk that undesirable roll axis oscillations would have been experienced on a production Comanche aircraft. (Ref. 3)

However, the designed Production Main Rotor System, Figures 128 and 129, also had other problems to solve. The higher loads resulted in Sikorsky seeking relief from traditional required static load factor criteria, e.g., V-n diagrams. Sikorsky had little experience in developing a single main rotor soft in-plane bearing less rotor with large hinge offset. However, Sikorsky had experience with designing, developing, and testing a coaxial rotor, the XH-59A (Ref. 21). The XH-59A helicopter was an Advancing Blade Concept (ABC) technology demonstrator developed by Sikorsky. It was successfully demonstrated for the Army/Navy/NASA. More recently, it was verified in their X2 demonstrator. Thus, it could fly at high-speed, as well as vertical take-off/land.

A derivative of it was initially proposed for LHX as a coaxial compound helicopter, Figure 117. However, the Army required LHX to be a conventional helicopter. Sikorsky successfully developed later, the X-2 Technology Demonstrator. (Ref. 6) It was also the basis for development of the Sikorsky S-97 Raider as a high-speed Future Long-Range Assault Aircraft (FLRAA) as a potential replacement for the UH-60 Black Hawk. The S-97 made its maiden flight in 2018. However, the Army recently selected the Bell V-280 tiltrotor for FLRAA because of its longer range and higher speed, which I believe was the correct solution.

A smaller X-2 derived version is the Sikorsky-Boeing SB-1 Defiant, which is the Sikorsky Aircraft and Boeing entry for the United States Army's Future Attack and Reconnaissance Aircraft (FARA) Program to replace the retired OH-58D Kiowa Warrior. Based on my previous experiences in combat and rotorcraft design and development I believe the coaxial compound helicopter may be the best FARA solution.

The last project the Sikorsky and Boeing companies teamed up for was the RAH-66 Comanche, which started in the 1980s and cost $7 billion before being canceled in 2004.

The First Team said that factors outside their control, like budget cuts, "requirement creep," and a long development period, caused problems with the Comanche and not team dysfunctionality. I believe this was partially true.

However, Comanche lessons learned also site lack of a proven Main Rotor System, an inadequate Systems Engineering approach, and a lack of productive use of Integrated Product Teams (IPTs) for Integrated Product and Process Development (IPPD). I believe IPPD and the use of IPTs, as required by the Secretary of the Army, was the right way to go.

In fact, my 1999 AHS Nikolsky Lecture, "Technology for Rotorcraft Affordability Through Integrated Product/Process Development (IPPD)," showed how to use IPPD for Concurrent Engineering, cost effectiveness, and Total Quality Management (TQM). (Ref. 4). Following my Nikolsky Lecture in 199, Dr. Ken Rosen, then Chief Engineer, Sikorsky, came up to me and said he thought the lecture was good, but it should be something industry does. I agreed whole-heartedly. However, it was something the First Team didn't understand or implement successfully for the RAH-66.

IPPD will be a foundation in my Book 3: *Development of a Graduate Program in Aerospace Systems Design.* Also, weight control and effective use of planned value was not sufficiently used or understood by the First Team.

Under the Comanche program, each company mostly built different parts of the aircraft. For the Joint Multi-Role (JMR) Demonstration Program, employees from both companies worked better together. Up to 2013, Sikorsky and partners had spent approximately $250 million on X2, Figure 113, and Raider, Figure 114. The Sikorsky-Boeing Team had confidence in the SB-1 Defiant and paid for more than half of its design costs.

On 5 December 2022, the US Army selected the rival Bell V-280 Valor as the winner of the Future Long-Range Assault Aircraft (FLRAA) Program; however, I feel Defiant may be best for FARA, Figures 132A and B.

Figure 159A. Sikorsky X-2 Demonstrator in first Flight   Figure 159B. SB-1 Defiant in Flight

Summary RAH-66 Comanche MEP Lessons Learned

RAH-66 Comanche Mission Equipment Package (MEP) Lessons Learned were largely based upon the requirements and Space Weight and Power (SWaP) without sufficient system integration. The Comanche MEP Architecture is illustrated in Figure 160. These considerations will be addressed. There was no consideration to use legacy boxes to fulfill the needs of the MEP.

The alternative form factor for RAH-66 Communications, Navigation, Identification (CNI) system was the Integrated Communications, Navigation, Identification Avionics (ICNIA). ICNIA consisted of two integrated avionics racks, two audio control units, two communication control panels, and all communication antennas.

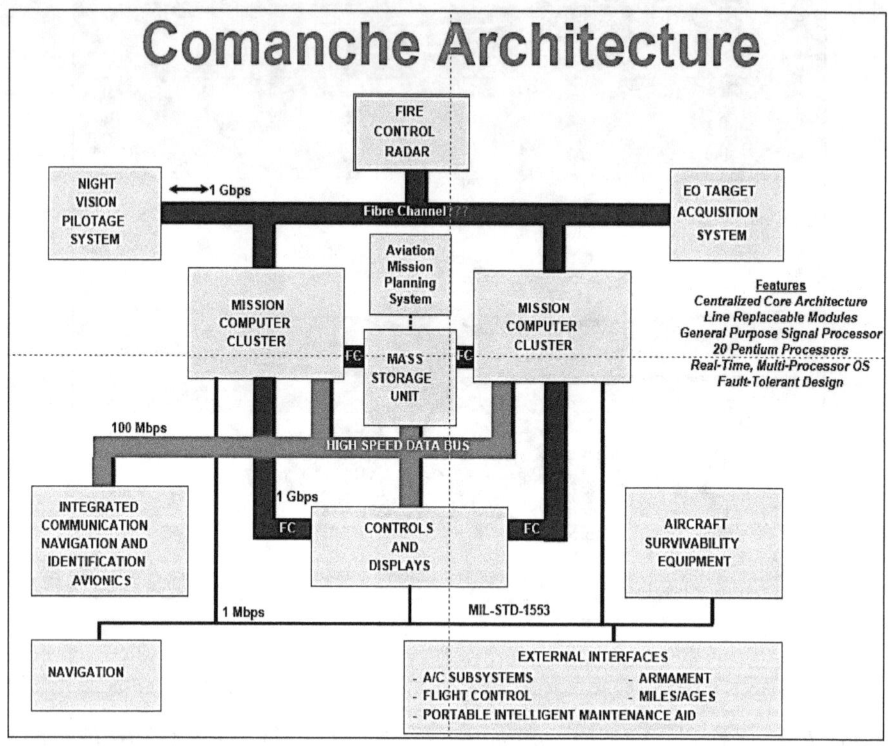

Figure 160. Comanche Architecture

The two avionics racks were populated with Standard Electronic Modules-Format E (SEM-E) modules. Figure 161. There were two each, antennas on the aircraft, for Very High Frequency-Frequency Modulation (VHF-FM), Very High Frequency-Amplitude Modulation (VHF-AM). Ultra-High Frequency (UHF-AM), and IFF (Identify Friend or Foe) (Modes 1, 2, 3A/C, and 4) were divided between upper and lower aircraft locations.

Rack Components

Figure 161. The two Comanche Avionics Racks

There was one transmit antenna and one receive antenna for SATCOM, one EPLRS, WNW, and Link-16 antenna. The Link-16 antenna was shared with the upper IFF antenna. The IFF upper antenna always had priority. The ICNIA system could receive any 4 UHF/VHF functions, 2 UHF/VHF guard channels, EPLRS, Link-16, and WNW simultaneously, provided the antenna allocation was not exceeded for that band. Figure 136 is the RAH-66 ICNIA EMD Schedule supporting LUT, ITC, and 1 – 3.

•Config	Added Functions	Added Hardware	Added Software
•Basic • 5/04 • LUT	4 chan clear voice IFF Transpond	2 Integrated Racks 2 Audio Units 2 Control Panels	Clear U/V IFF Transpond Basic infrastructure
•Digital • 8/05 • ITC	SINCGARS/HQ Tactical Internet	None	SINCGARS/Secure U/V IDM/TI Control Crypto
•Block 1 • 9/06 •IOTE	EPLRS SATCOM DAMA JTRS SCA Compliance	Updated RF LRMs External SATCOM PA	EPLRS DAMA Full Diagnostics
•Block 2 •10/07 • •	Link 16 WNW	Additional LRMs L-Band Integrated PA JTIDS Preprocessor WNW PA and Antenna	Link 16 Updated infrastructure WNW
•Block 3 • 6/08 •	IBS	None	IBS Updated infrastructure

Figure 162. Comanche ICNIA EMD Configurations

ICNIA had many external dependencies that needed management as part of its development cycle. The dependencies were usually in the form of software waveforms that came from other government agencies and provided as GFI. The two offices that were to give the software to the IPT were the JTRS JPO for radio waveforms, and the IDM APM for the IDM core software.

There were issues with obtaining the software from both offices. The JTRS JPO software was immature, and as a result, documentation was lacking. This made it difficult for the ICNIA vendor to design its hardware. This was an important issue. Documentation and software should be available before the hardware design phase.

The IDM GFI software was available in many cases, however; the company that handled exporting to the target processors did not staff properly, resulting in schedule slips. Also, since source code was not provided there was no way to help with the effort. There were also cases of dependencies at the IDM level where dependencies existed for other capabilities resulting in development delays.

Antenna placement plagued the program due to SWaP and co-site issues. These were exacerbated by the Low Observable requirements of the aircraft. There seemed to be no good solution for the problem. Directional stability problems were identified during flight tests requiring an empennage redesign. It was determined that the best solution involved placing end plated on the horizontal stabilizer. This addition resulted in being an antenna farm as well as solving the weak directional stability issues. Antenna placement should be completed prior to airframe final design. It is difficult to generate real estate on the airframe once the air vehicle design is locked.

UAS Integration Comanche's requirement for integrated Unmanned Aerial Vehicle System (UAVS) control was to provide Level 2 UAVS Control in Block I and Level 4 UAVS Control in Block II as listed below:

Block I Communication and Interoperability

Comanche must be capable of controlling and tasking (Level 2 control) UAVs (Objective is Level 4 control of UAVs), Figure 163.

Block II Communication and Interoperability

Comanche must be capable of controlling dynamically tasking (Level 4 control) Tactical UAVs and their payloads (Objective is Level 4 control of DoD/Interagency non-tactical UAVs objective and Unmanned Ground Vehicles (UGVs) using common data links). The RAH-66 requires interoperability with the family of Tactical UAVs (UGVs desired) operating in the 2008 and beyond timeline. This interoperability will entail Level 4 control and provide the aircrew with the ability to dynamically re-task the vehicle and payload of the UAV/UGV.

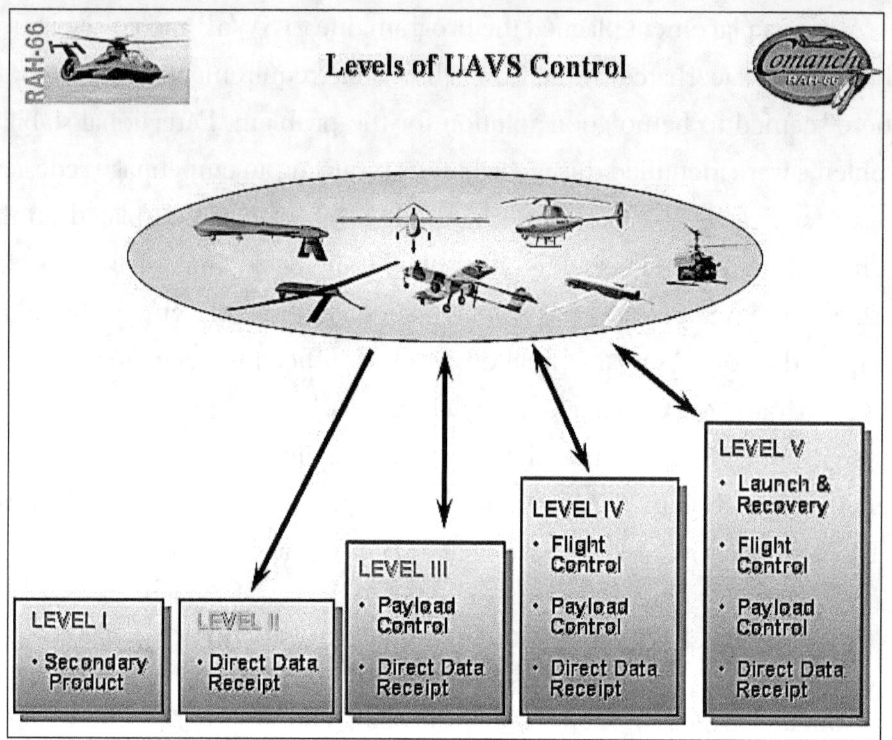

Figure 163. Levels of UAVS Control 21

The technical solution to meet the Block I Level 2 UAVS Control requirement was Integrated Communications, Navigation, Identification, and Avionics (ICNIA) enabled Joint Variable Message Format (JVMF) messaging. As the currently fielded Brigade and below Class IV UAVS, Shadow 200 was the defined team-mate for Comanche.

Comanche was to team with the Shadow system by JVMF enabled commands to the Shadow 200 Ground Control Station (GCS) and receipt of spot reports and status messages from the GCS. This capability was not truly Level 2 control since there was no capability for direct communication with the Shadow 200 air vehicle. This was a limitation of the Shadow 200 system. The Shadow air vehicle is not currently equipped with onboard processing or communications capability (i.e., Tactical Common Data Link (TCDL)) to support Level 2 control.

The technical solution to meet the Block II Level 4 UAVS Control requirement was TCDL. At program termination, TCDL was the current Joint accepted high bandwidth communications standard for UAVS control. Comanche's TCDL design was to employ a mast mounted directional antenna to provide high bandwidth data transmission at long ranges (See Figures 164 and 165).

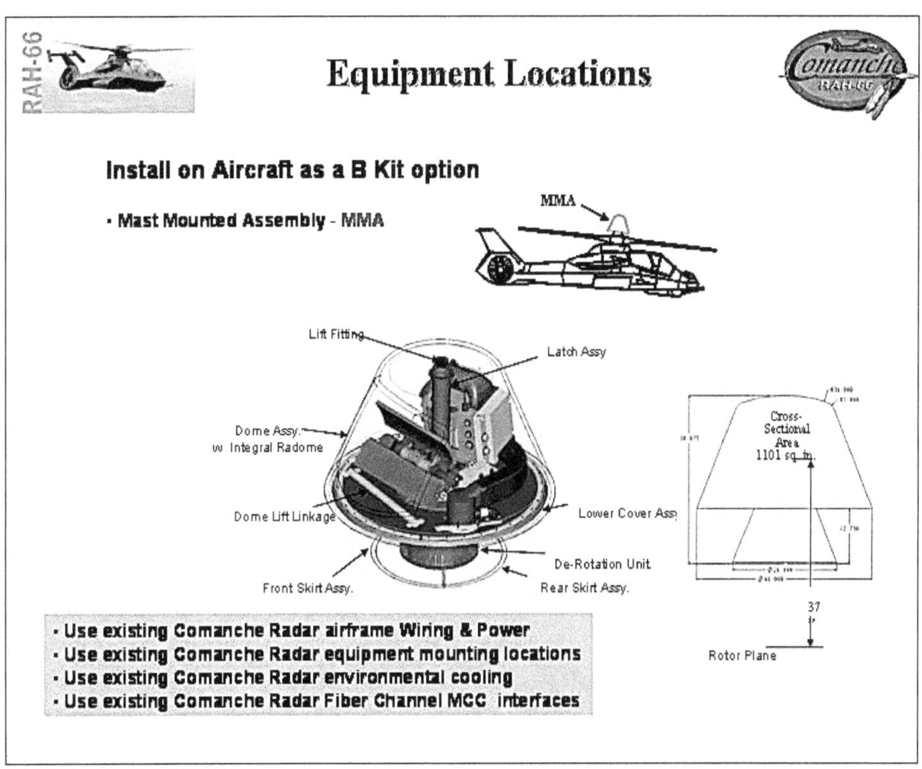

Fig 164. TCDL Equipment Location

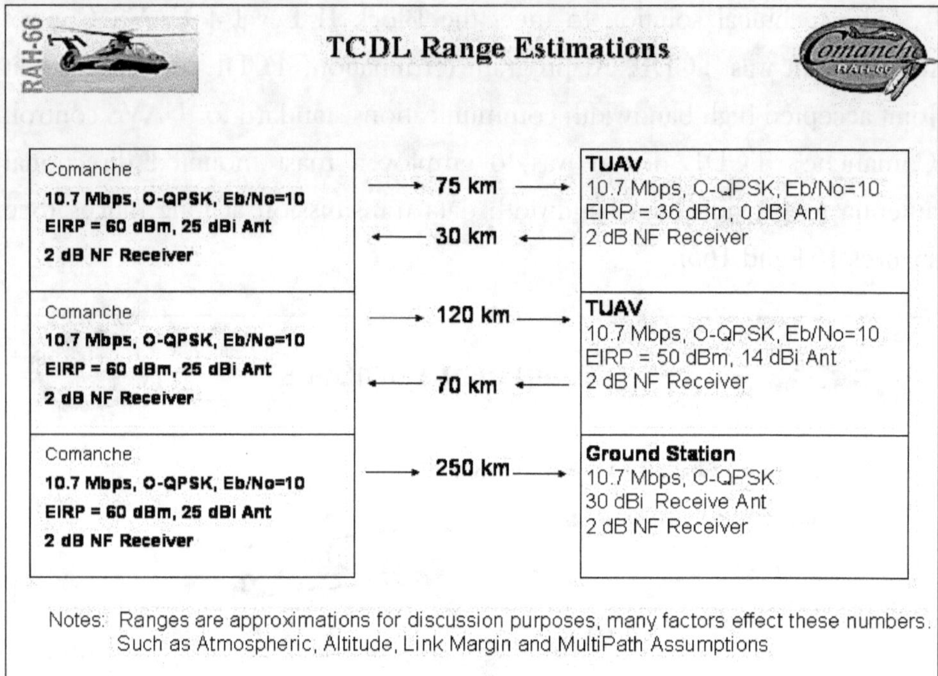

Figure 165. TCDL Range Estimations

The TCDL antenna array would be interchangeable with the Comanche Radar (CR) Mast Mounted Assembly (MMA). The TCDL Electronics Unit (TEU) would also be interchangeable with the Radar Electronics Unit (REU). Common internal wiring would support both systems. TCDL "kits" were to be fielded to 1/3 of the Comanche operational fleet. The most promising proposal for the TEU included an internal UAVS video storage, search, transmission/relay, and playback capability.

At program termination, the requirement for UAVS control requirement was tied to three UAV systems:

1. Shadow 200,

2. Fire Scout (FCS Class IV UAVS) and the

3. Unmanned Combat Armed Rotorcraft (UCAR).

These systems were identified as the primary probable "teammates" for Comanche during manned-unmanned (MUM) teaming operations. The

UAVS Interoperability IPT had set up communication with the PMO for each system.

Priority Information Exchange Requirements (PIER) Working Groups. PIER Working Groups were critical to successfully executing Comanche-UAV interoperability. Comprised of representatives from both Program Offices and the Comanche TRADOC Systems Manager (TSM), these groups worked to establish Memorandums of Agreement (MOAs) to define the actions required by both programs to successfully meet their requirements. This process uncovered many requirement and funding issues within the Shadow program that made Comanche Shadow teaming a challenge.

A Complete Digital Model of the RAH-66 Comanche is illustrated in Figure 166.

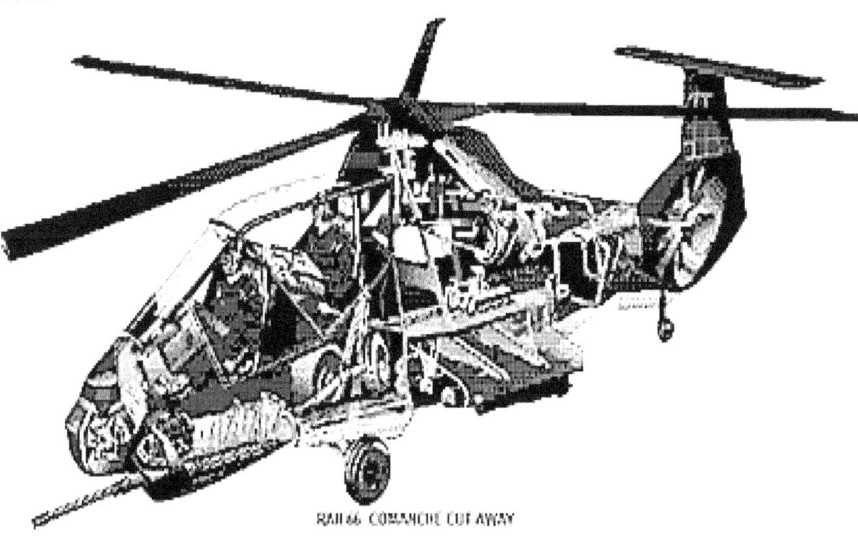

Figure 166 Comanche Complete Digital Mode

Endnotes/References

1. Schrage, D.P., "Army Aircraft Requirements in the 1990s: A View Forward", AIAA-83-2434, AIAA Aircraft Design, Systems and Technology Meeting, October 17-19, 1983, Fort Worth, TX

2. Singley, G.T., et al, "Army Light Family of Rotorcraft (LHX) Concept Formulation, AIAA-83-2552, AIAA Aircraft Design, Systems and Technology Meeting, October 17-19, Fort Worth, TX

3. Government Comanche Lessons Learned and Recommendations, 2005-2006

4. Schrage, D.P., 1999 AHS Nikolsky Lecture, "Technology for Rotorcraft Affordability Through Integrated Product/Process Development (IPPD), AHS Forum, 1999.

5. X2 Technology - A New Perspective on Helicopters, AHS Forum 2006 Presentation

6. Linden, A.W. and the Comanche Team, "The RAH-66 Comanche Helicopter: Technical Accomplishment, Program Frustration", American Institute of Aeronautics and Astronautics, Inc. 2021,

Chapter Eight

PREPARATION FOR BECOMING A ROTORCRAFT DESIGN PROFESSOR

Developing a Graduate Program in Aerospace Systems Design, Moving to Atlanta, and a Third Career
A Review of My Careers to Date

My first career was documented, mostly in Book 1. It included growing up, becoming a good athlete, and graduating from West Point. It followed with my becoming an active-duty military officer and then being assigned as a tactical nuclear weapons battery commander in an Honest John Missile battalion in Germany during the Cold War, 1968-69. This was followed by my attending rotary wing flight school and becoming an Army Aviator at Fort Walters, TX, and Fort Rucker, AL. I was then assigned to a combat tour in the Vietnam War as a helicopter lift ship platoon leader, air mission commander and gunship platoon leader in the 162nd Assault Helicopter Company (AHC) operating in the Mekong Delta and Cambodia. I was then assigned as the Asst S-3 and later S-3 for the 13th Combat Aviation Battalion, overseeing operations in the Mekong Delta. I then orchestrated the transfer of the Soc Trang Army Airfield in November 1970 to the VNAF and USAF as part of Helicopter Vietnamization. Next was the transfer and moving of the 13th CAB to Can Tho Army Airfield. This was followed over Christmas in 1970 with a one-week Rest and Relaxation (R&R) with my wife Nancy in Honolulu, Hawaii. I spent January 1971 as the S-3 13th CAB at Can Tho Army Airfield before deploying back to the USA in early February 1971.

My return from Vietnam in 1971 is the first chapter in Book 2. After reuniting with my family, I then attended the Field Artillery Advanced Course at Fort Sill, OK. This was followed by two years of graduate school in Aerospace Engineering at Georgia Tech, which is addressed in Chapter two of Book 2. I received a master's degree, MS, in Aerospace Engineering at Georgia Tech. I was then assigned to the Army Aviation Systems Command (AVSCOM) in St. Louis, MO, in June 1974. I was still on active duty as an Army Captain, Aviator and Aerospace Engineer.

In my first four years at AVSCOM, 1974-1978, I worked as an Aerospace Engineer specializing in Aeroelasticity, Dynamics, and Vibration (ADV). I served as a troubleshooter on the Army Utility Tactical Transport Aircraft System (UTTAS) and Advanced Attack Helicopter (AAH) Programs and on their Source Selection and Evaluation Boards (SSEBs). I also served on the AH-1 Cobra Improved Main Rotor Blade (IMRB) SSEB. In addition, during these four years, I worked on and received two advanced degrees: a Master of Arts (MA) in Business Administration from Webster College; and a Doctor of Science (DSc) in Mechanical Engineering from Washington University, in St. Louis, MO. My DSc thesis was to develop a new approach for analyzing rotor blade loads and stability. I was given the American Institute of Aeronautics and Astronautics (AIAA) Young Engineer Award from the St. Louis Chapter for these accomplishments. I was then promoted to active-duty Army Major in January 1978 and became the acting Aeromechanics Branch Chief. My follow-on military assignment was to be an instructor in the Department of Mechanics, United States Military Academy

(USMA) West Point, NY. However, in Spring 1978 I made the decision to leave active duty and become a reserves officer which became effective on July 1, 1978,

With the establishment of AVRADCOM, 1977-1983, I could see the opportunity to transfer from active duty to the US Army Reserves, which I did on July 1, 1978. I then was given a GS-14 Civil Service position as the Aeromechanics Branch Chief in the Development and Airworthiness Qualification Directorate, AVRADCOM. In parallel, COL Rogers, Director, Department of Mechanics, USMA, offered me the position as a Mobilization Designee (Mob Des), Assistant Professor in the Department of Mechanics at USMA two weeks a year.

In addition, I was also offered a position as a Military Academy Liaison Officer (MALO) with the Directorate for Admissions, USMA, for the St. Louis Area. In this capacity, I helped recruit interested and qualified high school students who wanted to attend USMA. I continued to serve in these positions until 1996, when I retired with the rank of Colonel, USAR.

In 1979 I was promoted to GS-15 as the Director of the Structures and Aeromechanics Division. I oversaw the technical development of the Near-Term Scout Helicopter (NTSH), which was part of the Army Helicopter Improvement Program (AHIP) which led to the OH-58D Kiowa Helicopter.

I also served as the Technical Chief for the AHIP SSEB and led the Airworthiness and Technical Development of the Army CH-47D Chinook Modernization Program.

In 1981 I was selected for Senior Executive Service (SES), Level 3, and became the Director of Advanced Systems (DAS) and the AVRADCOM Associate Technical Director for Science and Technology (S&T). In this capacity, I led the Concept Formulation Program for the Light Helicopter Experimental (LHX), which led to the Comanche Helicopter Development. As the Associate Director of S&T, I oversaw the AVRADCOM Research and Technology Labs (RTLs). In addition, in 1983 I served for six months as the Interim Chief Scientist for the Army Combined Arms Center (CAC) at Fort Leavenworth, KS under LTG Jack Merrit and worked with BG Colin

Powell, Director of the Combined Arms Combat Development Activity (CACDA).

As described in the Preface and Background Information in Book 2, AVSCOM was changed to AVRADCOM and then back to AVSCOM from 1977 to 1983. Also, the new AVSCOM command's two primary mission concerns were readiness (immediate) and research and development (eventual). This was a reversal of priorities from AVRADCOM, and the new AVSCOM would be led by a readiness officer, MG Orlando Gonzales, who had little if any, research and development experience. I could see that the Technical Director's role would be primarily to solve field problems. It was time to leave Civil Service for Academia.

While solving field problems was an important role for Army Aviation, it was not my desire, as I had established my name and reputation in research, design, and development. I had been approached by the Chair, and Associate Chair of the School of Aerospace Engineering, Dr. Arnold Ducoffe and Dr. Robin Gray, respectively, in 1982-83 to become the Rotorcraft Design Professor and the Associate Director for the Georgia Tech Center of Excellence in Rotary Wing Aircraft Technology (CERWAT), also called Rotorcraft Center of Excellence (RCOE) and later, the Vertical Lift Research Center of Excellence (VLRCOE).This solicitation was initiated by Dr. Norman R. Augustine, who was Assistant Secretary of the Army and in 1975 became Under Secretary of the Army, and later Acting Secretary of the Army. It focused on sponsoring long-term rotorcraft research at respected universities, and they were awarded in 1982 to Georgia Tech, the University of Maryland, and RPI. I told Drs. Ducoffe and Gray that I wasn't sure if I was ready to leave my government senior executive position.

I told them that I would help them find someone if I didn't take the position. I did approach Ray Prouty, a well-known rotorcraft expert and author in Rotor and Wing Magazine, to see if he was interested. However, he told me that after he joined MDHC, he didn't have sufficient time to draw retirement.

In the late 1970s, before the Army RCOE solicitation had been issued, I had encouraged Drs. Ducoffe and Gray to submit an unsolicited proposal to the Army Development and Readiness Command (DARCOM) with an emphasis on rotorcraft design and technology assessment. I imagined that DARCOM would forward it to AVRADCOM. I told them this should be based on the lessons learned from my participation in all the Army Aviation Design and Development Programs in the 1970s, which are addressed in this book, Book 2.

They submitted an unsolicited proposal to DARCOM. However, it had gotten lost in somebody's desk or filing cabinet at DARCOM, which made Dr. Ducoffe particularly upset. However, I told him that this wasn't unusual in a bureaucracy, which is typical of government in which most of the important decisions are often not made or are lost by government officials.

The good news was that when the Army RCOE solicitation came out in 1980, Georgia Tech had draft material with a focus on rotorcraft design for submittal. They ended up being the top university selected for the Army RCOE Program.

In the meantime, I had also been approached by Dr. Bob Lynn, VP of Engineering, Bell Helicopter Textron (BHT), to become their Director of Technology, which could lead to VP of Engineering upon Bob Lynn's retirement in a few years. However, I thought that being named a full professor in rotorcraft design at Georgia Tech was a great opportunity and the preferred choice for me and my family, who wanted to spend more time together.

In January 1984, I left civil service at AVRADCOM/AVSCOM. I became the Rotorcraft Design Professor in the School of Aerospace Engineering at Georgia Tech, and Associate Director for the Army-Sponsored Center of Excellence in Rotary Wing Aircraft Technology (CERWAT).

Finally, in Book 2, I have tried to set the record straight and answer the question, "Why has it been so hard for the Army to develop and field new Aviation Systems." I have shown it wasn't a high priority with readiness (immediate) and research and development (eventual) for the new AVSCOM in 1984. It is still a major issue today where the FVL Program didn't start for over 30 years and, hopefully, will complete the development of the FLRAA, FARA, and FTUAS in the 2030s. Fortunately, Dr. Bill Lewis, a former colleague, and my PhD student at Georgia Tech, was there to initiate the FVL Program. Hopefully, previous lessons learned in this book and elsewhere can continue to be applied to the next generations of rotorcraft.

Book 3 will focus on my 35 years as Rotorcraft Design Professor and Director of the Georgia Tech CERWAT/RCOE/VLRCOE, which were renewed every three or five years from 1982 to 2025.

Establishment of Rotorcraft Design as the Key for Creating Next Generations of Army Aviation Systems

Lessons Learned as an Aeroelasticity Dynamics and Vibrations (ADV) along with Design and Development of the UTTAS and AAH Prototypes, the OH-58D Kiowa, the CH-47D Modernization and the LHX Concept Exploration.

The impact of design changes during the development of the UTTAS and AAH are covered in Chapter 3. In Chapter 4 while pursuing a D.Sc. degree in Mechanical Engineering at Washington U. (St. Louis, MO) I developed an innovative tool for evaluating rotor systems designs for rotor loads and stability. It was called an eigenvalue and modal decoupling analysis using Floquet Theory and was applied to the tail rotors of the two UTTAS candidate aircraft, the Sikorsky YUH-60A and the Boeing YUH-61A. In addition, I became heavily involved in the redesign of the Hughes YAH-64A tail. In addition, while working as an Engineer in the AVSCOM Aeromechanics from 1974-1978 I became involved with Systems Research Integration Office (SRIO) located at AVSCOM in St. Louis which was focusing on the early, more conceptual phase of preliminary design and developed design and performance codes SSP-1 and SSP-2 based on momentum theory. When I went to Georgia Tech in 1984, we combined

SSP-1 and SSP-2 into the Georgia Tech Preliminary Design Program (GTPDP), which was used in my rotorcraft design courses for the first ten years, until 1994. This resulted in Georgia Tech winning over 90% of the American Helicopter Society (AHS) Student Design Competitions (SDCs), as illustrated in Figure 167 for a variety of Vertical Take-Off and Landing (VTOL) aircraft.

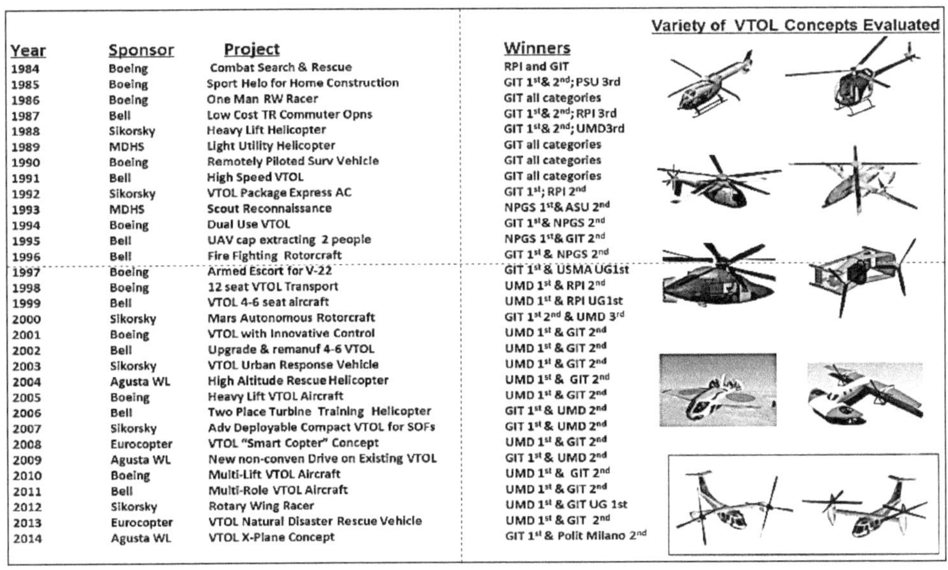

Year	Sponsor	Project	Winners
1984	Boeing	Combat Search & Rescue	RPI and GIT
1985	Boeing	Sport Helo for Home Construction	GIT 1st & 2nd; PSU 3rd
1986	Boeing	One Man RW Racer	GIT all categories
1987	Bell	Low Cost TR Commuter Opns	GIT 1st & 2nd; RPI 3rd
1988	Sikorsky	Heavy Lift Helicopter	GIT 1st & 2nd; UMD3rd
1989	MDHS	Light Utility Helicopter	GIT all categories
1990	Boeing	Remotely Piloted Surv Vehicle	GIT all categories
1991	Bell	High Speed VTOL	GIT all categories
1992	Sikorsky	VTOL Package Express AC	GIT 1st; RPI 2nd
1993	MDHS	Scout Reconnaissance	NPGS 1st & ASU 2nd
1994	Boeing	Dual Use VTOL	GIT 1st & NPGS 2nd
1995	Bell	UAV cap extracting 2 people	NPGS 1st & GIT 2nd
1996	Bell	Fire Fighting Rotorcraft	GIT 1st & NPGS 2nd
1997	Boeing	Armed Escort for V-22	GIT 1st & USMA UG1st
1998	Boeing	12 seat VTOL Transport	UMD 1st & RPI 2nd
1999	Bell	VTOL 4-6 seat aircraft	UMD 1st & RPI UG1st
2000	Sikorsky	Mars Autonomous Rotorcraft	GIT 1st 2nd & UMD 3rd
2001	Boeing	VTOL with Innovative Control	UMD 1st & GIT 2nd
2002	Bell	Upgrade & remanuf 4-6 VTOL	UMD 1st & GIT 2nd
2003	Sikorsky	VTOL Urban Response Vehicle	UMD 1st & GIT 2nd
2004	Agusta WL	High Altitude Rescue Helicopter	UMD 1st & GIT 2nd
2005	Boeing	Heavy Lift VTOL Aircraft	UMD 1st & GIT 2nd
2006	Bell	Two Place Turbine Training Helicopter	GIT 1st & UMD 2nd
2007	Sikorsky	Adv Deployable Compact VTOL for SOFs	GIT 1st & UMD 2nd
2008	Eurocopter	VTOL "Smart Copter" Concept	UMD 1st & GIT 2nd
2009	Agusta WL	New non-conven Drive on Existing VTOL	GIT 1st & UMD 2nd
2010	Boeing	Multi-Lift VTOL Aircraft	UMD 1st & GIT 2nd
2011	Bell	Multi-Role VTOL Aircraft	UMD 1st & GIT 2nd
2012	Sikorsky	Rotary Wing Racer	UMD 1st & GIT UG 1st
2013	Eurocopter	VTOL Natural Disaster Rescue Vehicle	UMD 1st & GIT 2nd
2014	Agusta WL	VTOL X-Plane Concept	GIT 1st & Polit Milano 2nd

Figure 167. AHS SDCs over 20 years, 1984-2014

The University of Maryland (UM) became so frustrated that they couldn't win they brought the designer of Russian helicopters, Dr. Marat Tishchenko, to the USA. In 1998, he joined the University of Maryland's Rotorcraft Center to try and help them in the AHS Student Design Competitions. Tishchenko served the UM Rotorcraft Center as a visiting professor for a period of 3–6 months, and then returned almost every year thereafter. The University of Maryland won a first place prize in the graduate category that first year and repeatedly through 2005. However, Georgia Tech won two of the three first place prizes in the Mars Autonomous Rotorcraft Competition in 2000. I used to tease the UM faculty that they let the Russians be their designers.

However, by the early 1990s we at Georgia Tech initiated our graduate program in Aerospace Systems Design and the Aerospace Systems Design Laboratory (ASDL), Figure 168, which included fixed wing, rotary wing, and spacecraft systems. It also focused on Integrated Product and Process Development (IPPD) and is now recognized as the world's leading academic Aerospace Systems Design Center. Book 3 will focus on how the Georgia Tech Graduate Program in Aerospace Systems Design and the establishment of the Aerospace Systems Design Laboratory (ASDL) and the Space Systems Design Laboratory (SSDL) were developed and became world renown. Beginning in the early 1990s the author and Dr. Dimitri Mavris, my former postdoc fellow, expanded the rotorcraft design program into a graduate program in Aerospace Systems Design.

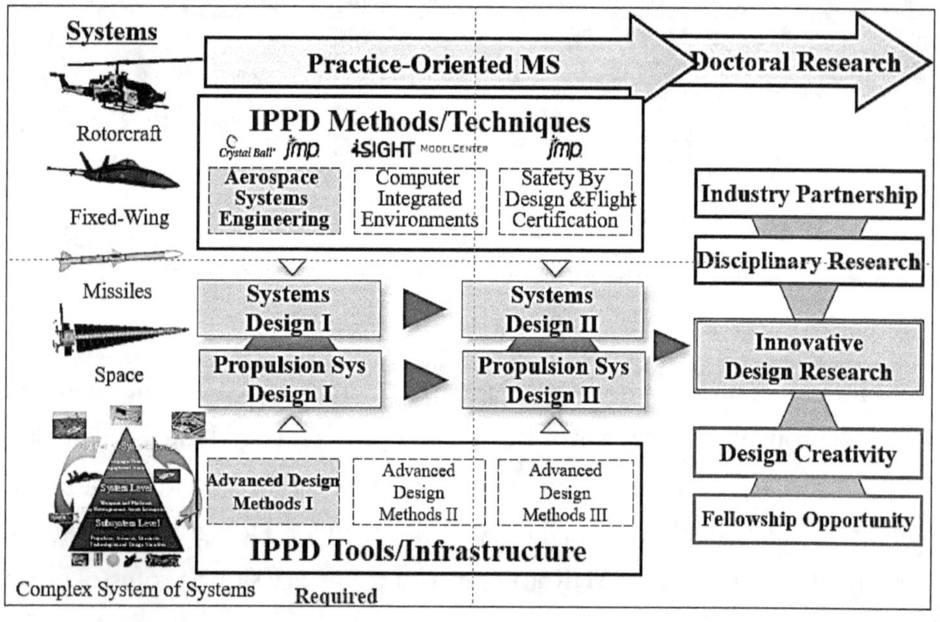

Figure 168. Georgia Tech Graduate Program in Aerospace Systems Design

This program was based on the rotorcraft design experiences I had at AVRADCOM with the design and development of the UTTAS, AAH, OH-58D Kiowa, the CH-47D Modernization and the LHX Concept Formulation.

Army Aviation Loss of seasoned Government Rotorcraft Design Specialists for Future Vertical Lift Aircraft

It is evident from the Future Vertical Lift (FVL) Program that there is a lack of Government Rotorcraft Design Specialists. This is especially true for the Future Attack and Reconnaissance Aircraft (FARA) which was cancelled based on the inexperience and incompetence of the industry design teams, along with the lack of seasoned Army design engineers. With this problem I have included an Epilogue which describes how the FVL designs could be evaluated based on my rotorcraft design experience. It also provides an introduction to hybrid electric solutions which could make the FVL and future Army aircraft better solutions in the future.

EPILOGUE

HOW LESSONS LEARNED SHOULD BE UNDERSTOOD AND APPLIED FOR CONCEPT DESIGN AND ANALYSIS (CDA) FOR FVL & OTHER PROGRAMS

Introduction to Lessons Learned from FARA's Cancellation & Other Programs

The chapters in Book 2 up to now have addressed and documented problems and lessons learned during the development of the 3rd Generation of Army Aviation Systems in Figure 2. This is based on my knowledge and direct involvement in all of these Army Aviation Systems developments, as well as my own combat experience that led to their development. It is also based on my thirty-five years of teaching rotorcraft design at Georgia Tech and as a consultant to government and industry. As mentioned in Chapter 8 beginning in the early 1990s the author and Dr. Dimitri Mavris, my former postdoc fellow, expanded the rotorcraft design program into a graduate program in Aerospace Systems Design, the largest in the world which will be the central theme and example in Book 3 in this trilogy.

Review of the FARA Program & Overly Constrained Problem (Ref. 1)

An Example Preliminary Design Study for FVL Risk Assessment and Selection will be illustrated. The dilemma facing FARA selection needs to be understood by government design engineers and decision-makers in a timely manner. This section provides how experienced VTOL design experts familiar with Army Aviation experience can provide this understanding in a timely manner. In July 2021, during the US Army Industry Days at Ft. Rucker, Alabama, the Future Vertical Lift (FVL) Project Manager for the Future Attack Reconnaissance Aircraft (FARA) made some remarkable comments. He was in the Army's Program Executive Office for Aviation, PEO(A), which took over leadership of the FARA project from the Army Futures Command (AFC) that year. **It was clear that the Army Futures Command or PEO Aviation didn't have the necessary technical expertise**. Aviation Week's Steve Trimble published the story, "Physics-Busting Requirements Challenge US Army FARA Program," on Aug. 3, 2021. The article starts by stating, "The US Army's project manager now says the original set of performance requirements for a future rotorcraft that represents the aviation branch's top modernization priority are not compatible with the laws of physics."

As noted in the VFS Director Mike Hirschberg's Commentary, "What FVL Needs to Succeed" (Ref. 1. Vertiflite, Nov/Dec 2018): "*W*ith FARA, the Army wants to field a new system in less time that is smaller, faster, longer range, more affordable, more capable, more lethal, more connected, more survivable, and more autonomous than current systems. To solve the tension between these conflicting desires, designers need to iterate the design sensitivities with operational analysis to show the pros and cons of each attribute, alone and in concert. In order to fly competitive prototypes (not just demonstrators) four years from [then], the Army must prioritize its metrics and potentially make hard trade-off decisions. Now the time has come for Army leadership to make those hard decisions." In this article, he identifies that the six key attributes of the FARA Competitive Prototypes are:

1. First flight: in fiscal 2023/2024
2. Maximum speed: at least 180 kt (333 km/h).
3. Engine: a single 3,000-shp (2,240-kW) GE T901 being developed in the Improved Turbine Engine Program (ITEP)
4. Rotor: 40 ft (12.2 m) diameter.
5. Max gross weight: 14,000 lb (6,350 kg) \
6. Mission: range, endurance, capability as specified

As mentioned, the Futures Command and PEO (Aviation)didn't have the technical expertise to conduct the necessary tradeoffs to meet these requirements. Also, the AMRDEC Aviation Development Directorate (ADD) was rolled into the Technology Directorate. It also appears that engineers in ADD and its Research Labs also didn't have the necessary technical expertise or experience to work with the FARA industry engineers in making these necessary tradeoffs.

The Two FARA Candidate Aircraft Shown in Figures 169 & 170.

Artist illustrations of the Bell Invictus are shown in Figure 169A with a ducted fan anti-torque device, while Figure 169B is shown with a tail rotor anti-torque device. The main rotor system is articulated with four main rotor blades very similar to the Bell 525 RELENTLESS which has five main rotor blades, Figure 169C. The Fly By Wire (FBW) System will be similar for both aircraft.

The Bell INVICTUS in Figure 169A appears similar to the planned RAH 66 Comanche except it has an articulated rotor system instead of a Bearing less Main Rotor (BMR) system. Also, a small wing is on both Bell

INVICTUS aircraft, Figures 169A and 169B, which is intended to provide some lift in high-speed flight to unload the main rotor and be able to attain the Army's required dash speed of 180 knots. However, Bell's claim of 190-200 kts is very questionable without the necessary blade loading, $C_T/6$, and a larger wing which will be discussed later in the Epilogue.

Figure 169A. Bell INVICTUS w/Ducted Fan.   Figure 169B. Bell INVICTUS w/Tail Rotor

A summary of the Bell INVICTUS is provided, along with a picture of the Bell Relentless in Figure 169C.

Single Main Rotor Helicopter with Ducted Tail Rotor

- Leverages technologies and lessons learned developing 525 Relentless
- Lift-sharing wing for high-speed maneuverability
- Supplemental Power Unit increases performance during high power demands
- Robust articulated main rotor enabling high speed flight
- Fly-by-wire flight control system
- Armed with a 20 mm cannon, integrated munitions launcher with ability to integrate air-launched effects, and future weapons, as well as current inventory of munitions
- Modular Open Systems Approach (MOSA) enabled

Figure 169C. The Bell FARA RELENTLESS Summary and Bell 525 INVICTUS Picture. The Lockheed-Sikorsky FARA candidate, S-97 Raider and X-2 coaxial compounds, Figs. 170A-C

Fig.170A. Lockheed-Sikorsky S-97 Raider Flight   *Fig. 170B. X2 Demonstrator ~ 230-250*

The summary of the S-97 Raider is shown in Figure 170C. Attitude control in low-speed NOE flight is critical, as is high-speed blade loading for air-to-air combat in high-speed flight, something essential for Comanche but ignored so far for FARA.

Figure 170C. Summary Sikorsky S-97 Raider Flight Attributes

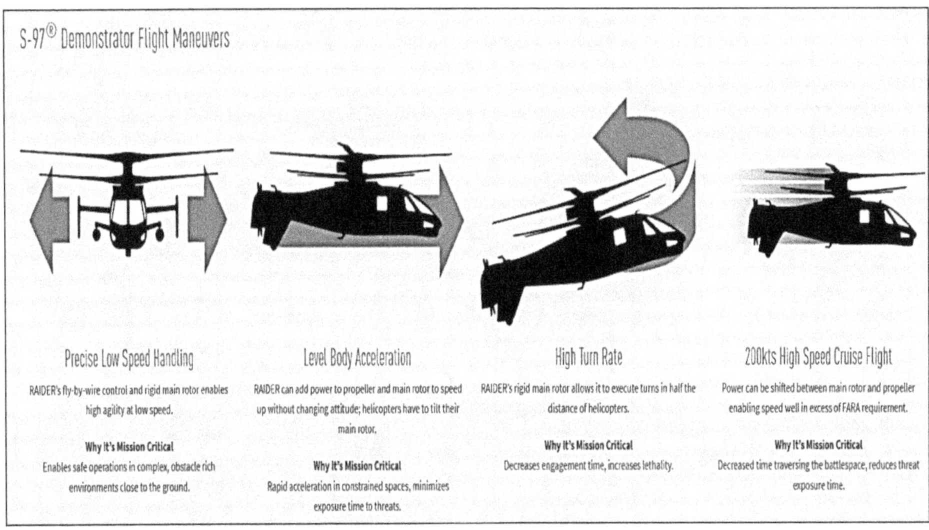

Figure 170D. Attitude Control and high Blade Loading are Very Important, as I learned flying Nap of Earth (NOE) in Vietnam

Based on my experience flying combat helicopters, UH-1C, UH-1H, OH-58A, and OH-6A, often in NOE, and possibly in air-t0-air combat I feel these capabilities while currently desired, should be required.

These capabilities of a coaxial compound for a High-speed High Maneuverable Rotorcraft (HSMR) were explored in the 1993 Georgia Tech AHS Award Winning SDC. (Ref. 3) It is illustrated in Figure 171A as the only feasible design from the potential design concepts considered in Figure 171B.

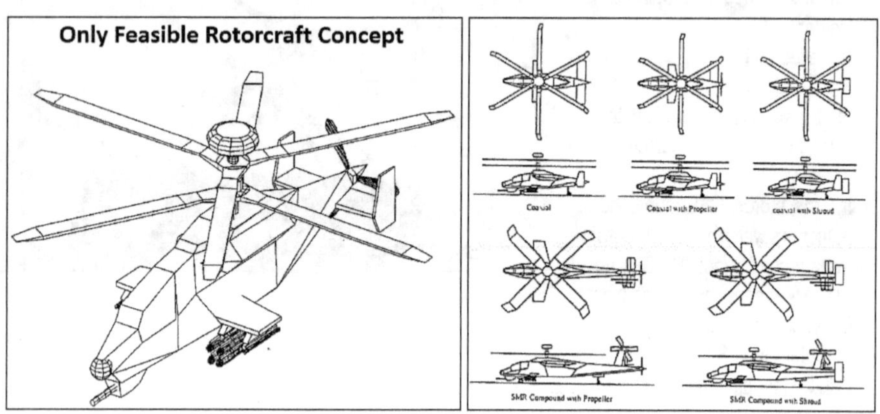

Figure 171A. Only Feasible Rotorcraft Concept. Figure 171B. Potential Feasible Concepts

The 1993 Tenth AHS SDC Requirements for a HSHMR are identified in the Table below.

Performance, Maneuver & Handling Qualities Requirements (4k ft, 95 deg F)	
Range	430 km
Reserve	30 min
Speed	*200 knots at IRP(Concept Driver - Compound or TR)
Vertical Rate of Climb	*800 fpm at IRP (Concept Driver - low download)
Payload	Primary - 1274 lbs
Sustained/Trans Manu LFs	*2.0g @ V_{mp}/4.0g @ 160 knots (Concept Driver-Coax)
Ferry Range/Ferry Reserve	1260 nm/100nm
Multiple Engines/Onboard APU	at least 2
Disk Loading	*< 15.0 lbs/ft^2 (Concept Driver - Compound Helo or TR)
MR/TR Norm Oper Tip Speeds	never exceed - 725 fps/never exceed - 650 fps
Autorotation t/k index	0.8 sec or greater
Crew	2
Cannon	30 mm
Handling Qualities	MIL-H-8501A (Pitch/Roll Control Power vs Damping Good}

Table 1. 1993 AHS SDC Design Requirements

The underlined design requirements with asterisk were most critical, especially were the Sustained/Transient Maneuvers Load Factors (LFs) of 2.0g's sustained at V_{MCP} and transient 4.0g's at 160 kt. While the Georgia Tech Team finished second to the Naval Post Graduate School (NPGS) with a Single Main Rotor Compound (SMRC) it was evident that the Transient Maneuver of 4.0g's at 160 kts couldn't be easily achieved.

I later ran into one of the NPGS graduate students, an Army Aviator, who was on their first-place team. He bragged that they were on the winning team and had beaten us for once. I asked him how their team met the Transient Maneuver Requirement with their SMRC. He said he was sure they did, but later reported back to me that they hadn't analyzed it. Therefore, I provided a little tutorial to explain why they couldn't analyze it. The lift distribution for the SMR and Coaxial Rotor illustrated along with their Rotor Thrust Loadings are shown Figures 172A and B.

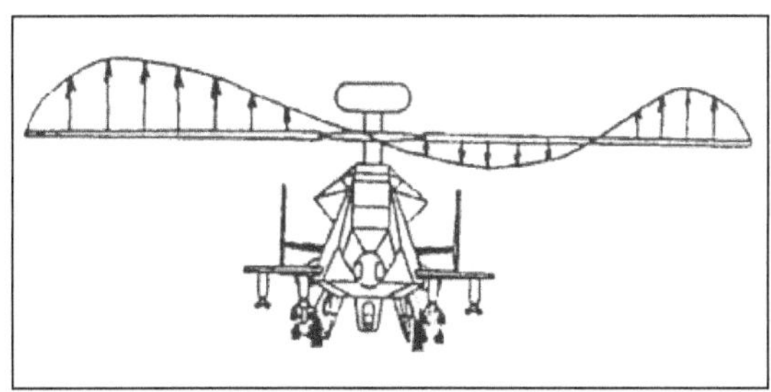

Single Main Rotor

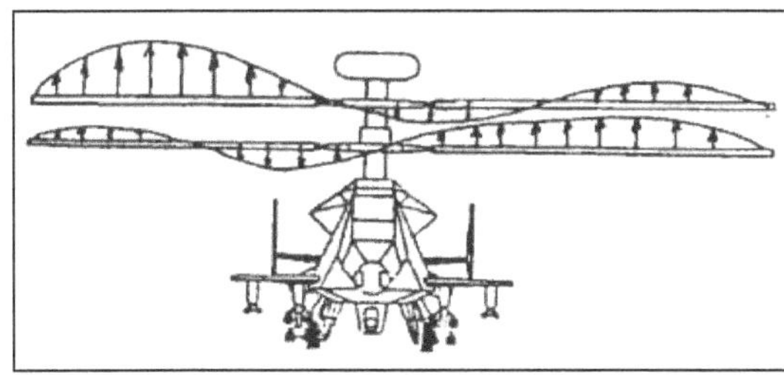

Coaxial Rotor

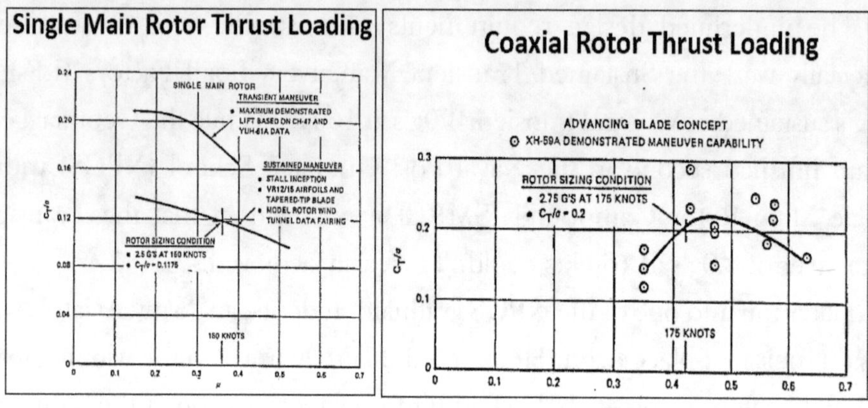

Fig. 172A. Single MR Thrust Loading vs. Airspeed. Fig. 172B Coaxial Rotor Thrust Loading

In an attempt to understand why the SMRC yields such high blade chords or radii, the mathematical expression linking the chord sizing to the rotor blade loading, $C_T/6$, and the lift, L_w, must be reviewed. This expression is presented below:

$$c = \frac{(4 \cdot W - L_w)}{\left(\dfrac{C_T}{\sigma}\right) \cdot \rho \cdot R \cdot V_T^2 \cdot b}$$

Inspection of this equation shows immediately that in order to achieve reasonable chord sizes for the critical transient maneuver requirement, a rotor system offering the highest possible value of $C_T/6$ at the given speed must be obtained. Since the density, ρ, is fixed and the rotor radius and tip speed are selected based on disc loading requirements (downwash considerations) -- the only other parameters that can be altered to achieve this effect on the number of blades (solidity) and the amount of lift provided by the wing (proportional to wing area, incident angle, etc.).

As far as the maximum $C_T/6$ is concerned, the coaxial rotor, as can be seen in Figure 172B, can provide according to the XH-59A flight test data, a maximum sustained $C_T/6$ of 0.21 at 180 kt and an assumed (very conservative) transient $C_T/6$ of 0.25. This selected value is justified based on typical transient load behaviors when compared to the sustained loads. Figures 172 A&B. For the SMRC, the only additional lift than the rotor would have to come from the wing. As illustrated in Figure 173 the size of the wing area must be greatly increased in radius or chord and can never practically meet the Transient Maneuver Requirement of 4g's at 160kt.

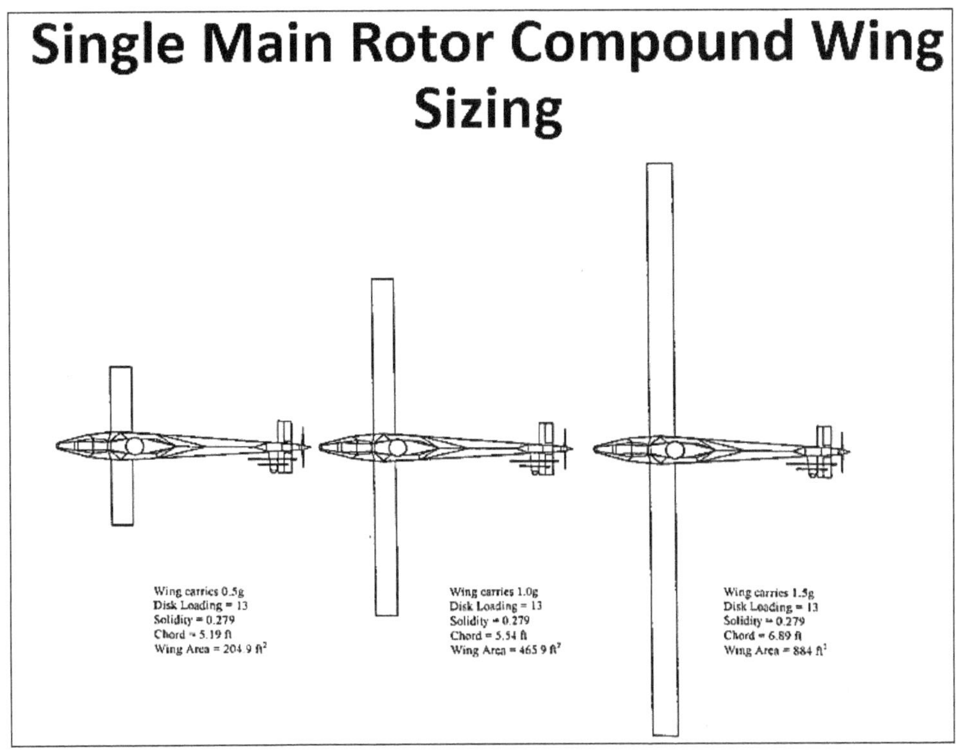

Figure 173. Unrealistic Wing Sizing in an Attempt for 4g's Transient Manu at 160kt

In summary, based on my experience in helping in evaluation and development of the UTTAS (YUH-60A and YUH-61A), the AAH (YAH-63A and YAH-64A), and the LHX/RAH-66 none of them could meet their high-speed cruise or dash speed requirements. I believe the risk for a FARA Sikorsky S-97 Raider is much less than it is for the Bell 360 Invictus. The Army 180 kt dash speed is questionable for the Bell 360 Invictus with little maneuverability capability; while the Sikorsky S-97 will have difficulty achieving 250 kt dash speed it should have maneuverability up to 220- 230kts.

If air-to-air (ATA) combat is still a critical Army Requirement the Sikorsky S-97 Raider should be the more capable aircraft.

An Example Preliminary Design Study for FARA Risk Assessment and Selection

A. Sizing, Performance, Technology Pull and Push and Laws of Physics (Ref. 3) The requirements, design, and performance for an aircraft are initially determined through sizing and synthesis with technologies applied based on trade studies and the laws of physics. The overall design and Systematic Operations Analysis process is illustrated in Figure 174.

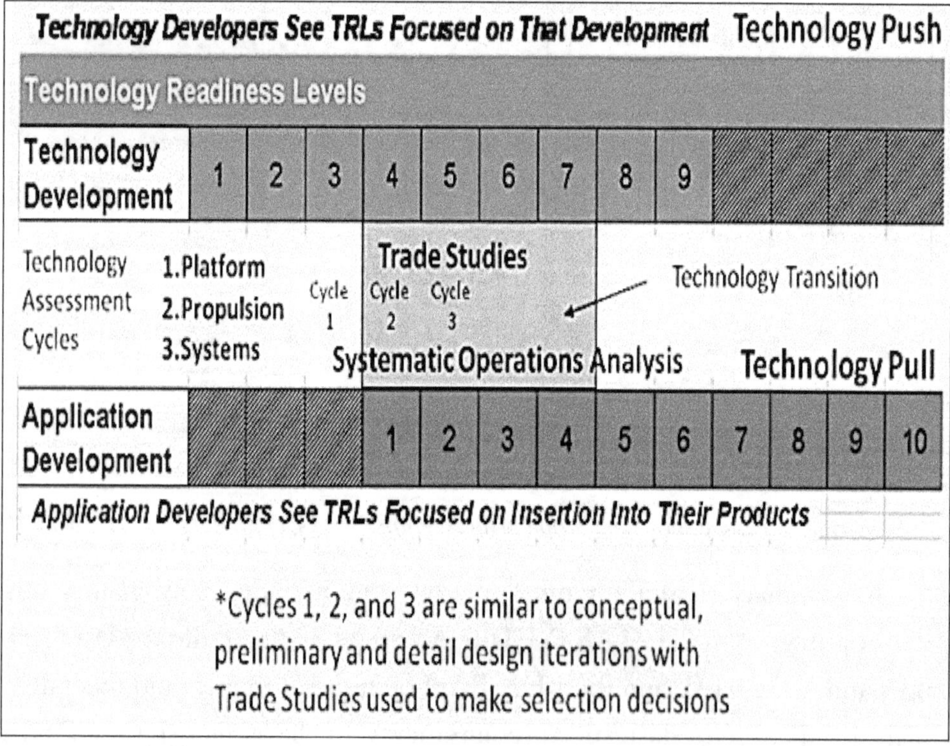

Figure 174. How the Six Key Attributes and Constraints Need to be Addressed.

This recommended approach was discussed in the Preface and Background and should be used in evaluating the Future Heavy Lift (FHL)/High Speed VTOL Aircraft described in Chapter 7.

Sizing and Synthesis for VTOL Aircraft

The Advanced Vehicle Design Sizing and Synthesis for VTOL Aircraft developed by the author is illustrated in Figures 175A&B. I have successfully taught this approach in Rotorcraft Design for 35 years, 1984-2019.

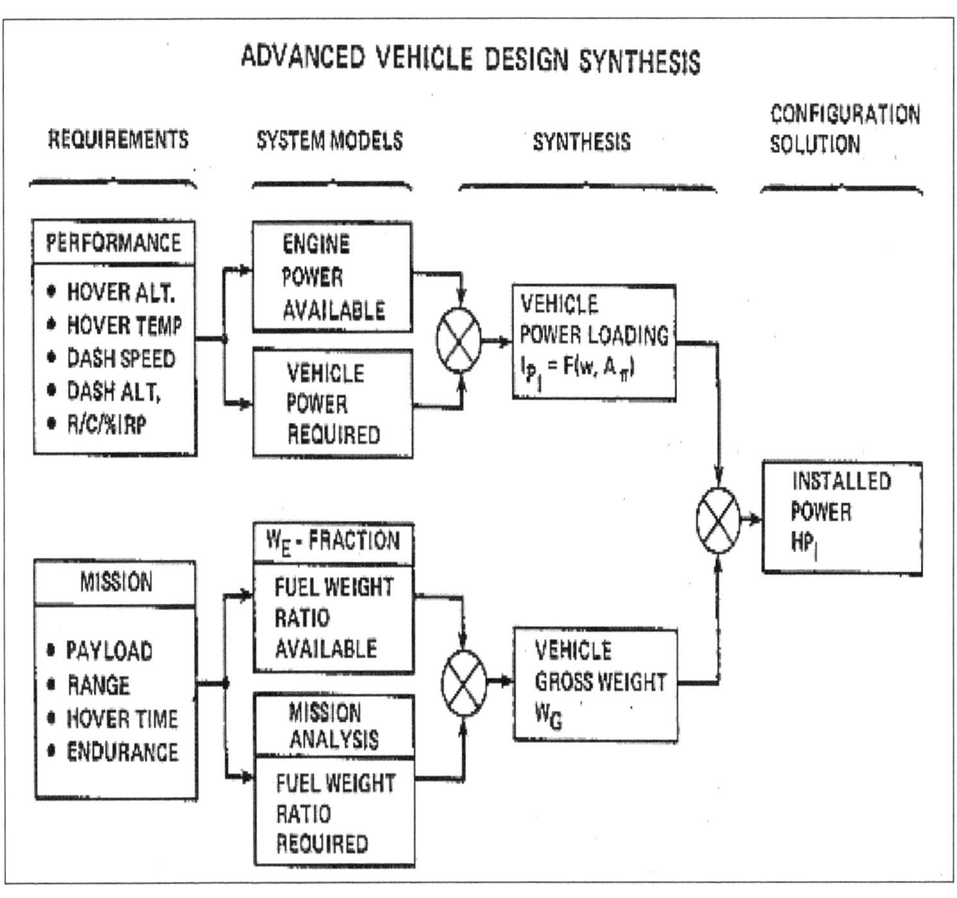

Figure 175A. Sizing and Synthesis for VTOL Aircraft

The Total Horsepower Required (THP) equation and Power Balance are illustrated in Figure 175B. Every engineer in industry and government involved in vertical lift design should understand these basic relationships. Unfortunately, it is difficult to find five engineers in industry and government who have this understanding.

Total Horsepower Required(THP) for Generic Subsonic Fixed Wing and Rotorcraft:

Parasite Power   Wing Induced Power   Rotor Induced and Profile Power

$$THP = \frac{\sigma f V^3}{146000\,\eta_p} + \frac{0.332}{\sigma e \eta_p}\left[(1-K_L)\frac{W_G}{b}\right]^2 \frac{1}{V} + \frac{1}{\eta_H}\left[K_L^{3/2} K_u \frac{ihp_H}{\sqrt{\sigma}} + K_\mu \sigma rhp_H\right]$$

$$K_u = \frac{U_o}{V}$$ Induced Rotor Correction Factor

$$K_\mu = 1 + 3\mu^2 + C\mu^4$$ Profile Rotor Correction Factor

$$ihp_H = \frac{0.0938\,\omega^{3/2} R^2}{B\sqrt{\sigma}}$$ Induced Rotor Power Equation

$$rhp_H = \frac{\sigma A_B}{1.85} V_{T_t}^{3/2} \delta_m$$ Profile Rotor Power Equation

Horsepower Available(HP_A) as a Function of Altitude, Temp, Time, and No. of Engines:

$$HP_A = HP'\left(\frac{N-1}{N}\right)(1-K_{h1}h_1)(1-K_t \,\mathrm{delta}\, T_S)\left(1+\frac{e^{-0.0173t}}{4}\right)$$

HP' = HP SLS
K_{h1} = Altitude Lapse Rate
K_t = Temperature Lapse Rate
t= Time in minutes

THP = HP_A for Hover, Forward Speed, Maneuver

Critical Power Loading(THP/GW) sizes the Engine

Figure 175B. THP & Key Power Contributions (induced, parasite & profile powers) for Power Balance

A fuel balance of Required, RFR, and Available, RFA, is used to determine a gross weight, Wo, as illustrated in Figure 175C. The minimum gross weight is determined by the intersection of the lowest disk loadings, W_2.

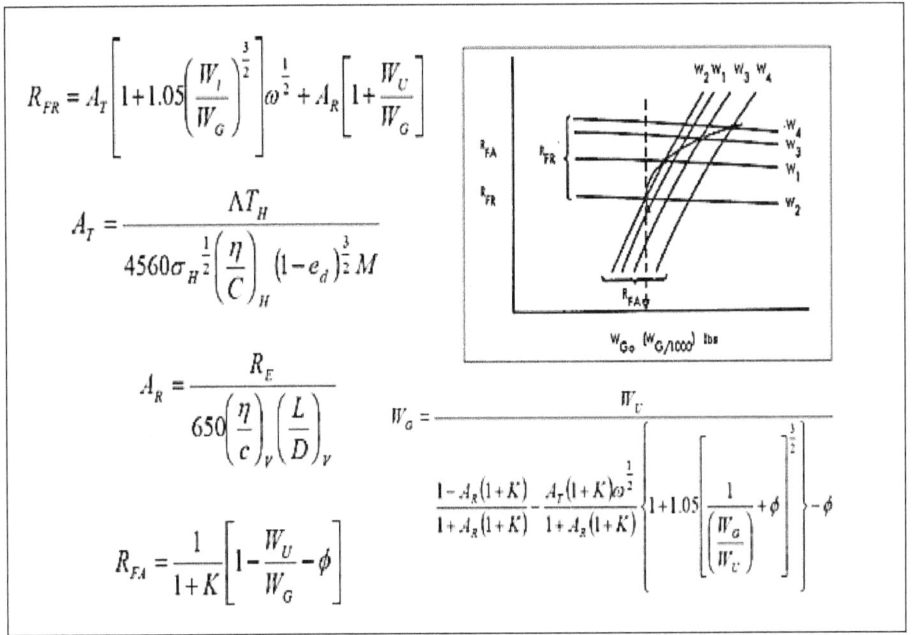

$$R_{FR} = A_T \left[1 + 1.05 \left(\frac{W_I}{W_G} \right)^{\frac{3}{2}} \right] \omega^{\frac{1}{2}} + A_R \left[1 + \frac{W_U}{W_G} \right]$$

$$A_T = \frac{\Lambda T_H}{4560 \sigma_H^{\frac{1}{2}} \left(\dfrac{\eta}{C} \right)_H (1 - e_d)^{\frac{3}{2}} M}$$

$$A_R = \frac{R_E}{650 \left(\dfrac{\eta}{c} \right)_V \left(\dfrac{L}{D} \right)_V}$$

$$W_G = \frac{W_U}{\dfrac{1 - A_R(1+K)}{1 + A_R(1+K)} - \dfrac{A_T(1+K)\omega^{\frac{1}{2}}}{1 + A_R(1+K)} \left\{ 1 + 1.05 \left[\dfrac{1}{\left(\dfrac{W_G}{W_U} \right)} + \phi \right]^{\frac{3}{2}} \right\} - \phi}$$

$$R_{FA} = \frac{1}{1+K} \left[1 - \frac{W_U}{W_G} - \phi \right]$$

Figure 175C. Minimum Gross Weight Achieved through a Fuel Balance

The Fuel Balance Method can be used to evaluate and assess the six FARA attributes identified. For instance, the Power Required is broken down in Figure 176A, a Breakdown of Power Required versus airspeed in Figure 176B, and a Balance with Power Available in Figure 176C.

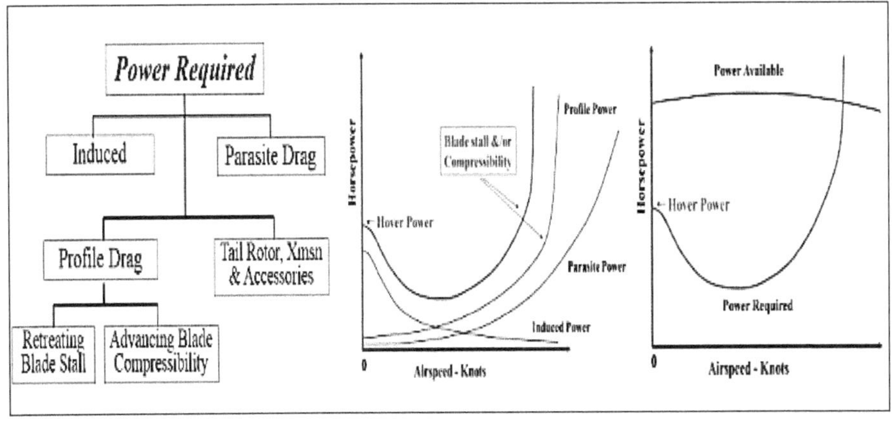

Fig. 176A. Power Required Fig.176B. Power Breakdown Fig.176C. Power Required & Available

Additional plots that can be calculated for physics-based modeling are illustrated in Figures 177A, 177B, and 177C, along with Figures 178A& B.

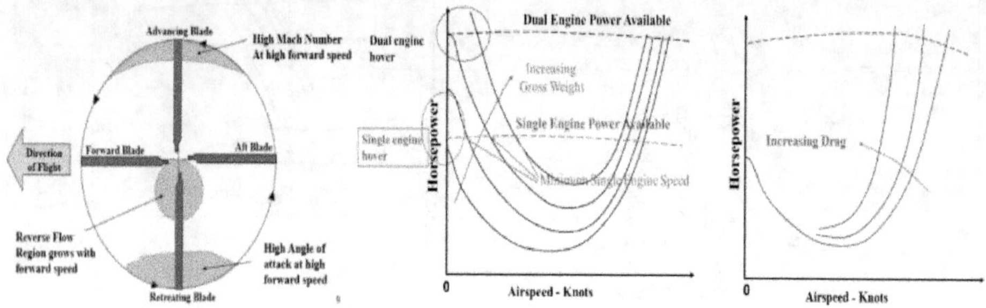

Fig. 177A. Asymmetric Velocity Forward Flight Fig.177B. Impact Increasing GW Fig. 177C. Impact of Increasing Drag

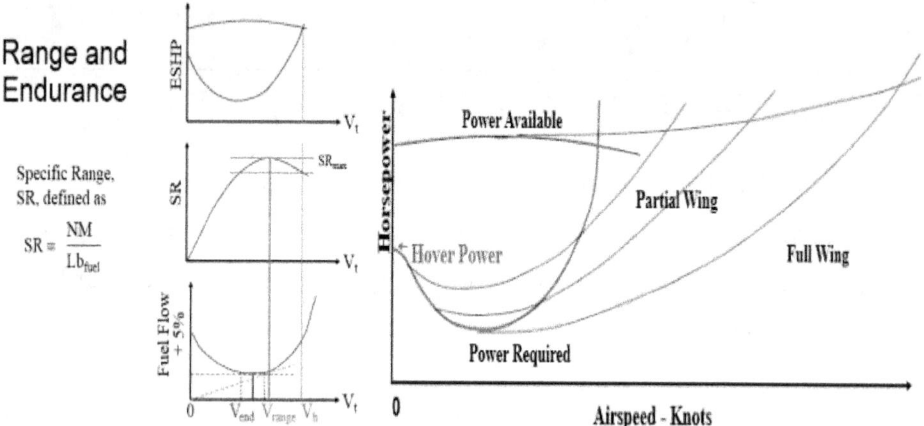

Specific Range, SR, defined as

$$SR = \frac{NM}{Lb_{fuel}}$$

Fig. 178A. Mission Requirements Drive Range and Endurance. Fig. 178B. Wing Flattens the Power Required Curve

Compound Helicopters Sizing, Synthesis, and Performance

The RF Fuel Balance Method is the basis for most fixed-wing and rotary-wing aircraft sizing and performance in industry and government. Georgia Tech has developed it and used it in the AHS Student Design Competitions starting in 1985 and dominating the first ten years, winning 80% of the AHS SDCs. All types of VTOL aircraft were designed. In addition, Georgia Tech has had several contracts with the Air Force, Army and Navy on Combat

Search and Rescue (CSAR) VTOL aircraft, including coaxial compounds, tiltrotor aircraft and for the evaluation of the Piasecki Speed Hawk, Fig.179.

Figure 179. Compound Helicopters Evaluated by Georgia Tech Using the RF Method

An Approach for Overcoming the Overly Constrained FARA Problem

The Overly Constrained FARA Problem is partially due to the lack of sufficient power available and not using technology currently being incorporated in hybrid-electric solutions for Civil VTOL Aircraft. An initial study of potential compound helicopters has been conducted at Georgia Tech for overcoming these Constraints.

Task 1: Build up the mission elements for a representative future compound attack helicopter (CAH) mission representative of FVL

Capability Sets 1 and 3, resulting in mission gross weights of 10,000, 15,000, and 20,000 lbs. with max continuous power (MCP) cruise speeds of 185 kt, 200 kt, and 215 kt for a Generic Mission for an Advanced Single Main Rotor Compound (ASMRC) Study. The generic mission utilized in Phase 2 is based on the NASA Ames studies for the "Design of a Slowed-Rotor

Compound Helicopter for Future Service Missions" by C. Silva, H. Yeo, and W. Johnson. The resulting mission is illustrated in Figure 180.

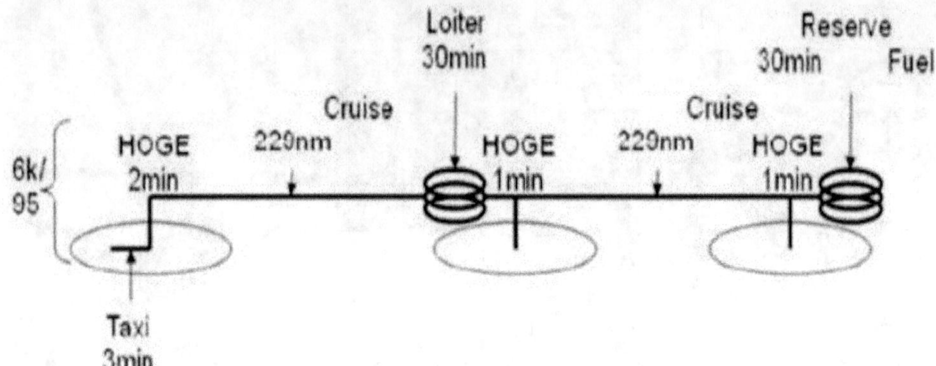

Figure 180. Design Mission Profile for Design of a Slowed-Rotor Compound Helicopter Task 2: Identify design parameters from the Phase 1 selected ASMRC for Phase 2

The selected ASMRC configuration is similar in arrangement to the AH56A Cheyenne (lifting wing, conventional tail rotor or ducted propeller or fan, and an open propeller aft of the tail rotor), as notionally depicted in Figure 181.

Figure 181. Notional Clean ASMRC

Previously, a Combat Search And Rescue (CSAR) compound helicopter study was conducted at Georgia Tech for the USAF using a similar sizing and performance fuel and power balance method, RF, as used in this study. The objective of the CSAR study was to parametrically demonstrate the impact of helicopter design and technology variables on the requirement capabilities in a typical CSAR mission. During the CSAR study, confidence in the methodology was gained through calibration and benchmarking, where it was sought to compare CSAR vehicle concepts using the RF Method and make distinctions, and hopefully eliminations of alternatives. Many combinations were examined, but only four vehicle concepts, as illustrated in Figure 182, are shown here to make the necessary notable distinctions between what was found to be the main alternatives. The four alternatives are depicted below; the third concept from the left included a wing, separate tail rotor, an open propeller aft of it and with a slowed rotor, it was the recommended as an ASARC concept configuration for evaluation.

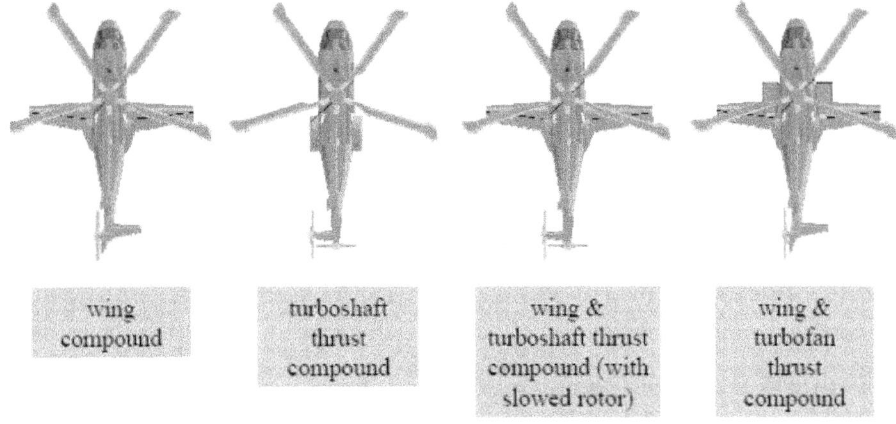

Figure 182 – Several of the CSAR Vehicle Concepts Analyzed

Also, from this CSAR study, the trade-off between weight and performance was indistinguishable between the different Aux Propulsion Driver alternatives. Therefore, the AH-56A Cheyenne configuration with an open tail rotor and an open propeller was selected as the baseline configuration for this study. The use of a slowed rotor is essential for high-speed rotorcraft. In a recent internal design study, design parameters such as empty weight ratio and equivalent flat plate drag area were assumed, and payload available for the Single Main Rotor Compound (SMRH) at the

Hover-out-of-Ground Effect (HOGE), the HOGE 6K 95deg F condition, which was used for the other configurations, the Single Main Rotor Compound (SMRC) Baseline and the ASMRC Baseline. In Phase 2 these parameters were built up based on emerging technologies that will reach a TRL 6 in the next five years for the ASMRC. The Improved Technology Engine (ITE), as described in Figures 183, 184 and Table 1, was approximately represented as a fixed engine for weight, shaft horsepower and fuel consumption in this study. Payload was adjusted as necessary to have mission gross weights at 10,000, 15,000 and 20,000 lb.

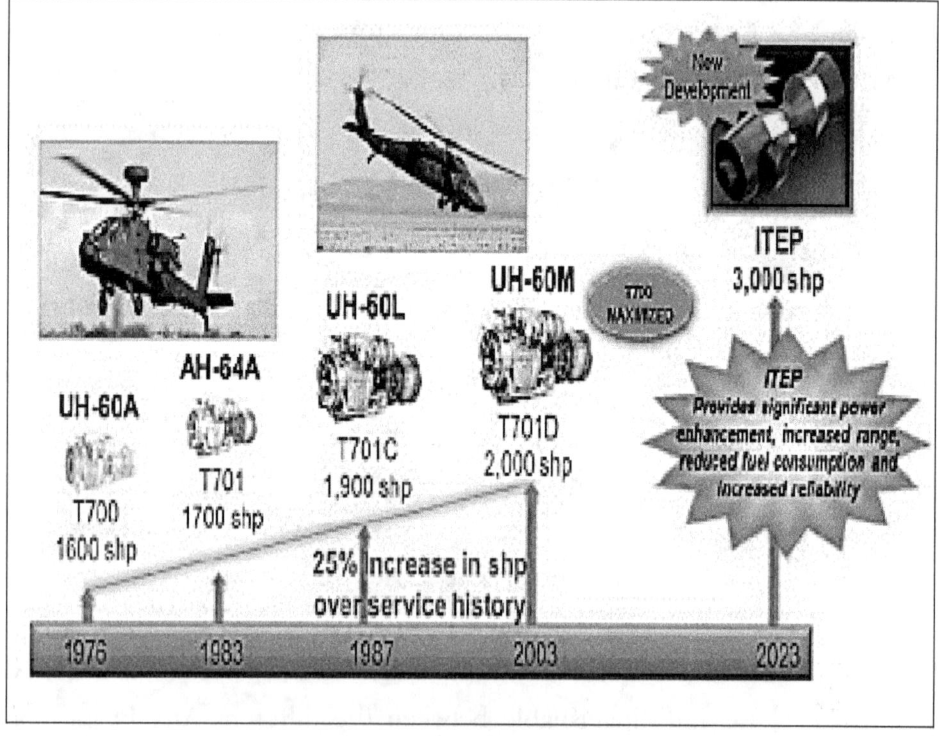

Figure 183. T700 Development and the Leap to the ITE

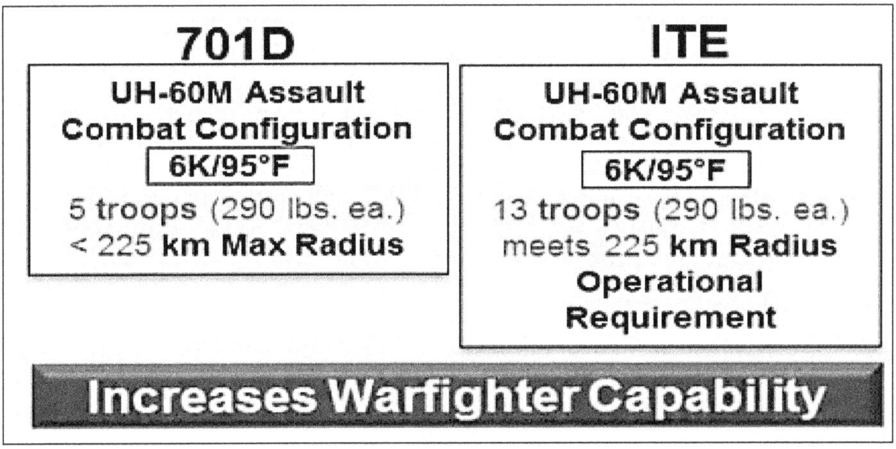

Figure 184. Worldwide Mission Capability While Meeting Operational Range and Payload Requirements

Power	Objective: 2050 SHP at 6,000 ft., 95°F, static, Maximum Rated Power (10 Min) Threshold: 1850 SHP at 6,000 ft., 95°F, static, Maximum Rated Power (10 Min)
Specific Fuel Consumption	Objective: ≤.352 lbs./hp-hr @ 1450 SHP when operated at 6,000 ft., 95°F conditions ≤.347 lbs./hp-hr @ SL/ISA STD Intermediate Rated Power baseline. Threshold: ≤.409 lbs./hp-hr @ 1450 SHP when operated at 6,000 ft., 95°F conditions. ≤.370 lbs./hp-hr @ SL/ISA STD Intermediate Rated Power baseline.
Weight (Dry)	Objective: ≤ 456 lbs. Threshold: ≤ 500 lbs.
Engine Growth Capability	Threshold: 25% inherent power growth capability within the current engine mounting, nacelle and installation envelope.

Table 1. Major ITE Engine Performance Requirements

Simplified Engine Power Methodology The engine is based on the ITE estimated improvements published by GE for their T-901 engines, compared to their T-701 engines. Using the well-documented performance values for the 701 engines, we then multiplied these values by the improvement factor from the T-901 engine improvement estimates. For example, their performance goal is an increase in power of 50% over the 701 series, so we multiplied the sea level standard MCP specification for the 701 by 1.5. Engine power for environmental conditions other than sea level standard was calculated from the equation.

*PowerAvailable = Power at SL Standard * (1 − 0.195 * hoverAlt 10000) * (1− 0.005 * (hoverTemp − ISATemp@Alt)*

Parasite Drag is an important consideration for rotorcraft because of its very strong influence on total power required in high-speed, level flight. Figure 161 provides historical values for rotorcraft and fixed-wing aircraft parasite drag (gross weight over equivalent flat plate area, FE) as related to gross weight. Clean airframe design was assumed for the Phase 2 SMRC designs; therefore, the "Clean Helicopters" trend line of Figure 185 was adopted for the SMRC designs in Phase 2. Reading from this trend line for our 10,000, 15,000, and 20,000 pounds gross weight designs, the parasite drag values are:

- GW/fe = 790, thus fe = 10,000/790 = 12.7 ft2 for the 10,000 lb GW design; GW/fe = 950, thus fe = 15,000/950 = 16.3 ft2 for the 15,000 lb GW design; GW/fe = 1050, thus fe = 20,000/1050 = 19.0 ft2 for the 20,000 lb GW design

To achieve this level of drag "cleanliness" will require, as a minimum, retractable landing gear, internal or conformal carriage of stores, and conformal mounting of sensors. The placement of the propeller at the rear of the Phase 2 designs will contribute to the achievement of lower parasite drag because the flow into the propeller disc will "clean-up" low-pressure flow areas at the rear of the fuselage. In aircraft industry design practice, low drag design also depends on knowledgeable customers and a strong-willed program manager.

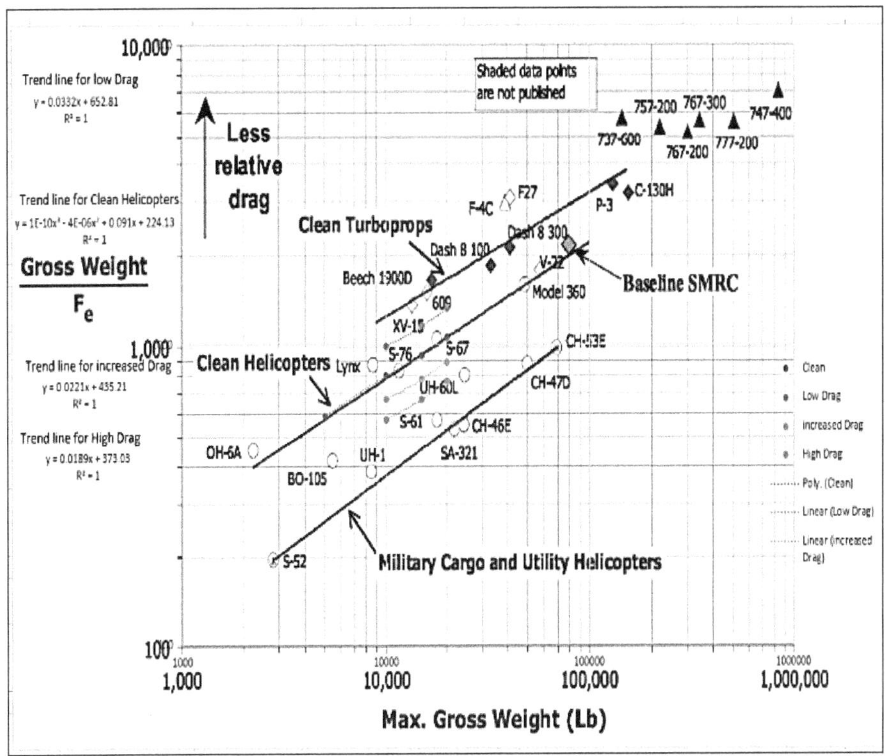

Figure 185: Rotorcraft Parasite Drag vs. GW Comparison

Advancing Blade Tip Mach Number Limit

In order to minimize main rotor profile drag rise due to compressibility effects on the rotor blades in high-speed flight, it is useful to slow the rotor rotational speed as a function of atmospheric temperature and flight airspeed. This limits the advancing blade tip Mach number (MAT) to a value below drag divergence (MDD): $\Omega R = MAT\ VS - V$ Keeping MAT below MDD also has the important advantage of lower rotor vibratory loads in high-speed flight. The Boeing VR-15 is a good choice for a rotor blade airfoil in the blade tip region. It is considered an "advanced" airfoil but is "low risk." The MDD of the VR-15 near zero lift coefficient, which is where one would expect the advancing blade tip region in high-speed cruise to be operating (especially for a compound helicopter, with the rotor substantially unloaded in thrust by the wing and auxiliary propulsion), is about 0.825 at Reynolds numbers typical for a helicopter rotor blade. The VR-15 airfoil section characteristics were assumed for the blade tip region of the main rotor for the Phase 2 designs.

The sizing methodology for Phase 2 was modified to incorporate a MAT limit which slows the main rotor rotational speed to keep MAT at or below an input MDD, which was set to 0.81 to provide some "cushion" below 0.825. This value is the MAT limit used in the compound helicopter design in other studies. If rotor rotational speed reduction is achieved by means of a multi-speed transmission, there are penalties in weight, complexity, and cost, and possibly also in R&M, but no penalty in engine performance.

On the other hand, if rotational speed reduction is achieved by means of slowing engine output shaft RPM, the multi-speed transmission penalties are avoided. But, there is a penalty in loss of power available at maximum continuous power rating. The Phase 2 designs use a full-RPM design tip speed of 650 ft/sec, and the design sizing mission is flown at 6K'/95; thus, rotor rotational tip speed was limited as follows:

Airspeed, knots	Max Rotor Rotational Tip Speed, ft/sec	Percent of Design Tip Speed (650 ft/sec)
185	623	96
200	598	92
215	572	88

Aircraft turboshaft engine output shaft RPM can be slowed to the 80 to 85 percent range without undue loss in engine performance. Therefore, this approach was chosen for the Phase 2 designs rather than the multi-speed transmission method.

Wing Lift Sharing & Tail Rotor Offload Scheduling

In order to be able to trim the rotor and reduce vibratory loads, the main rotor will not be completely unloaded, as illustrated in Figures 186A and B. It will provide 20% of the lift at the vehicle mission weight at the design cruise speed. The offloading of the rotor begins at 30.4 knots (35 MPH), where the wing begins to provide lift, and rotor thrust is reduced linearly until it provides 20% of the lift at the design airspeed of 185, 200, or 215 knots. The equilibrium diagram in Figure 175 above (where propeller thrust, TP, is represented by the equation: $TP = D - KLWG\alpha$), illustrates how propeller thrust and main rotor thrust can be scheduled for pitch attitude control, a strong feature of an SMRC helicopter over conventional helicopters where airframe pitch attitude changes during acceleration and decelerations. Thus, the auxiliary propeller can provide up to 100% of the propulsive force for all airspeeds in trimmed-level flight. The auxiliary propeller and separate tail rotor will provide excellent low-airspeed maneuverability and agility compared to ducted propellers, fans, and NOTAR.

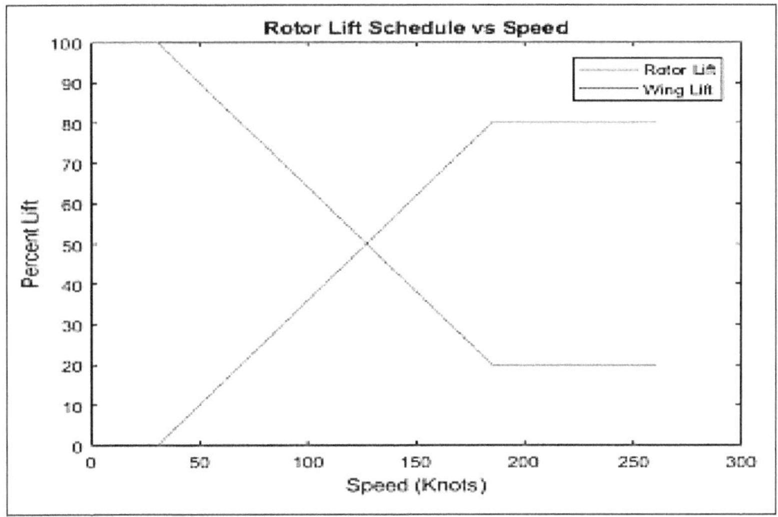

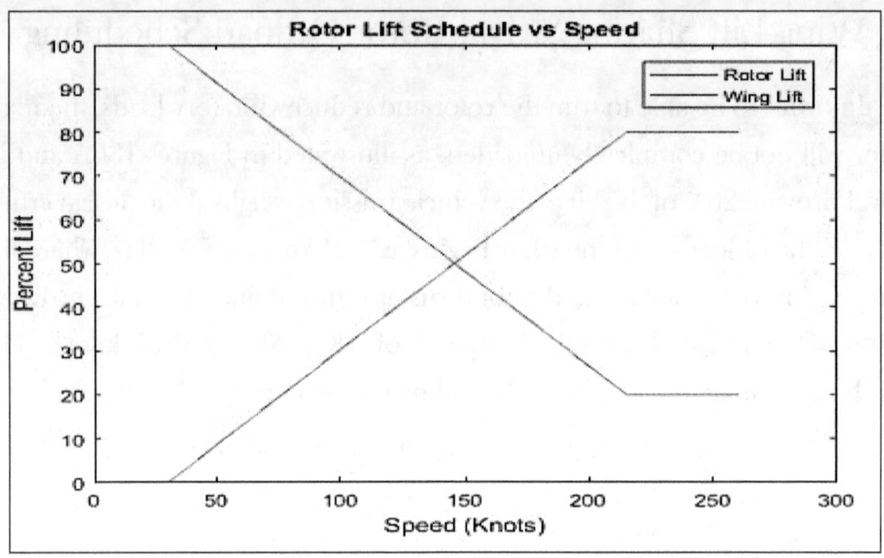

Fig. 186A: Rotor Lift Share Schedule (Design Airspeed of 185 knots). Fig.186B. Design AS of 200 knots

A more advanced rotor–wing lift share schedule could be utilized to minimize power across all airspeeds. In a similar fashion, the tail rotor is linearly offloaded starting at 56.5 knots (65 MPH) until it is fully offloaded at the design airspeed. The vertical fin and rudder are used to produce counter-torque moment as the tail rotor is offloaded.

Empty Weight Estimation

Empty weight fraction was estimated using the Shinn weight equations from the 1981 Advanced Scout Helicopter study and the JVX Technical Assessment (JTA) study. The Shinn weight equations, as do most weight estimating methods, use a multiple linear regression general formula which has the form: W=kAxByC z....etc., where W = weight of the respective component (e.g., main rotor blades) A,B,C = aircraft design parameters that have some effect on the weight of the component (e.g., gross weight, rotor radius, solidity, horsepower, etc.) k, x, y, z ....= constants determined by the multiple linear regression computer program.

Initial Application of Hybrid-Electric Power Effort

An overview of the emergence of Hybrid-Electric Propulsion (HEP) technology for rotorcraft is included. A brief description of how it could possibly be included for a compound helicopter is summarized here. If the power required to hover or to meet the design speed was not sufficient for a given number of ITE engines, then the additional power could be supplied via a HEP system. The electric motors could be sized to make up the difference in required horsepower, either hover or high-speed, or both. The battery is sized based on the required horsepower and the mission hover time, plus one minute (5 minutes total) for the mission in Figure 180. For the hybrid configurations, the transmission is sized to meet the greater of the horsepower required for cruise or hover at 6K 95F, multiplied by 1.5, or the maximum rated power of the engines at sea level standard. For the standard configurations (15,000 lb.) with two ITE engines, the transmission is sized for the power of the two ITE engines, at sea level standard, without the need for auxiliary HEP.

Power Required and Available Curves

Sensitivity Analysis Payload fraction is used to determine the best combination of disk loading, main rotor solidity, and necessary use of maximum rated power (MRP, a10 minute rating) or intermediate rated power (IRP, a 30-minute rating) versus the use of MCP. Also, the possible use of HEP is initially investigated. These considerations are especially true in view of the Army's need for the ITE to help carry thirteen troops (+crew) at 290 lb each for a 225 km mission radius, as illustrated in Figure 180.

In accordance with the compound helicopter sizing section in AMCP 706-201, Helicopter Engineering, Part One: Preliminary Design, rotor geometric solidity, bc/πR, is determined based on the following rationale.

The broad requirements for the rotor are to operate efficiently in hover at moderate speeds, to provide its share of the required maneuvering capability, and, at high speeds, to produce as much lift as possible without producing excessive vibration or loads. Once the disk loading is chosen, either on the basis of using all of the installed power to hover or on the basis of maximum allowable downwash velocity, several other rotor parameters must be defined:

- Tip speed
- Solidity
- Twist
- Airfoil section

The main rotor tip speed was established by the combination of the maximum forward speed and the maximum allowable advancing tip Mach number. For this study, tip speed was controlled by slowing the rotor to tip speed Mach number not to exceed 0.81, as discussed in the Advancing Blade Tip Mach Number Limit section. The twist and airfoil selections were based on projected future compound helicopter rotors.

As stated in AMCP 706-201, the rotor solidity is determined by one or the other of two requirements—hover performance or maneuver capability. For a given disk loading and tip speed, the solidity helps to define the hover Figure of Merit (although the Figure of Merit is also a function of CT/σ, twist, and tip Mach number). For steady turns, the rotor must supply whatever lift the wing cannot. The solidity that meets this requirement is illustrated in the equation below:

$$\sigma_{opt_{max}} = \frac{nW_g - L_{w_{max}}}{\rho(\Omega R)^2 A \left(\dfrac{C_T}{\sigma}\right)_{max}} \text{, d'less}$$

where

$$n = \text{load factor for turn, dimensionless}$$
$$W_g = \text{helicopter gross weight, lb}$$
$$L_{w_{max}} = \text{maximum lift of wing, lb}$$

Rather than trying to find a maximum optimum solidity for maneuvering flight, a parametric variation of solidity values, 0.07, 0.09, 0.11, and 0.13, is provided in the Power Required and Available plots included in the Appendices. A solidity of 0.09 was selected to emphasize hover efficiency in the results presented in this main part of the report.

The Power Required and Available Sensitivity Analysis plots for 10,000 lb,15,000 lb., and 20,000 lb. mission gross weights are presented below, along with Empty Weight (EW) fraction charts.

Figures 159, 160, and 161 show the power curve for a range of disk loadings with a vehicle that weighs 10,000 lb. Depending on the design disk loading, the 10,000 lb. GW ASMRC can meet the hover and maximum airspeed requirements with a single ITE engine running at the ten-minute MRP or the thirty-minute intermediate power rating (IRP). A higher solidity of 0.13 was also assessed to account somewhat for maneuvering flight and blade loads. The key takeaway when viewing Figures 187A to C is that the power required at hover and the maximum forward airspeed depends heavily on the Disk Loading. Constraining FARA to a set disk loading by mandating a 14,000-pound maximum gross weight and a 40-foot diameter rotor removes the trade space available to the engineers. Figures 187A, 187B and 187C show that with an appropriate disk loading, a 10,000-pound vehicle can meet the 6K95F hover requirement and the max speed requirement with a single ITEP when considering an SMRC design. If the gross weight is increased to 15,000 pounds (1,000 pounds greater than the FARA max gross weight), then no disk loading could meet the requirement on a single ITEP. If a single-engine design is an Army priority, it must allow the engineers to vary both the max gross weight and the disk loading or reexamine the performance requirements.

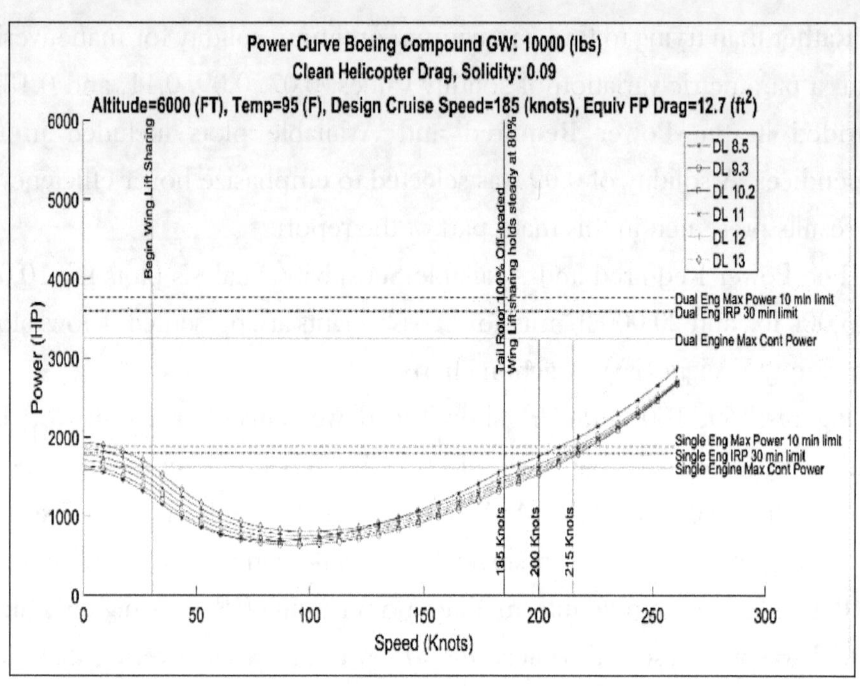

Figure 187A. 10,000 lb GW, 185 Knots Design Maximum Cruise Airspeed

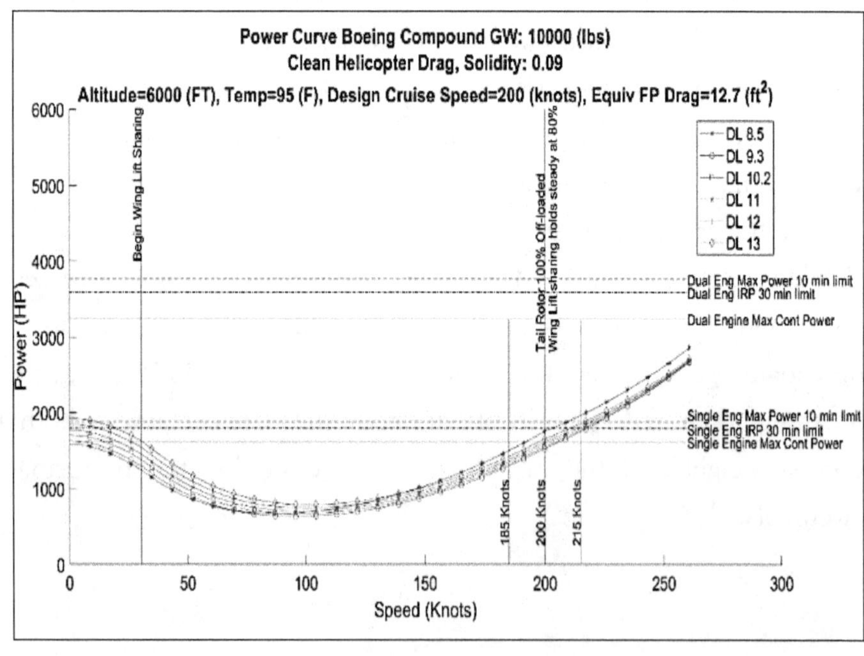

Figure 187B. 10,000 lb. GW, 200 Knots Design Maximum Cruise Airspeed

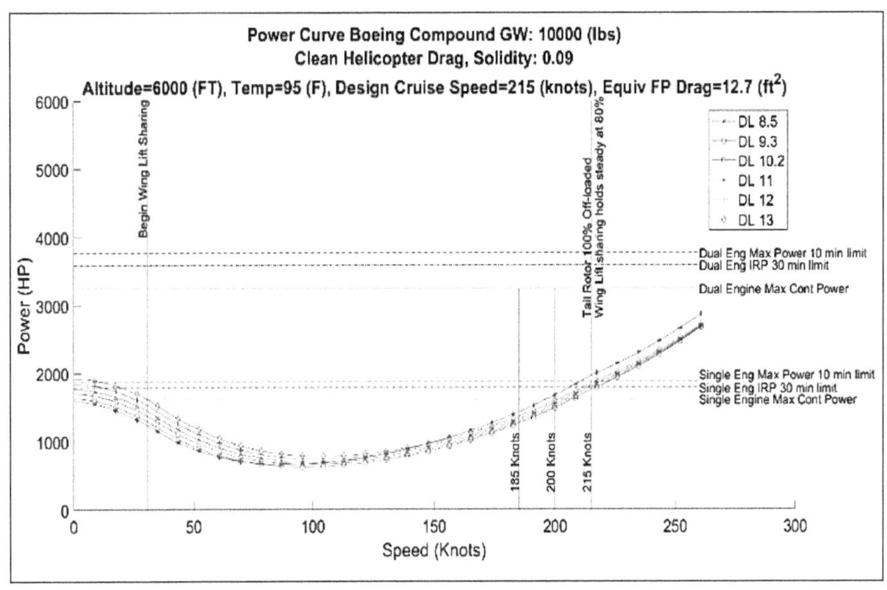

Figure 187C. 10,000 lb GW, 215 Knots Design Maximum Cruise Airspeed

Two ITE engines provide the appropriate amount of power available for the 15,000 lb. GW ASMRC for all disk loadings and design airspeeds and have margin for growth in gross weight or parasite drag, as illustrated in Figures 188A and B and for 20,000lbs in Figure 188C.

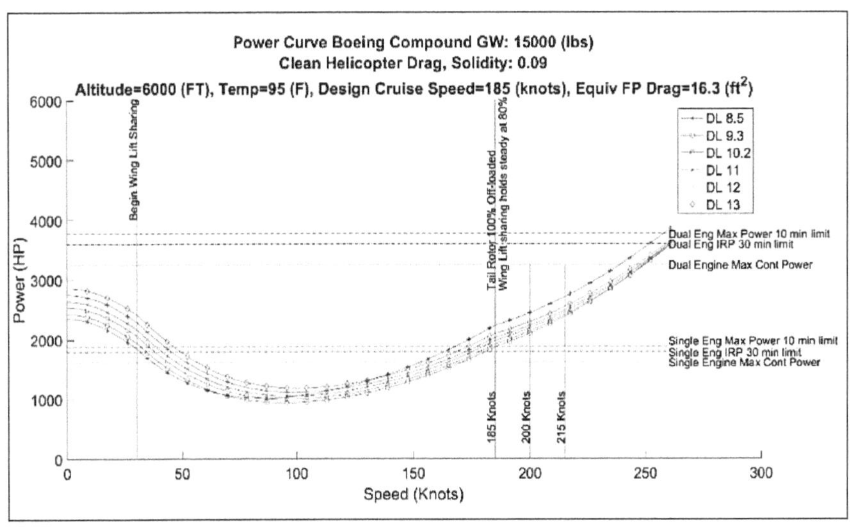

Figure 188A. 15,000 lb GW, 185 Knots Design Maximum Cruise Airspeed

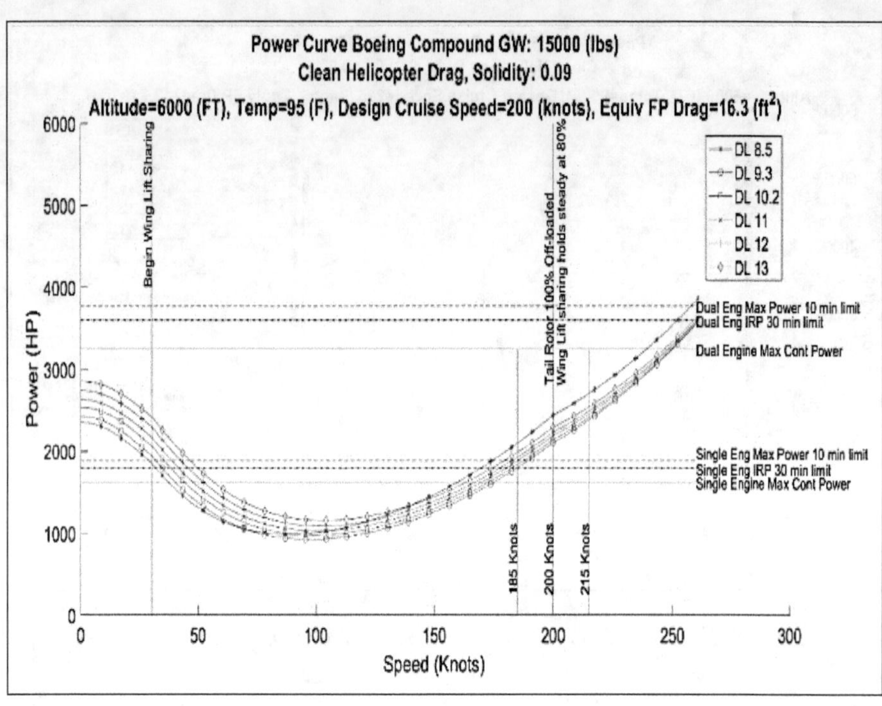

Figure 188B. 15,000 lb. GW, 200 Knots Design Maximum Cruise Airspeed

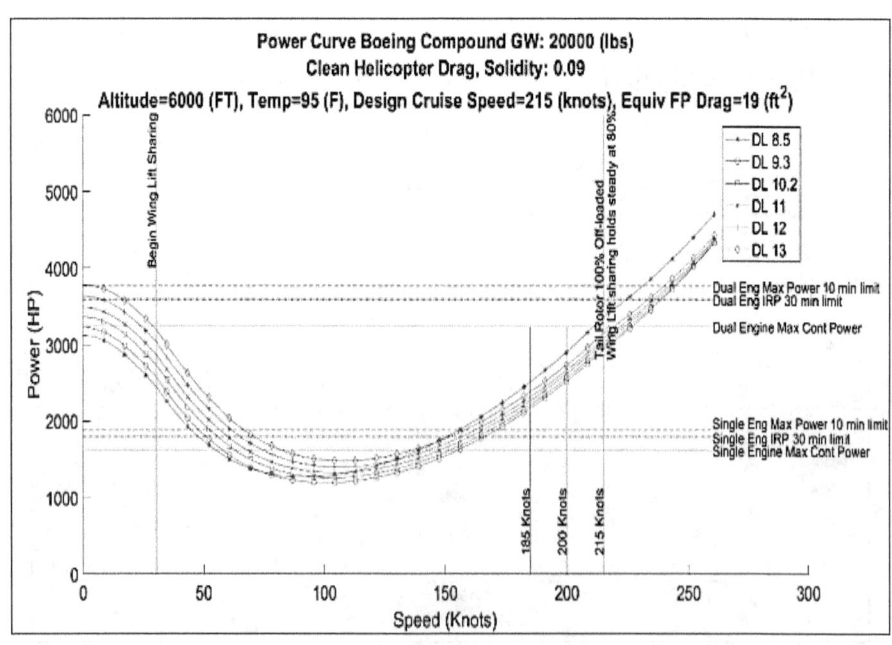

Figure 188C. 20,000 lb. GW, 215 Knots Design Maximum Cruise Airspeed

There is little change in the power required curves for each design maximum cruise airspeed. The major reason for this small difference between these design airspeeds is the lift scheduling and tail rotor offloading. The design airspeed has a more significant impact on payload since higher airspeeds require more power, and thus more fuel is required for the mission. This results in less weight available (in the fixed gross weight) to be devoted to the payload. An example of the impact on Empty Weight Fraction (EW/GW) of design gross weight is presented in Figure 189 for the high airspeed flight condition of 185 kt.

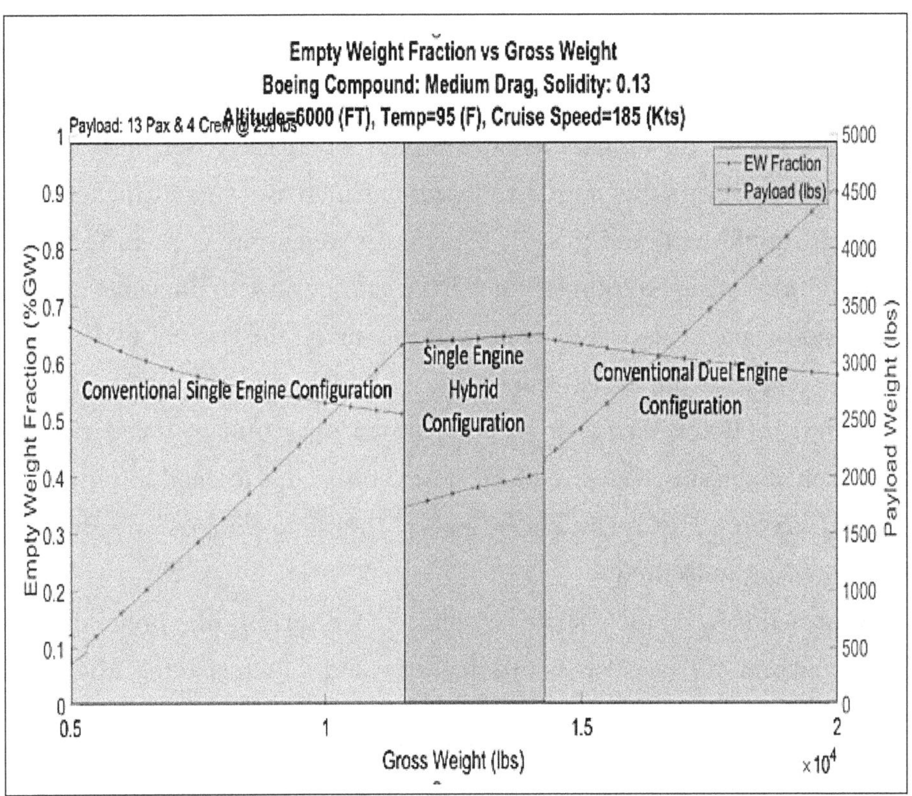

Figure 189: Configuration for Empty Weight Fraction and Payload vs. Gross Weight, Disk Loading of 8.5

The optimal power train configuration (single ITEP, HEP, or dual ITEP), which results in an aircraft with the lowest gross weight and thus the lowest cost, changes depending on the gross weight of the vehicle. As seen in

Figures 189, at very low gross weights, the conventional single-engine configuration provides sufficient power and has the lowest empty weight fraction. As the gross weight increases, the single-engine no longer provides enough power for the 6K 95F mission requirement. As the power required continues to increase with an increased gross weight, a conventional dual-engine configuration becomes lighter than the hybrid configuration of similar total power.

For both conventional single and dual-engine configurations, the empty weight fraction decreases as the gross weight increases because the drive section's weight is a function of horsepower. Since engine power is fixed, the drive section becomes a smaller portion of the total vehicle weight as the gross weight is increased. The empty weight fraction increases slightly in the hybrid configuration as gross weight increases because the battery weight and motor weight increase with the power requirement increase due to higher gross weight. It should be noted that the exact gross weight break point for which the power train configuration is best will be very sensitive to the battery energy density and capacity. Using the values in this study, the current FARA gross weight is very near the break point between a HEP and a conventional dual ITEP design. If the Army desires the lower operating cost and reduced complexity that a single-engine ITEP design offers, it will need to reduce the required maximum gross weight by several thousand pounds or reduce the performance requirements.

Figures 190 A, B and C show how the shaded regions (the optimal power train configuration) vary across disk loadings and design speeds. The trends discussed above hold true for all disk loadings that were investigated. The differences in results among the different disc loadings are largely due to the weights of the drive section and of the rotor section, which are functions of power and thus are impacted by disk loading and rotor diameter, the latter being a fall-out of the chosen disk loading. The payload is equal to the design gross weight minus the fuel weight and the empty weight.

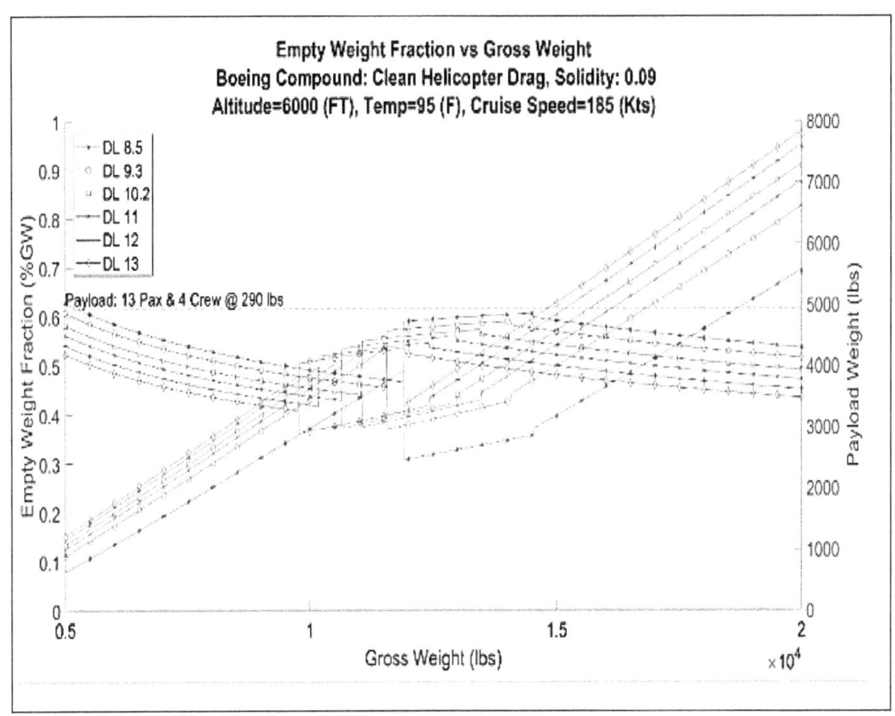

Figure 190A: Payload & Empty Weight Fraction vs. GW, Design Airspeed of 185 Knots

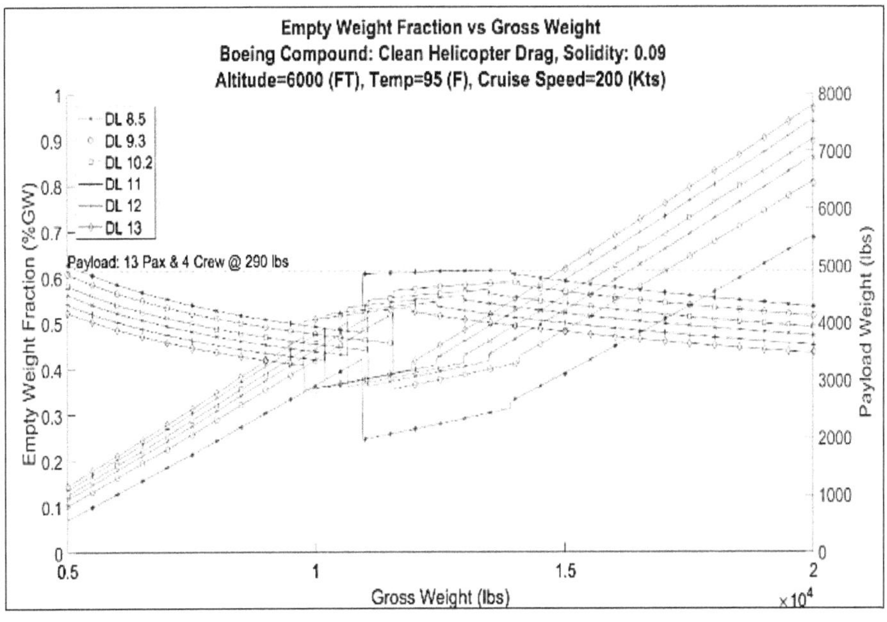

Figure 190B. Payload & Empty Weight Fraction vs. GW, Design Airspeed of 200 Knots

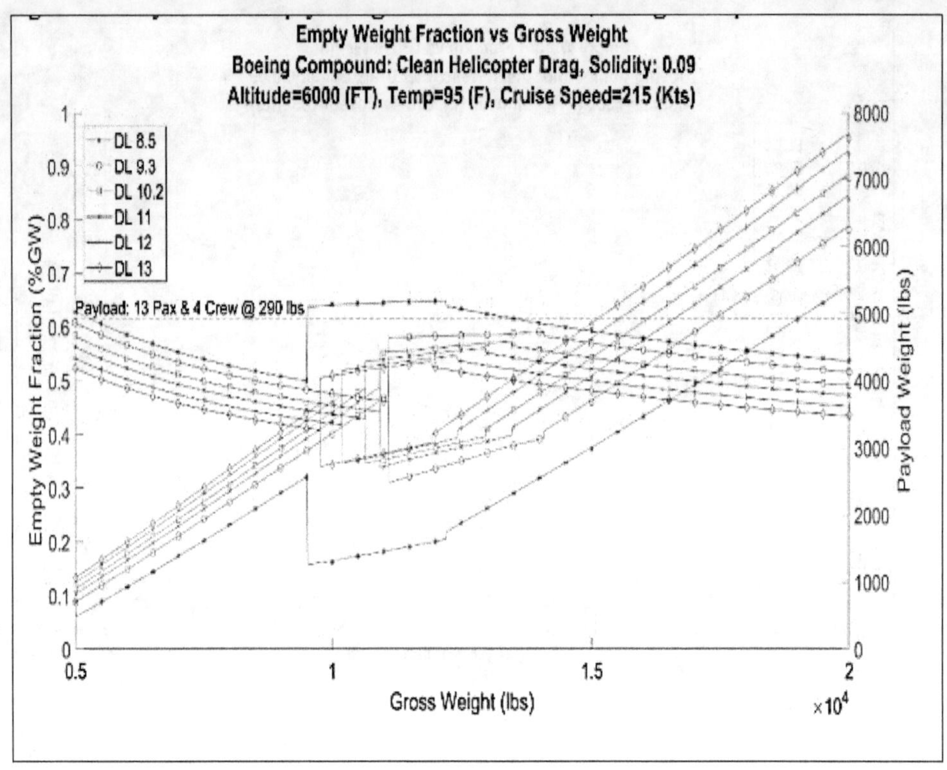

Figure 190C. Payload & EW Fraction vs. GW, Design Airspeed of 215 Knots

There are only slight changes in the required fuel between the 185 and 200 knot design airspeeds as they are not far from the best range airspeed. However, once the design airspeed is pushed to 215 knots, we see a larger change in power required for the lower disk loadings (e.g., 8.5) as this airspeed is much further from the best range airspeed, resulting in higher fuel requirement, and thus less available payload.

The effect of advanced technology factors for weight reduction would be to lower the empty weight and thus increase the payload weights for all aircraft gross weights. In summary, operation at MRP offers a hover and high airspeed solution at all three gross weights and all three maximum airspeed requirements; however, the rapid advancement in HEP may offer, in the 2030-35 timeframe, a viable alternative to operating the engine(s) at MRP.

Task 1: Air Vehicle Configuration Concepts

A general overview of the vehicle configuration is illustrated in Figure 191. The design team focused on generating a conceptual layout for the three different design weight classes, 10,000 lb., 15,000 lb., and 20,000 lb., for the SMRC helicopter. The basic aircraft geometries were created in PTC's CREO Parametric 2.0. Geometric variables such as rotor radii, wingspan, wing chord, and vertical and horizontal tail areas are derived from the sizing and performance routine. Each of the three SMRC helicopter configurations consists of a single main rotor, a lifting wing, a counter-torque tail rotor mounted atop a vertical tail, a horizontal tail, and a pusher propeller.

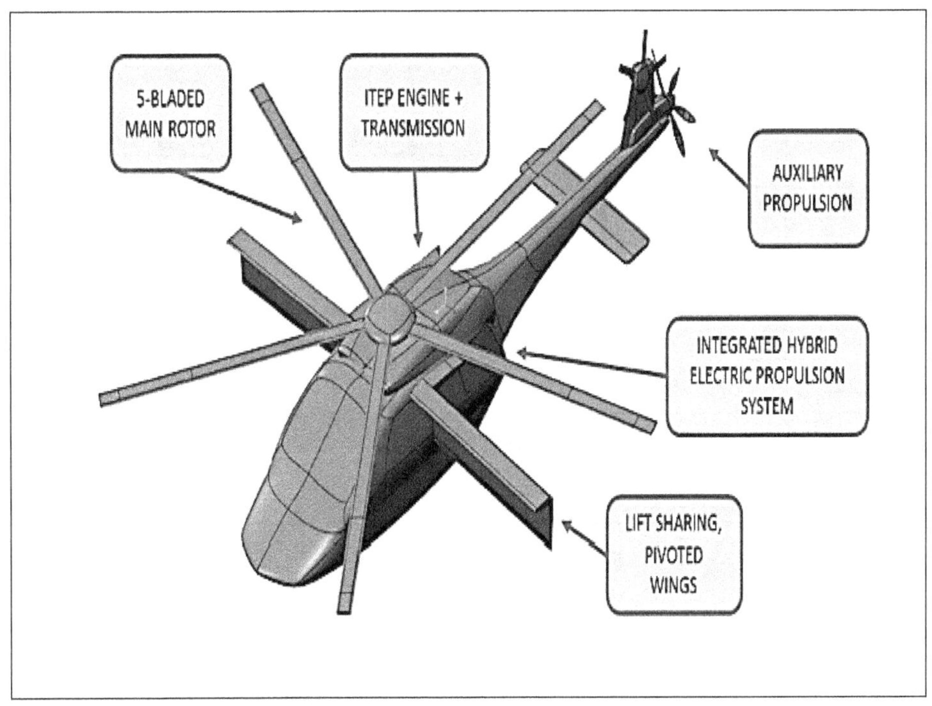

Figure 191. Vehicle Configuration Schematic

The engine(s) are mounted on top of the fuselage and supply power to the main rotor, tail rotor, and propeller through the transmission. The five-bladed main rotor incorporates moderate linear twist and tapered, swept blade tips.

Task 2: Conduct the initial vehicle sizing and final design parameter selection for the selected ASMRC concept.

From the Power Required and Available Sensitivity Analyses plots for 10,000 lb., 15,000 lb., and 20,000 lb. mission gross weight (Figures 187A, B and C), and the Payload and Empty Weight versus Gross Weight plots (Figures 188A, B and C) for the three different cruise speeds: 185 kt, 200 kt, and 215 kt), a Point Design for each weight category was identified and is included in Table 2. These baseline point designs include nineteen baseline technologies, and a main rotor solidity of 0.13 to initially address maneuvering flight. The incorporation of advanced technologies, particularly drag and empty weight reduction, will be addressed in the next sections.

GW	lb	20000	15000	10000
Propulsion	--	2 ITE Engines	2 ITE Engines	1 ITE Engine
Disk Loading	lb/ft^2	10.2	13	12
Radius	ft	25.0	19.2	16.3
Solidity	--	0.13	0.13	0.13
V_{tip}(Hover)	ft/s	650	650	650
V_{tip}(Cruise)	ft/s	572	572	598
V Cruise (max continuous)	Kts	213	215+	200+
Equiv FP Drag	ft^2	19.0	16.3	12.7
Empty Weight Fraction	--	0.466	0.510	0.450
Download Factor (ed)	% GW	0.078	0.093	0.088
Payload	lb	7010	4500	3550
Fuel	lb	3670	2850	1950

Table 2: Point Designs with No Advanced Technology Factors Applied

The point designs were chosen by prioritizing first, less aircraft complexity (conventional design wins over a HEP design) and second, payload. At the 10,000 lb. class, a disk loading of 12 lb./ft2 was selected. This kept the design with a single ITE engine with no HEP needed. The payload of 3550 lb. does not meet the desired FVL aircraft payload value of 4930 lb. for thirteen passengers and four crew at 290 lb. each. This 10,000 lb. aircraft would be able to carry eight passengers and four crew, or ten passengers and two crew. The 15,000 lb. class vehicle maximizes payload with a disk loading of 13 lb./ft2 and does not require HEP. It also misses the thirteen passengers and four crew desired payload. The 20,000 lb. vehicle maximizes payload with a disk loading of 10.2 lb./ft2 and can carry twenty passengers and four crew.

Advanced Technologies for Airframe Weight Reduction

Several trend lines for parasite drag for the SMRC configurations are presented in Figure 185, for use in the parasite drag variation study. Advanced technologies for parasite drag reduction studied in Reference 5 (Wilkerson, J.B., and Smith, R.L., *Aircraft System Analysis of Technology Benefits to Civil Transport Rotorcraft*, NASA/CR-2009-214594) were used to identify the projected trends in Table 3 below.

Reference 5 also included Active Flow Control (AFC). Active Flow Control is the use of small, distributed, airframe surface orifices which energize local flow conditions by means of very low, or zero, oscillating mass flow, in order to delay boundary layer separation. AFC has been experimentally proven to reduce drag, and several possible applications have been explored in model-scale wind tunnel tests, and at least one full-scale flight test. Research test results reported in Reference 5 indicate application of AFC to lifting wings can limit wing profile drag coefficient (based on wing planform area) to 0.008, up to a wing lift coefficient of at least 1.0. This result was analyzed for our SMRC helicopter designs as follows. Assuming the parasite drag breakdown of the SMRC helicopter in Reference 5 is typical, the fraction of aircraft parasite drag due to the wing profile drag of our Phase 2 design is 20% of the total parasite drag:

wing fe = 0.2 x 12.7 = 2.54 ft2 for the 10,000 lb. GW design;

wing fe = 0.2 x 16.3 = 3.26 ft2 for the 15,000 lb GW design;

and wing fe = 0.2 x 19.0 = 3.8 ft2 for the 20,000 lb GW design.

With the application of AFC to the wing of the Phase 2 design, the wing profile drag coefficient is assumed constant at 0.008, thus:

(wing fe)AFC = 0.008 AW,

where AW is the platform area of the wing. Then the parasite drag reduction due to the application of AFC to the wing was calculated as:

(Δ fe)W = 0.008 AW – 2.54 ft2 for GW = 10,000 lb.

(Δ fe)W = 0.008 AW – 3.26 ft2 for GW = 15,000 lb.; and

(Δ fe)W = 0.008 AW – 3.80 ft2 for GW = 20,000 lb.

Research results also reported in Reference 5 indicate that by applying AFC in a very tailored manner to the main rotor pylon and the upper fuselage and wing area near the pylon, a 50% reduction in pylon and rotor hub drag could be achieved. This would probably require analytical and experimental development of the specific AFC configuration, including a wind tunnel test series.

It is typical in helicopter parasite drag breakdowns that the drag of the main rotor hub and pylon is approximately 50% of the total parasite drag of the aircraft. However, since the SMRC helicopter has a wing, we assumed that the main rotor hub and pylon constitute only 40% of the total parasite drag. Then, with the indicated 50% drag reduction with AFC, the reduction in parasite drags for each design gross weight, due to the application of AFC to the main rotor pylon and the upper fuselage and wing area near the pylon, was calculated as:

Δ fe)H & P = -(0.5 x 0.4 x 12.7) = -2.54 ft2 for GW = 10,000 lb,

(Δ fe)H & P = -(0.5 x 0.4 x 16.3) = -3.26 ft2 for GW = 15,000 lb, and

(Δ fe)H & P = -(0.5 x 0.4 x 19.0) = -3.80 ft2 for GW = 20,000 lb

GW	Low	Clean	Medium	High
10,000	10.2	12.7	15.2	17.8
15,000	13.0	16.3	19.6	22.8
20,000	15.2	19.0	22.8	26.6

Table 3. Summary of Parasite Drag Estimates: High, Medium, and Low

Hover-out-of-Ground Effect (HOGE) Download Reduction

As part of the study of advanced technologies for the Phase 2 SMRC helicopter design, providing a swiveling capability for the wing was considered, for the purpose of reducing the HOGE download due to vertical drag produced by the rotor downwash on the airframe. Specifically, part of the wing was assumed to swivel about a span-wise axis so that it is at an angle of attack to the rotor downwash such that its lift coefficient is zero. The wing would be locked against swiveling except in hover and very low airspeeds.

It was conservatively assumed that the profile drag of the wing, when swiveled into the rotor downwash, is approximately 10% of the vertical drag of the wing in the normal position, since, in the latter case, it is like a flat plate in crossflow. Further, it was estimated that the wing in the normal "down" position would be 50% of the total aircraft download in HOGE. Then, if the percent download of the complete aircraft is (%DV)1, (%DV)2 = (0.5) (%DV)1 + [(0.5)(%DV)1 - (0.9)(0.5)(%DV)1] = (0.5)(%DV)1 + (0.05)(%DV)1 = (0.55)(%DV)1

The design team considered the incorporation of a wing pivoting mechanism allowing 75-85% of the wing area to be pivoted downward, to reduce hover download, and thus reduce hover power required. Figure 174 shows the aircraft in hover configuration with the wing pivoted down. Further reduction in hover download could be incorporated by the pivoting of the entire wing about a span-wise axis, which would reduce up to 90% of the project wing area. A detailed analysis of the pivoting mechanism would be needed, along with the sensitivity of hover download to wing projected area, to justify the added weight and complexity of pivoting the wing. There could also be the issue of hindrance of passengers' ingress/egress.

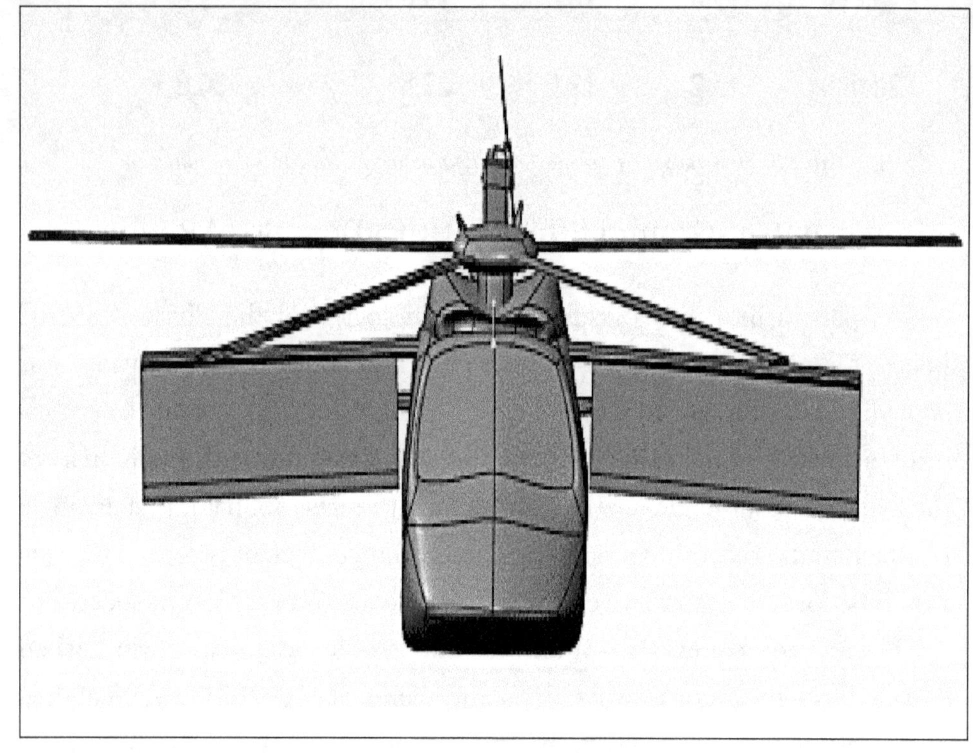

Figure 192: Hover Configuration

Task 4: Compare the performance characteristics of the selected ASMRC CAH qualitatively with other potential SMRC CAH concepts.

A subjective evaluation of VTOL aircraft concepts versus desirable features was initially developed by the Army Aviation Flight Dynamics Directorate (AFDD) a number of years ago. As part of the current study, this was updated to provide a qualitative evaluation of the SMRC against other VTOL aircraft concepts, Table 4.

VTOL Concept / Features / Examples	Helicopter	Compound (Tail Rotor + Aux Prop)	Compound (Advancing Blade Concept)	Compound (Vectrd Thrst- Open Prop)	Compound (Vectrd Thrst- Ducted Prop)	Tilt Rotor	Tilt Wing	Canard Rotor Wing	Fan-in-Wing	Jet Lift
Examples	AH-64, RAH-66	AH-56	Sikorsky XH-59	Sikorsky AAFSS	Piasecki VTDP	XV-3, XV-15, V-22	CL-84, TW-68, XC-142	Boeing CRW	XV-5, Grumman ACAS	AV-8B
Hover / Loiter / Ground Ops										
Efficiency (Endurance)	Best	Good	Good	Good	Good	Good	Fair	Low	Low	Poor
Downwash / Temperature	Low	Low	Low	Low	Low	Moderate	Marginal	Marginal	High	Extreme
Ground Ops Protection	Optional	Open Prop	No Anti-Torque	Open Prop	Ducted Prop	—	Optional	No Anti-Torque	—	Jet Exhaust
Cruise / Dash										
Max Speed (Dash)	Low	> Helicopter	> Helicopter	> Helicopter	> Helicopter	Good	> Tilt Rotor	High	High	Highest
Cruise Efficiency (Range)	Lowest	Low	Low	Low	Low	Good	Good	Good	Good	Highest
Maneuverability / Agility										
Low Speed	Good	Good	Good	Good	Good	Good	Fair	Limited	Limited	Limited
High Speed	Limited	Good	Good	Good	Good	Good	Good	Good	Good	Excellent
Conversion	—	—	—	Limited Agility	Limited Agility	Benign	Lmtd Corridor	Lmtd Corridor	Lmtd Corridor	Benign
Attitude vs Accel (Low Speed)	Coupled	Independent	Coupled	Coupled	Coupled	Independent	Independent	Independent	Independent	Independent
Reverse Thrust Capability	—	All Speeds	—	High Speed	High Speed	—	High Speed	—	—	—
Yaw Control	Symmetrical	Symmetrical	Symmetrical	Symmetrical	Limited / Unsymmetrical	Symmetrical	Symmetrical	Limited in Low Speed	Symmetrical	Symmetrical
Survivability										
RCS (Low Speed Mode)	Rotating Blade	Rotating Blade	Rotating Blade	Rotating Blade	Rotating Blade	Rotating Blade	Rotating Blade	Rotating Blade	Doors / Louvers	Short Inlet
RCS (High Speed Mode)	Rotating Blade	Rotating Blade	Rotating Blade	Rotating Blade	Rotating Blade	Rotating Blade	Rotating Blade	Good	Good	Short Inlet
IR (Engine Exhaust)	Lowest Power	Low Power	Low Power	Low Power	Low Power	Low Power	Medium Power	High Power	High Power	Short Exhaust
Cost / Supportability										
Vibration Environment	Highest	High	Very High	High	High	Moderate	—	Moderate	Low	Low
Empty Weight	Low	Increased	Increased	Increased	Increased	Increased	Increased	Increased	High	Moderate
Complexity (Min # of Thrusters)	2 Fixed	3 Fixed	2 Fixed	1 Fixed + 1 Vectored	1 Fixed + 1 Vectored	2 Vectored	1 Fixed + 2 Vectored	Convertible Engine	3 Vectored	1 Thruster + 8 Nozzles

Key to Table: **Advantage** Neutral *Disadvantage*

Table 4. Subjective Evaluation of VTOL Aircraft Concepts Using the Army Aviation Matrix

Summary and Conclusions

This independent assessment has considered performance and design parameter tradeoffs for a SMRC helicopter as a potential Future Vertical Lift (FVL) aircraft. This has been a two-phase effort. An Advanced VTOL Vehicle Synthesis Environment, Figure 175, was used based on a simple, but visible and adequate fuel and power balance method, called the RF Method, which has the following key elements:

- Analytical design methods which obviate the need for trial and error are presented for initial VTOL Aircraft Sizing and Performance.

- Simultaneous solutions of analytical expressions representing both weight and aerodynamic performances are developed.

- Two sub-methods are presented and used: The first, which requires some a priori knowledge of the empty-to-gross weight ratio (φ), equivalent range, and HOGE payload required of the configuration type, is useful for configuration "sensitivity" (trade-off) studies. The second, which uses an analytical description of the empty weight ratio, φ, as a function of design parameters, mission phases, and power required for each phase, is useful for initial vehicle sizing and design parameter selection (i.e., rotor diameter, disk loading, solidity, etc.) convert some of the Survivability and Cost/Supportability sub-evaluation criteria in Table 4 from red to at least blue or green.

Overview of Emerging Hybrid-Electric Propulsion (HEP) for Rotorcraft (Ref. 6)

FARA and future VTOL aircraft need to move to HEP. This Study was unique for FARA, as it was able to provide a mission for conducting technology tradeoffs for empty weight, gross weight, propulsion (both ITE and HEP), and maximum cruise and dash speeds.

There have been substantial initiatives, both in industry and government, over the past few years to introduce hybrid-electric power for aircraft applications. HEP has been utilized in automobiles for a number of years, with the Toyota Prius being an excellent example. While Uber Elevate had pushed strongly for electric VTOL (eVTOL) air taxi demonstrations, there has also been a substantial growing effort by NASA and industry in pursuing HEP for all aircraft applications, both conventional fixed-wing, as well as VTOL aircraft.

At the recent International 5th Transformative Vertical Flight Workshop in San Francisco, CA, Empirical Systems Aerospace, Inc. presented their emerging results from a NASA Phase II SBIR entitled "Hybrid-Electric Propulsion System Conceptual Design Challenges." An overview of their efforts, including applications for a compound helicopter and a tiltrotor aircraft, is illustrated in Figure 192.

A challenge: adapting Design Tools for Hybrid Air Vehicles is illustrated in Figure 193.

A challenge: the impact of Redundant Capability Requirements is illustrated in Figure 194.

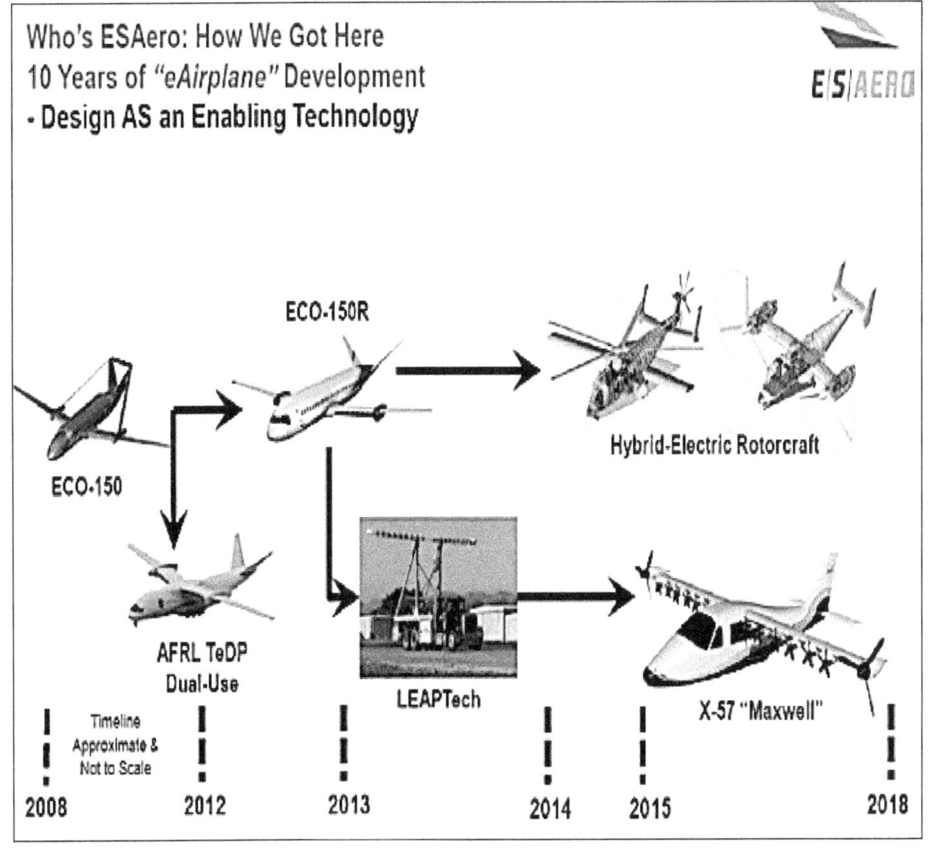

Figure 193. Overview of ES Aero, including applications for a compound helo & tiltrotor aircraft

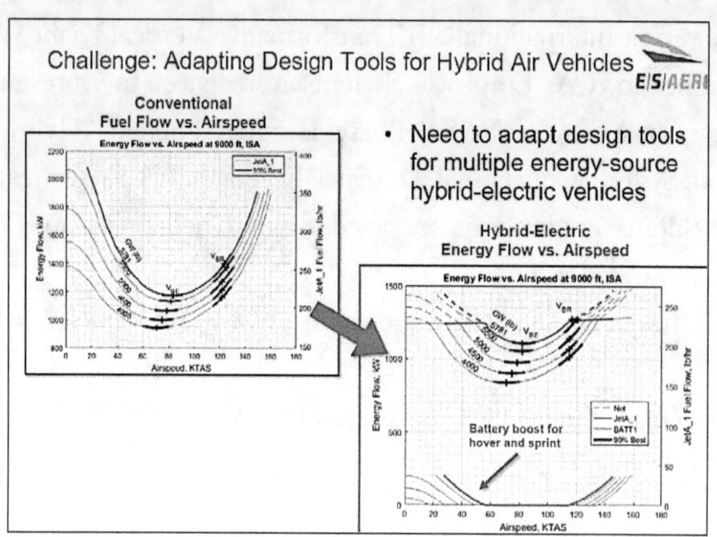

Figure 193. Challenge: Adapting Design Tools for Hybrid Air Vehicles

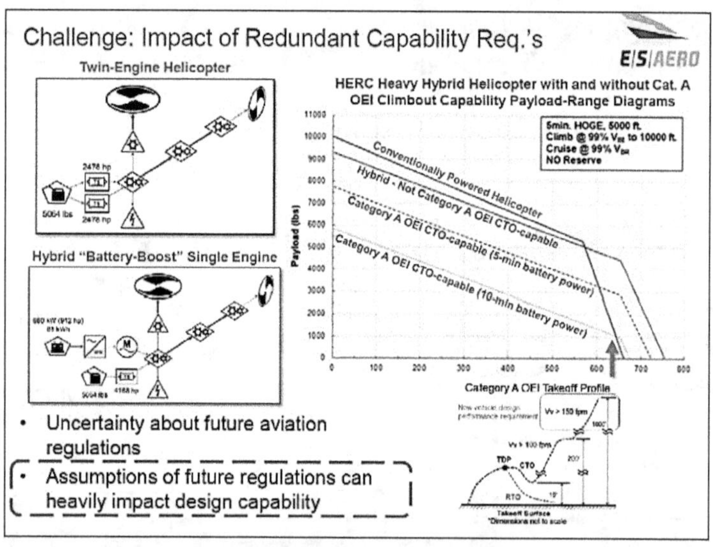

Figure 194. Challenge: Impact of Redundant Capability Requirements

Rolls Royce Allison is a propulsion company that is aggressively pursuing hybrid-electric propulsion for upgrades to existing aircraft, as well as for new aircraft. At the HAI Expo in Las Vegas, NV, in February 2018, Dr. Mike M. Mekhiche, Global Head, Rolls Royce Electrical, gave a seminar on Electrifying Aerospace which addressed Electric Propulsion Benefits, and how Hybrid-Electric Propulsion Transforms Aircraft Design Space and Creates a Strong Value Proposition, including for VTOL and rotorcraft.

Parallel Hybrid Propulsion System

A parallel hybrid-electric system was conceptually studied. The propulsion arrangement consists of the engine(s) coupled with an electric motor generator, connected via a main transmission, as shown in Figure 178. The motor generator is powered by a battery pack having enough capacity to supply five minutes more power during hover flight or high-speed flight. The amount of added power supplied is the larger of the power required above the MRP, at either hover or high-speed dash. The battery pack can be recharged during less demanding flight regimes, such as during loiter and cruise, through a fast-charging Power Control Module (PCM). The battery weight was calculated using a current technology battery energy density of 150 Wh/kg. The electric motor weight was scaled based on current aviation-grade motors, using multiples of fifteen kw brushless motors connected in parallel, with an overall power-to-weight ratio of 2.78 kW/kg. The overall electric power generation system efficiency was assumed to be 88%. With these assumptions, it appears that additional power for limited times could be incorporated, with an acceptable weight penalty.

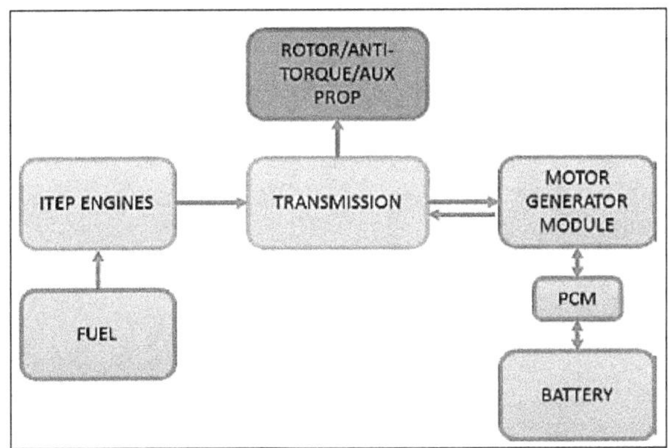

Figure 195. Parallel Hybrid System Schematic

Endnotes

1. Hirschberg, M., "Why the US Army Needs Ten Years to Begin Fielding A New Recon Helicopter." 2020

2. Future Attack Reconnaissance Aircraft (FARA), May 17, 2020, Global Security.com

3. 1993 AHS Student Design Competition, Hight Speed High Maneuverable Rotorcraft (HSHMR)

4. Schrage, D.P., Rotorcraft Design Notes, Georgia Institute of Technology, 1985-2019

5. Wilkerson, J.B., and Smith, R.L., *Aircraft System Analysis of Technology Benefits to Civil Transport Rotorcraft*, NASA/CR-2009-214594)

6. Danis, R., "Hybrid-Electric Propulsion System Conceptual Design Challenges," Presentation at AHS International 5th Transformative Vertical Flight Workshop, San Francisco, CA, January 2018

ACKNOWLEDGMENTS

First of all, I would like to acknowledge my family for their support and dedication to me in helping document my full lifetime career in writing this trilogy. My wife, Nancy, has been with me since we started dating in high school in 1961 and has shared unique experiences and stories in this trilogy for the past 62 years. Her support and the support from our children, Steven, Susan, Michael, and Alex have made completing this book possible.

I would also like to acknowledge my mentors and friends who have encouraged me to write the second book of this trilogy. This acknowledgment includes Charles C. "Charlie" Crawford, who allowed me to join his Airworthiness Qualification and Development Directorate in the Army Aviation Systems Command (AVSCOM). Bob Wolfe and Ron Gormant, the SSEB Technical Chiefs at AVSCOM and the SSEBs. BG Story Stevens and Dick Lewis with AVRADCOM. Dr. Robin Gray, Dr. Al Pierce, and Dr. Arnold Ducoffe with the Georgia Tech School of Aerospace Engineering. All were instrumental in my education and career development. AE

Finally, I would like to thank my publisher, Frank Eastland, and his team, Teresa Evans, Raeghan Rebstock, and Bob Laning, at Publish Authority, for all their help in bringing my memoir to reality.

ABOUT THE AUTHOR

Dr. Daniel P. Schrage grew up in the Midwest as the son of two teachers and graduated from USMA West Point in 1967. He has advanced degrees in Aerospace Engineering (AE), MS Georgia Tech, 1974; Business Administration, Webster College. 1975; and a DSc. Mechanical Engineering (ME), Washington U. (St. Louis) in 1978.

Dr. Schrage has had three careers and excelled in each. In his first career, he commanded an Honest John Nuclear Missile Battery during the Cold War in Europe, 1968-69. He was an Army Aviation Air Mission Commander in South Vietnam and commanded lift ship and gunship platoons. He then served as the S-3 3th Combat Aviation Battalion and ran all operations in the Mekong Delta. He orchestrated the transfer of the first Army Aviation Airfield, Soc Trang, to the Vietnamese Air Force (VNAF) under the Helicopter Vietnamization Program, 1970- 71. In his second career, he was an Aerospace Engineer, Manager, and Senior Executive for Army Aviation Development, 1974-78. He led the technical development of the next generation of Army Aviation Systems as the youngest Senior Executive Servant (SES) in the Army Material Command (AMC). In his third career, he was a Full Professor and Director of Research for the Army Vertical Lift Research Center of Excellence (VLRCOE) at Georgia Tech. He retired as a Professor Emeritus. His careers followed and documented the changes in U.S. warfare, technology development, and academic transition in a changing world. He is a Technical Fellow of both the American Institute of Aeronautics and Astronautics (AIAA) and the Vertical Flight Society. He and his wife reside in Atlanta, GA.

You can find more information about the author and his books on his website at DanielSchrage.com.